MYXOMATOSIS

MYXOMATOSIS

BY

FRANK FENNER

M.B.E., M.D., F.R.S.

*Professor of Microbiology
John Curtin School of Medical Research
Australian National University
Canberra*

AND

F. N. RATCLIFFE

O.B.E., B.A.

*formerly Officer-in-Charge, Wildlife Survey Section,
Commonwealth Scientific & Industrial
Research Organization, Canberra*

CAMBRIDGE

AT THE UNIVERSITY PRESS

1965

CAMBRIDGE UNIVERSITY PRESS
Cambridge, New York, Melbourne, Madrid, Cape Town, Singapore, São Paulo, Delhi

Cambridge University Press
The Edinburgh Building, Cambridge CB2 8RU, UK

Published in the United States of America by Cambridge University Press, New York

www.cambridge.org
Information on this title: www.cambridge.org/9780521112963

First published 1965
This digitally printed version 2009

A catalogue record for this publication is available from the British Library

Library of Congress Catalogue Card Number: 65–17207

ISBN 978-0-521-04991-7 hardback
ISBN 978-0-521-11296-3 paperback

*For our wives, Bobbie and Agnes
whose patient understanding contributed materially to
the preparation of this book*

CONTENTS

vii

CONTENTS

viii

LIST OF PLATES

FOREWORD

Few diseases of wild animals not transmissible to man have had such an impact on the general public as infectious myxomatosis of rabbits. Although first recognized in 1896, until 1950 this disease was known only to a few pathologists interested in viruses and tumours. In December 1950 it spread in dramatic fashion through the wild rabbit population of south-eastern Australia, and greatly reduced the serious rabbit problem in that part of the world. In 1952 it was introduced into wild rabbits in France, and from this focus spread over the greater part of Europe, where it proved extremely lethal to both wild and domestic rabbits.

In Australia the virus was introduced to assist in the control of the major animal pest of that continent, the rabbit, and its behaviour was followed with great interest by both scientists and the general public. In Europe attitudes to the rabbit are more variable, and public reactions varied according to outlook and interests; hunting organizations were appalled at the loss of the major game animal, foresters and farmers welcomed the destruction of a pest, and breeders of domestic rabbits sought methods of protection of their stock.

For the scientist interested in the ecology of infectious disease, the spread of myxoma virus through the wild rabbits of Australia and Europe offered a unique opportunity to observe the interaction of a very lethal parasite with a highly susceptible, very common and reasonably large mammalian host. In this book an attempt has been made to consolidate the results of the investigations on the virus and its host which preceded and followed this unique event, and to see what bearing they may have on our knowledge of infectious diseases.

F. N. Ratcliffe was responsible for writing chapters 2, 3, 4, 12, and 16 and F. Fenner the other chapters, each author having received criticisms and suggestions from the other. For the reader who is not interested in technical details of virology and pathology a short summary has been provided at the end of each chapter concerned with these topics.

F.F.
F.N.R.

Canberra
December 1963

ACKNOWLEDGEMENTS

The study of the epidemiology of a disease of wild animals in four continents has been made possible only through the help of many individuals and organizations. Financial help for the Australian investigations has been afforded on a generous scale by the Wool Research Trust Fund, the Rural Credits Development Fund of the Commonwealth Bank of Australia, and the two organizations to which we belong, the Australian National University and the Commonwealth Scientific and Industrial Research Organization. One of us (F. F.) is indebted to the Master and Fellows of Churchill College, Cambridge, for an Overseas Fellowship which facilitated the collection of data on myxomatosis in Europe, and the writing of this book.

We are also indebted to many individual scientists for collaboration and assistance. In Australia the following have been valued colleagues: Drs I. D. Marshall and Gwendolyn M. Woodroofe of the Department of Microbiology in the Australian National University; Dr R. Mykytowycz and Messrs A. L. Dyce, J. H. Calaby, K. Myers, W. E. Poole, B. V. Fennessy and E. J. Waterhouse of the Wildlife Survey Section (now the Division of Wildlife Research), C.S.I.R.O.; Mr G. W. Douglas of the Victorian Department of Crown Lands and Survey; and Mr D. J. Lee of the School of Public Health and Tropical Medicine, Sydney. Dr M. F. Day of the Division of Entomology, C.S.I.R.O., and Dr W. R. Sobey of the Division of Animal Genetics, C.S.I.R.O., have helped greatly with certain aspects of the work and Dr E. H. Mercer of the Electron Microscopy Unit, Australian National University, has allowed us to use some of his unpublished electron micrographs. Mr H. V. Thompson assisted us with information on myxomatosis in Europe, and Drs P. J. Chapple and H. Jacotot supplied strains of virus from Britain and France. Dr J. Giban and Dr R. Vittoz kindly supplied statistical data on the effects and incidence of myxomatosis in France. In America we are indebted particularly to Dr David C. Regnery, who gave us access to much unpublished material, to Dr Richard E. Shope and to Dr K.

McKercher. The late Dr H. B. Aragão was a valued and stimulating correspondent.

Some of those whose cooperation in the research work has already been acknowledged, especially Dr I. D. Marshall and Messrs A. L. Dyce and G. W. Douglas, gave helpful suggestions on certain parts of the text.

We are indebted to the following for the provision of photographs and figures: Sir Christopher Andrewes, Dr L. B. Bull, Dr H. Dalmat, Dr M. F. Day, Mr B. K. Filshie, Dr H. Jacotot, Dr L. Kilham, Dame Jean Macnamara, Dr E. H. Mercer, Professor A. A. Miles, Mr K. Myers, Dr R. Mykytowycz, Dr B. L. Padgett, Mr V. Paral, Dr R. F. Parker, Mr D. F. Nery-Guimaraes, Mr M. H. Siriez, Mr E. Slater, Dr W. R. Sobey and Mr H. V. Thompson.

We also wish to thank Her Majesty's Stationery Office, the Rockefeller Institute, and the editors and publishers of the following journals for their permission to use the photographs and figures ascribed to them in the individual captions and references: *Annals of Applied Biology, Annales de l'Institut Pasteur, Australian Journal of Biological Sciences, Australian Journal of Experimental Biology and Medical Science, Country Life Sydney, Journal of Bacteriology, Journal of Experimental Medicine, Journal of Hygiene, Journal of the National Cancer Institute, Proceedings of the Linnean Society of New South Wales, Proceedings of the Society of Experimental Biology and Medicine, Virology.*

CHAPTER I

HISTORY AND DISTRIBUTION
OF MYXOMATOSIS

In the years between 1867 and 1881 Pasteur and Koch formulated
and won general acceptance for the germ theory of infectious disease.
As a consequence Institutes of Hygiene, devoted to the task of apply-
ing this new-found knowledge to the prevention of the infectious
diseases of man and animals, were established throughout the western
world. In Italy these institutes were attached to the universities, and
the director of the Hygiene Institute of the University of Siena in
1895 was Professor Guiseppe Sanarelli. During that year Sanarelli was
invited by the Government of Uruguay to set up a Hygiene Institute
in Montevideo.

In the process of setting up this institute, Sanarelli acquired some
domestic European rabbits for the production of immune sera. The
source of the rabbits, and the methods used to replenish supplies of
them are unfortunately not known. It is possible that they were
imported from nearby Argentina or Brazil, for domestic rabbits had
been taken to South America long before. In 1896 a devastating
disease occurred in these rabbits, producing symptoms quite unlike
anything that Sanarelli or anyone else had seen in Europe. The
disease was infectious and highly lethal, and produced numerous
mucinous tumours in the skin of the infected animals. After briefly
reporting the outbreak at the IX International Congress for Hygiene
and Demography in Madrid in 1898, Sanarelli wrote an account of
his investigations (Sanarelli, 1898) in which he named the disease
infectious myxomatosis of rabbits. In this paper Sanarelli suggested
that myxomatosis was caused by a member of the newly defined
group of infectious agents, the 'filterable viruses'.

No further reports appear in the scientific literature until 1909,
when Splendore described the occurrence of the disease in European
rabbits bought in the local market in São Paulo, Brazil. According to
Moses (1911) and Aragão (1927), outbreaks of myxomatosis have

I

1

occurred sporadically in European rabbits maintained for various purposes in several places in Brazil. In South America the disease has also been reported in Argentina, where isolated outbreaks have occurred from time to time. In 1954 myxomatosis was deliberately introduced into wild *Oryctolagus* in Tierra del Fuego by the Chilean Government, and it has caused extensive outbreaks there and further north in Chile. In 1958 and in 1959 infections occurred in laboratory rabbits in the Gorgas Memorial Laboratory in Panama, and in 1960 cases were recognized in Colombia.

Myxomatosis was for many years regarded as a disease confined to South America, but in 1930 it broke out in commercial rabbitries of southern California (Kessel, Prouty & Meyer, 1931) and it has appeared sporadically in domestic rabbits in California ever since. Major outbreaks occurred in 1949 and 1959 among domestic rabbits bred commercially near San Diego. Outbreaks have also occurred at Sheridan, Manila, and Arcadia, California, in San Mateo County, California, and in Corvallis, Oregon (see Fig. 19).

The major impact of myxomatosis as a recognized infectious disease occurred, however, not in the Americas but in Australia and Europe. It was introduced to these continents in a conscious attempt to control the European rabbit, *Oryctolagus cuniculus*. The successful establishment of myxomatosis in Australia in 1950 was the culmination of a series of laboratory and field investigations extending over more than a decade. The preliminary laboratory experiments, designed to assess the dangers and potential of the disease employed as a method of biological control, and practically all the early field studies, were conducted (or sponsored) by Australia's major governmental scientific organization, the Commonwealth Scientific and Industrial Research Organization (C.S.I.R.O.), formerly the Council for Scientific and Industrial Research (C.S.I.R.)

In Europe, on the other hand, myxomatosis was introduced into France in 1952 by a private individual, Dr A. Delille, who inoculated two rabbits with virus obtained from Switzerland (Delille, 1953). In Australia, and with minor exceptions in Europe also, myxomatosis spread within a few years of its introduction until it became coextensive with the wild rabbit populations.

Thus myxomatosis is now a disease of four continents. In North America and most parts of South America it appears to be enzootic as a mild unrecognized infection of certain species of *Sylvilagus*, and recognition of the presence of the virus depends upon its transfer to colonies of domesticated European rabbits. In Europe and Australia it is now an enzootic and often lethal infection of wild European rabbits. In Tierra del Fuego and more northerly parts of Chile recurrent outbreaks of myxomatosis occur in wild *Oryctolagus*, but it is impossible to say whether these are due to repeated artificial introductions of the virus or whether myxomatosis is established there as an enzootic disease.

MYXOMATOSIS IN SOUTH AMERICA

We have seen that the existence of myxomatosis was first recognized when a novel highly lethal disease broke out among laboratory rabbits at the Institute of Hygiene in Montevideo, Uruguay, in 1896. Nearly half a century passed before the explanation of the 'spontaneous' outbreaks of myxomatosis that occur sporadically in various parts of South America was provided by Aragão (1942, 1943).

It was obvious that myxomatosis could not have been maintained in domestic rabbits, and it was known from laboratory experiments to be a disease of high host specificity. Aragão therefore examined again the local wild rabbit, *Sylvilagus brasiliensis*, which occurs very extensively in South America. The susceptibility of these animals had been tested by Moses (1911), who reported that 'wild rabbits (*Lepus brasiliensis*) were more resistant and they only rarely became infected'. Aragão, however, found that 40% of wild caught *S. brasiliensis* were easily infected with myxoma virus, and he was able to show that the remainder were resistant because of prior infection. The capture of a wild *S. brasiliensis* in the State of Rio with an ocular lesion from which myxomatosis was transmitted to domestic rabbits established beyond doubt that this animal was a natural reservoir of the virus.

In *S. brasiliensis* myxoma virus produces skin tumours which remain localized to the injection site. These tumours serve as the source of infection for mosquitoes, which transfer the virus mechanically when they feed on other rabbits.

Marshall (1961) has recently summarized available information on the distribution of myxomatosis in South and Central America, which will be discussed in some detail in chapter 15. In most of the areas where outbreaks have occurred in domestic rabbits maintained for food or fur production, or in biological laboratories, *S. brasiliensis* appears to be the reservoir host.

MYXOMATOSIS IN NORTH AMERICA

Myxomatosis was unknown in North America, except in the laboratory, until 1930, when Kessel, Prouty & Meyer (1931) recorded the occurrence of the disease in domestic (European) rabbits in San Diego (see also Kessel, Fisk & Prouty, 1934). This outbreak was thought to have originated in a shipment of diseased domestic rabbits sent from Baja California, Mexico, to San Diego, California (Vail & McKenny, 1943). While such importations may have occurred, subsequent investigations have made it quite clear that myxomatosis is also enzootic as a localized benign skin tumour in the Californian brush rabbit, *Sylvilagus bachmani* (Marshall & Regnery, 1960), which has a fairly extensive range along the west coast of the U.S.A.

Fenner & Marshall (1957) showed that two specimens of myxoma virus from California differed from the South American strains in the symptomatology of the disease they produced in *Oryctolagus* and in the size of the pocks they produced on the chorioallantoic membrane. Marshall, Regnery & Grodhaus (1963) subsequently showed that ten strains of myxoma virus recovered from different sources in California between 1949 and 1959 had the same characters as those described by Fenner & Marshall. In addition, there are clearcut antigenic differences between Californian and Brazilian strains of virus (see chapter 5). Strains from western United States appear to be homogeneous in antigenic constitution as well as in their other characters, but preliminary studies have revealed some regional differences in the soluble antigens and in certain biological characters of strains from South America. There appear to be at least two groups of myxoma viruses in America, the South American and the Californian, the former being rather heterogeneous.

MYXOMATOSIS IN AUSTRALIA

Aragão made two major contributions to the study and use of myxomatosis. He elucidated the natural history of the disease in Brazil,

PLATE I. Prophets and pioneers. *a*, The late Dr H. de Beaurepaire Aragão; *b*, the late Sir Charles Martin; *c*, Dr Lionel B. Bull; *d*, Dame Jean Macnamara; *e*, Dr Richard E. Shope; *f*, the late Dr Thomas M. Rivers.

and in 1919 he brought to the attention of the Australian Government, through Dr A. Breinl, then Director of the Australian Institute of Tropical Medicine in Townsville, the possibility of the deliberate use of myxomatosis for rabbit control. The first official response, from the newly established Institute of Science and Industry of the Commonwealth of Australia, was that ' . . . the trade in rabbits both fresh and frozen, either for local food or for export, has grown to be one of great importance, and popular sentiment here is opposed to the extermination of the rabbit by the use of some virulent organism'. The infected tissues sent by Aragão were kept under quarantine at the Commonwealth Serum Laboratories. The possibility of using myxomatosis for rabbit control was re-opened in 1924 by Dr H. R. Seddon, Director of Veterinary Research in the Department of Agriculture of New South Wales. He sought the virus from Dr Aragão, and after some difficulty due to transport delays, a viable sample arrived in Australia in November 1926.

In a letter acknowledging advice on the receipt of the virus, Aragão stated, 'I think that the rainy season is the best', his reason at that time being that spread would occur more readily if the ground and burrows were damp. No liberations of the virus were made, however, nor were field trials undertaken at this stage. The Department of Agriculture was only permitted to use the virus supplied by Dr Aragão for laboratory experiments, which confirmed the extreme lethality of the disease, and the difficulty of its transmission by airborne contagion and contaminated surroundings (White, 1929).

No further work was undertaken until 1934, when a Melbourne poliomyelitis specialist with an interest in rural problems, Dr (later Dame) Jean Macnamara, visited Dr Richard Shope at the Rockefeller Institute in New York. Shope was working on the relation between myxoma and fibroma, and therefore had in his laboratory many rabbits infected with myxomatosis. Unaware of Dr Aragão's previous efforts, but impressed by the possibilities of myxomatosis as a method of control of Australian wild rabbits, Dr Macnamara wrote a memorandum to the High Commissioner for Australia in London recommending that it should be tested. The Australian quarantine authorities were not prepared to allow the virus into the country without experimental evidence of its likely value in practice, and the Council for Scientific and Industrial Research accordingly arranged

5

for Sir Charles Martin to carry out investigations relating to the possible use of myxoma virus for rabbit control at the Institute of Animal Pathology at Cambridge. Martin's laboratory and colony experiments (Martin, 1936) led him to state that the virus should be suitable for the control of a population of rabbits in a circumscribed area, and that it appeared to be highly specific for the rabbit. He obtained no evidence that it would spread from one colony of rabbits to another.

On the strength of Martin's excellent report permission was given for the importation of the virus to Australia so that further investigations could be undertaken by the C.S.I.R. Division of Animal Health. The results of extensive inoculation tests (Bull & Dickinson, 1937) confirmed the specificity of the virus for rabbits, and dispelled fears that it might adversely affect either domestic or Australian native animals.

When Dr L. B. Bull started to develop his research programme in the years that followed, the natural history of myxomatosis had not been clarified, and the fact that it was essentially an insect-borne disease was still unknown. However, in addition to carrying out colony experiments with wild type rabbits, and investigations into techniques for disseminating the virus in the field, he paid increasing attention to the possibilities of insect transmission. It was shown in the laboratory that *Echidnophaga myrmecobii* (a native stickfast flea that had adapted itself to the introduced rabbit) and a variety of locally occurring mosquitoes could transmit the disease.

When the time came, and quarantine permission was granted, for field trials with the virus, there was some initial difficulty in obtaining suitable experimental sites, in part because of the hostility of those who had vested interests in the rabbit skin and carcass industry. Co-operation was forthcoming from the South Australia Government, however, and a series of trials was carried out first in the coastal regions and later in the semi-arid interior of that State. The most successful was an experiment carried out in a 90-acre enclosure in which the rabbits, living in thirteen large warrens, were infested with *Echidnophaga*. The rabbit population was virtually exterminated within two months of the introduction of myxoma virus. Much less successful results had attended other field experiments in areas where flea infestation did not occur (Bull & Mules, 1944). In the larger scale inland trials the results were, in general, disappointing. While warren colonies into which the disease was introduced were usually exterminated, the infection failed to spread from one warren to another. It was thought at the time that the activities of foxes and other predatory animals, to which sick rabbits fell easy victims, might be an important factor in the situation.

In summing up the work that terminated in 1943, Bull & Mules concluded

that myxomatosis cannot be used to control rabbit populations under most natural conditions in Australia with any promise of success. Nevertheless, it seems possible that in some parts of Australia under special conditions, including the presence of insect vectors in abundance and the absence of predatory animals, the disease could be used with some promise of temporary control of a rabbit population.

At this juncture the C.S.I.R. considered that if further trials with myxomatosis were to be done, they might properly be undertaken by the State authorities who had the responsibility for vermin control within their territories; but the Director-General of Health refused to release the virus from quarantine restrictions, and work on myxomatosis was discontinued for a second time.

For reasons not yet fully understood, but certainly in part due to manpower shortage and the failure to apply routine control measures, a very serious rabbit situation developed in Australia in the years immediately following the Second World War. In 1949 the Commonwealth Scientific and Industrial Research Organization (C.S.I.R.O. —the successor to the Council for Scientific and Industrial Research) established a Wildlife Survey Section under the leadership of F. N. Ratcliffe, which in addition to its responsibilities for the study of the Australian native fauna undertook to explore the possibilities of dealing with Australia's major animal pest, the rabbit, in a scientific fashion. Experiments with myxomatosis were undertaken again almost immediately. Stimulated by the insistence of Dame Jean Macnamara, in a newspaper controversy, that the early trials had been carried out in dry country and that adequate experiments in well-watered country had still to be done, and following Bull's suggestion that trials should be undertaken where there were abundant vectors, the 1950 trials were conducted in several sites in the Murray Valley. The results were similar to those obtained by Bull &

Mules in their South Australian liberations, the disease remaining localized and apparently dying out after it had been established in the rabbit population (Myers, 1954). However, in December 1950, myxomatosis flared up in one of the trial sites, and was reported almost simultaneously at several points along the river nearby; and in the following few weeks it spread with dramatic speed over south-eastern Australia. It has never since disappeared, and is now firmly established as an enzootic disease of the Australian wild rabbit, with epizootics developing periodically in association with local and seasonal vector activity. The subsequent history of myxomatosis in Australia will be related in detail in chapter 16.

With some reluctance, the quarantine restrictions imposed by the Director-General of Health were cancelled in 1951, when it was apparent that they could no longer apply. However, because of public anxiety about the use of fibroma virus vaccine for the protection of domestic rabbits raised for commercial purposes, the Director-General of Health in 1962 re-applied restrictions upon scientific laboratories studying fibroma and myxoma viruses, but not upon the distribution of myxoma virus for rabbit control.

MYXOMATOSIS IN EUROPE

At the time of Sir Charles Martin's experiments with the virus in Cambridge several attempts were made to introduce myxomatosis for the control of wild rabbits on Skokholm Island, off the Pembrokeshire coast, and in Denmark and Sweden (chapter 17). None of these led to the enzootic establishment of the disease, though the Swedish introduction gave rise to a localized epizootic of sufficient intensity to reduce the rabbits on one large estate to very low numbers. No further introductions were attempted in Europe until after the Australian experiences had demonstrated that myxomatosis could be a powerful instrument for the biological control of the rabbit.

Approaches were made to Australian authorities early in 1952 by French officials interested in forestry and agriculture, concerning the possible supply of myxoma virus for rabbit control. However, the actual introduction of the virus into Europe was made on 14 June 1952 when Dr P. F. Armand Delille, acting as a private individual, inoculated two wild-caught rabbits with virus obtained from the

Laboratoire de Bacteriologie, Lausanne; his intention being to destroy wild rabbits on his 600-acre walled estate at Maillebois near Paris (Delille, 1953). By the end of August virtually all rabbits on the estate were dead and outbreaks of myxomatosis had also occurred at St Ange and Sfraze, villages 8 and 45 km. distant from Maillebois. The first official identification of myxomatosis in France was made on a wild rabbit taken at Rambouillet in October 1952 (Jacotot & Vallée, 1953a).

There was little evidence of myxomatosis in the winter of 1952–53, but widespread and very destructive epizootics occurred in the summer of 1953 and by the end of that year the disease had been recorded from most parts of France, from Belgium, the Netherlands, Luxembourg, Germany, Spain and England. Occurrence of myxomatosis in England was confirmed on 13 October 1953 from an outbreak at Edenbridge, Kent (Thompson, 1954). The first natural outbreak of the disease in Sweden since the 1937–38 trials occurred as late as 1961 (Borg, 1962).

The history of myxomatosis in Europe and the changes observed there in host and parasite will be described in detail in chapter 17. The virus introduced into Europe was a strain which had been passed only a few times in domestic rabbits since originating from *S. brasiliensis* (see Fenner & Marshall (1957) on strain Brazil/Campinas/1949/1 (Lausanne)), whereas the virus used to initiate the disease in Australia, although originally derived from *S. brasiliensis* (Moses, 1911), had been passed many times in domestic rabbits. Both strains were of South American origin, and were of very high virulence for *O. cuniculus*.

CHAPTER 2

THE LEPORIDAE

The natural hosts of the myxoma virus are species of the New World genus *Sylvilagus*, belonging to the family Leporidae of the mammalian order Lagomorpha. The leporids are the animals commonly known as the rabbits and hares. The only other family in the order is the Ochotonidae—the pikas. These are small, short-eared, tail-less inhabitants of high mountain screes in North America, Asia and southeast Europe.

The Leporidae conform to a fairly standard pattern of morphology and general appearance (Fig. 1), and all members of the family were originally placed by Linnaeus in the one genus *Lepus*. Nine genera are now recognized, of which only three—*Lepus*, *Oryctolagus* and *Sylvilagus* —need to be considered here.

THE HARES AND JACKRABBITS

The genus *Lepus* comprises the hares and jackrabbits. They are surface living, usually open country animals, relying on speed rather than cover for safety, and are generally larger, longer legged and longer eared than the species of *Sylvilagus* and *Oryctolagus*. The young (rarely exceeding three to a litter) are born fully furred, open eyed and active, and are deposited in open 'forms', not in a specially constructed nest.

There are a round dozen species of *Lepus* in North America, two in the British Isles and western Europe, and several distributed over the rest of Eurasia and Africa. The two species found in Britain are the blue or mountain hare (*L. timidus*) confined to Ireland and the hill-country of Scotland, and the brown or European hare (*L. europaeus*) of England and the Scottish lowlands. The European hare has been introduced and successfully established in several countries, including the eastern United States, parts of South America, Australia and New Zealand.

10

The effect of the myxoma virus on species of the genus *Lepus* will be discussed in detail in a later chapter. Populations of both *L. europaeus* and *L. timidus* have been subject to the risk of natural infection during outbreaks of myxomatosis among local rabbits. While these Old World hares, as species, have a high degree of resistance to infection, occasional individuals of both of them have proved susceptible. There is an indigenous fibroma of *L. europaeus* in France and Italy, which has recently been shown to be caused by a virus related to myxoma virus (see chapter 7).

THE EUROPEAN AND AMERICAN RABBITS

The genus *Oryctolagus* contains one species only—*O. cuniculus*, the common or European rabbit. Because of its considerable northward and eastward spread during recent, and not so recent, historical times (this spread being in part deliberately assisted by man, and in part probably a normal exploitation of forest clearance and developing agriculture) it is impossible to define the original natural distribution of the species with precision. It included the regions bordering the western Mediterranean to the north and south, and extended into central Europe. Today it is found in all the countries from Portugal to western Poland, including the British Isles, with local populations in parts of the Ukraine, Norway and southern Sweden. Its introduction into countries outside Europe will be discussed later.

The young of *Oryctolagus* are born naked, helpless and blind, and are deposited in underground nests. The litter size normally ranges from two to eight (mostly three to five), with larger litters occasionally recorded. The European rabbit is exceptional among the leporids in being gregarious and a burrower. According to Thompson & Worden (1956)

all the names for the rabbit in central European languages are derived from the Latin *cuniculus*, which also signifies an underground passage. The Italian is *coniglio*, Spanish *conejo*, Portuguese *coelho*, German *kaninchen* and English *coney*. The terms *rabbet*, *rabytt* or *rabette* (whence the modern *rabbit*) were used in the fifteenth century for young rabbits only, and were apparently derived from the Walloon *rabett*.

Oryctolagus cuniculus is the only member of the family to have been domesticated. Most of the numerous fancy breeds that exist today seem to have been developed within the past hundred years. Some are much larger than the ancestral wild form; some have a different colour and type of coat; some have a conformation approximating to that of a hare; but all are derivatives of the European rabbit. In its domesticated form, as a hutch animal bred for meat and fur, and more importantly as a laboratory animal that has become indispensable to medical science, *O. cuniculus* has been introduced into virtually every country in the world.

Not only are species of the American genus *Sylvilagus* commonly called rabbits, but most have a close general resemblance to *Oryctolagus*. The descriptive term cottontail is applied to some species, but not to the exclusion of the commoner name. The South American forms now included within the species *S. brasiliensis*, which have vestigial button-like tails, are referred to in the literature as tapetis, or tapitis, a transcription of the Tupi Indians' name for the animals (Hershkovitz, 1950), but seem invariably to be called rabbits—or the Spanish or Portuguese equivalent.

With the doubtful exception of the pigmy rabbit (*S. idahoensis*) species of *Sylvilagus* are not burrowers, though they make ready use of, and may even take over, the burrows of other animals. The young at birth are in a slightly more advanced stage than the newborn *Oryctolagus*, with unopened eyes and practically no fur. The litter size approaches that of European rabbit, but seems on the average to be somewhat smaller. A well-formed nest is constructed. Apart from differences in burrowing and social behaviour, species of *Sylvilagus* and *Oryctolagus cuniculus* are generally similar in their habits.

The geographic ranges and relationships of the various forms of *Sylvilagus* are of great importance in the natural history of the myxoma and the related fibroma viruses, and form a rather complex picture. As has already been indicated, the genus is restricted to the Americas, its northern and southern limits being, roughly, the Canadian–U.S. border and a line drawn east of the Andes across southern Brazil and northern Argentina. The taxonomic status of one or two forms is still arguable, but twelve or thirteen full species are now generally recognized. Two of these (*S. floridanus* and *S. brasiliensis*) have very extensive ranges and are subdivided into a considerable number of named geographic races or subspecies; and with a third species, *S. bachmani*, they form the group with which myxoma and fibroma are naturally associated.

Although they include forms of considerable biological interest (e.g. the semi-aquatic marsh rabbit, *S. palustris*, of south-eastern U.S.A.), the other species of the genus need not concern us. The pattern of distribution of the different species of *Sylvilagus* is such that

(1) (2)

(3) (4) (5)

Fig. 1. Leporids of importance as hosts of viruses of the myxoma-fibroma subgroup of the poxviruses. (1) European hare (*Lepus europaeus*). (2) European rabbit (*Oryctolagus cuniculus*). (3) Eastern cottontail (*Sylvilagus floridanus*). (4) Tapeti or tropical forest rabbit (*Sylvilagus brasiliensis*). (5) Brush rabbit (*Sylvilagus bachmani*).

every region in the United States and Mexico has at least one: when two or more occur in a region, preference for different habitats results in a fairly effective ecological separation of the species.

Sylvilagus floridanus is known in the United States as the eastern cottontail. It ranges from just across the Canadian border in the north, over the eastern and central United States, through Mexico and Central America into the north of South America, where it inhabits the 'arid and semi-arid savannahs and scrublands of Caribbean coastal plains and highlands of Colombia and Venezuela', with some southward penetration into the valleys of the upper Orinoco and other rivers (Hershkovitz, 1950). That part of the eastern United States where fibroma occurs as an enzootic infection of this cottontail lies mostly within the range of the subspecies *S. floridanus mearnsii*.

Sylvilagus brasiliensis has recently (e.g. by Leopold, 1959) been referred to by the apt name of the tropical forest rabbit. Except for two races that have become adapted to the savannahs of the high Andean slopes, the species is essentially a forest dweller, and both the northern and southern limits of its range are determined by this ecological fact. To the north, *S. brasiliensis* extends through Central America up to the natural limit of the tropical forest in southern Mexico. In the south it stops at the line of transition between the forest and the open pampas. It is confined, on the west, by the high Andes. Thus it ranges over most of the northern two-thirds of South America, and is absent from all but the extreme north of Argentina, from the southernmost part of Brazil, from Uruguay and from Chile.

Before the classification of the Leporidae was revised in the light of modern taxonomic ideas, most of the distinctive local races of *S. brasiliensis* were given full specific rank. The form occurring in hilly country inland and north of Rio de Janeiro was given the name *minensis* (Thomas, 1901); and although it may not even merit the subspecific rank to which it was reduced by Thomas in 1913, the name lingered on. When Aragão, in 1942, published his evidence for myxomatosis existing as an enzootic infection of the local native rabbit population, he referred to the species as *Sylvilagus minensis*.

Sylvilagus brasiliensis and *S. floridanus* are the only representatives of the genus found in South America—in fact they are the only leporid species occurring naturally on the subcontinent, although the European rabbit and hare have been introduced to one or two South American countries and have become well established.

Sylvilagus bachmani, the brush rabbit, is restricted to the seaboard of western North America, from Oregon in the north to the tip of the peninsula of Baja California in Mexico. It is typically an inhabitant of the chaparral, a low scrubby plant association that extends inland onto the foothills of the coastal ranges. The role of *S. bachmani* as a natural host of myxoma remained undiscovered until 1960.

RABBITS AND MAN

The Leporidae must have been hunted and trapped by man, as a source of food, since the earliest times. The process still continues, and not only in rather primitive communities such as those of rural Mexico and South America. Particularly since their fur has been used as the raw material for the manufacture of felt hats, the commercial harvesting of wild rabbits has been organized on a large scale in, for example, Britain and Australia. In some countries where shooting is a very popular field sport, rabbits and hares enjoy the status of quite important game animals, notable examples being *Sylvilagus floridanus* in the United States and *Oryctolagus cuniculus* in France. (Recently, since the reduction of the rabbit by myxomatosis, the European hare has taken its place as one of the major game animals in France.)

Most members of the family are at any rate potential pests; and those species that are able to build up and maintain high populations in settled areas can be very destructive to crops and pastures. Jack-rabbits in western U.S.A. come into this category; but the leporid that has earned the greatest notoriety as a pest is, of course, the European rabbit—in Britain, and particularly in Australia, New Zealand and Chile, where it has been introduced.

When the rabbit is sufficiently common to have a significant effect on rural production, necessitating some form of control, it naturally invites organized commercial exploitation; and so sections of the community will almost inevitably develop a vested interest in its survival, and moreover its survival in reasonable abundance. This in turn results in the community as a whole having an ambivalent attitude to the animal, which can weaken the desire and drive for effective control, a point that will be referred to in later chapters.

THE DOMESTICATION AND WORLD-WIDE DISPERSAL OF *ORYCTOLAGUS CUNICULUS*

Mention has already been made of the domestication of the European rabbit, alone among the Leporidae. According to Zeuner (1963) the process was started by the Romans in the first century B.C. when Varro, a writer with an interest in natural history, suggested that rabbits should be put in *leporaria*—walled gardens originally used for holding hares destined for the table. The idea of the leporarium was perpetuated in the 'rabbit gardens' and 'rabbit islands' of medieval Europe, and in the enclosed 'warrens' that were maintained on English country estates until, presumably, wild rabbits had become sufficiently abundant to be readily available. This was semi-domestication at best. A domesticated rabbit is adjusted to a cage or hutch existence, and breeds readily and regularly in confinement. Zeuner (1963) states that true domestication was achieved between the fifth and tenth centuries, probably in France by monks seeking a supply of a Lenten delicacy—the newly born or unborn young of the rabbit were not regarded as meat. Moreover, he believes that domesticated rabbits spread through northern and eastern Europe if anything before the wild and semi-wild types.

At the end of the eighteenth and beginning of the nineteenth centuries—the period in which we are especially interested—two or three fancy varieties (e.g. the Angora) had been developed, and must have been maintained by inbreeding. There is reason to believe, however, that the majority of the ordinary run of hutch rabbits were still grey or 'agouti' coloured. Zeuner states that until about the middle of the sixteenth century 'the wild coloration was the rule', and that for a hundred years after that, or more—that is until they were bred extensively for their fur—'the market required wild-coloured rabbits'.

Thus at the time when the progenitors of overseas populations of *Oryctolagus* were being shipped around the world, we can be reasonably certain of two things: first that a random sample of hutch rabbits would mostly resemble the wild type in appearance, and second that (despite the not infrequent outbreeding that must have occurred in rural areas) a millennium of domestication, involving continuous close confinement, must have affected in some degree the ancestral ability of the species to adapt itself to a variable and often harsh natural environment.

The last and most important phase in the dispersal of *Oryctolagus cuniculus* from its European home was associated with the opening up and settlement of southern hemisphere lands during the past 150 years or so. The rabbits taken aboard the sailing ships (and subse-

16

quently given or sold to settlers, or liberated with the idea of supplementing the local food resources of some distant, probably struggling, white community) would normally have been domesticated animals. There is a lingering tradition of 'wild' importations, e.g. to the mainland of Chile during the early years of this century (Thompson & Worden, 1956); but we know of only one authenticated case of wild rabbits being transported overseas. This is the shipment that arrived in Australia in 1859 on the clipper *Lightning*, the significance of which in the history of Australia's rabbit infestation will be discussed in the next chapter.

The *Lightning* shipment was arranged by a Victorian landholder, Thomas Austin, 'an ardent acclimatizer who also introduced sparrows, (and who) made several attempts to introduce rabbits before his efforts were crowned with undreamed of success' (Kiddle, 1961). According to the records, the shipment as landed comprised '24 wild rabbits' (Holland, 1923). There is some conflict, in detail, between accounts of the background of this importation, which are based on Austin family recollections (Austin, 1955; Kiddle, 1961); but the essential facts are clear and unquestioned. Thomas Austin had stipulated that *wild* rabbits were to be sent out to him: his relative in England, who had the task of collecting them, was unable to get enough wild rabbits in the time available, and made up the shipment with a few hutch-bred rabbits. The majority of the animals that sailed for Australia, however, were genuinely wild rabbits.

This information about the *Lightning* shipment, trivial though it may seem, lends support to the assumption that the vast majority of European rabbits shipped around the world were of the domesticated type. Hutch rabbits would have been readily available in any British or European port or town, whereas to obtain wild ones a special effort would have been called for, such as those primarily concerned with loading or looking after the animals would be most unlikely to make.

Three *Oryctolagus* populations established in the southern hemisphere—in Chile, Australia and New Zealand—soon attained great economic importance. In Chile, rabbits have spread along the coastal belt from the extreme south to some hundreds of miles beyond Santiago, and have penetrated into some Andean valleys. The only reasonably complete information available relates to Tierra del Fuego. The infestation on that island apparently originated from hutch rabbits released in or about 1910, possibly augmented by later liberations and escapes (Thompson & Worden, 1956). By the 1940's, the productivity of the rabbit-infested land had become seriously affected; and the introduction of myxomatosis in 1954 was spectacularly successful.

The rabbit in Australia will be discussed in chapter 3. Allowing for the much smaller size of the country and the more restricted range of habitat types, the picture in New Zealand is closely comparable. As in Australia, introductions and liberations prior to the middle of the nineteenth century gave rise to no startling developments. Wodzicki (1950) states that 'strangely enough, the rabbits at this time remained localized, did not increase rapidly and even disappeared from many localities in later years'. Most of the New Zealand introductions—including the earliest, which occurred before 1838—were of rabbits obtained from Australia. The successful establishment and 'colonizing spread' of the rabbit in New Zealand dates from the mid 1860's, taking a decade or so to build up to full momentum. Wodzicki mentions further introductions at about this time; and one is tempted to suggest that some of these included wild-type rabbits, which by then would be available in Australia, and that such an infusion into the New Zealand stock accounted, in part, for the animals' rather sudden revelation of aggressive adaptability. Be this as it may, the rest of the story parallels Australian history—except for the failure of myxomatosis—with infestation becoming general wherever conditions favoured the rabbit's survival; loss in productivity and, locally, a serious degradation of pasture vegetation; the need for organized and costly control, and the development of a skin and carcass trade with a vested interest in the persistence of a commercially exploitable rabbit population.

In addition to Australia, New Zealand and Chile, rabbits have been liberated on offshore and oceanic islands in both hemispheres, too numerous to list. They range from subantarctic islands like Macquarie Island and Kerguelen to Pacific islands lying within the tropical belt. While some of the island populations died out for no obvious reason, the surviving majority provide a remarkable demonstration of the adaptability of *Oryctolagus cuniculus* to climatic extremes.

The history of the introduced rabbits on Pacific islands has been reviewed and discussed by Watson (1961). The main lesson to be learned from it is that unless conditions are only just sufficiently favourable to permit survival (e.g. on Phoenix Island within 4° of the equator, which lacks surface water), or unless environmental factors can be relied on to produce catastrophic mortalities at intervals (e.g. on Manana, an islet off Oahu in the Hawaiian group, where droughts cause periodic 'die-offs'), the rabbit will before very long completely destroy the vegetation on which it depends, or at best find itself surviving in precarious balance with a greatly altered and degraded flora.

The tendency of the rabbit to destroy or impoverish its environment is likely to be revealed most obviously and rapidly on small islands, but it has been manifest, in some degree, wherever the species has been introduced and established. In an area where predators occur in reasonable variety and abundance, and there is competition from herbivores of comparable biological efficiency, one might expect *Oryctolagus cuniculus* to be subject to a natural control; but such a situation is no longer to be found in developed lands. Thus where man does not intervene, the food-supply ceiling provides the only ultimate check on the population. The periodic or persistent over-exploitation of the ground flora, which provides the rabbits' food supply, inevitably leads to qualitative and quantitative changes in its composition. All too frequently these changes are rendered irreversible by water or wind erosion, following on a reduction in the density of the protective plant cover.

CHAPTER 3

THE RABBIT IN AUSTRALIA

THE HUMAN BACKGROUND

As a background to the story of the colonization of Australia by *Oryctolagus cuniculus*, it is useful to know something about its colonization by European man. To begin with, one may recall that 'what the continent was like, even in outline, was not known until fifteen years after the First Fleet arrived. That it was a continent at all, and not a cluster of islands, was not known.... What the inland territory was like was not known' (Scott, 1936).

The First Fleet arrived in Botany Bay, and Sydney was established as a convict settlement, in 1788. Within the next 50 years, settlements that were to develop into the other five State capitals were founded—Hobart (1804) and Brisbane (1824) as penal settlements; Perth (1829), Melbourne (1835–36) and Adelaide (1836) as free ones. All the eastern half of the continent was included in New South Wales until 1851, the name Australia (for the country as a whole) coming into use during the second decade of the nineteenth century.

In the early years of the colony, the home government actually tried to prevent the spread of settlement in New South Wales. The stockowners in the colonial community were not greatly concerned about land policies framed in London (as they were to demonstrate on several occasions), but they were effectively restrained by the difficult nature of the ranges that hemmed in the area of settlement around Sydney. In 1813 a route through the Blue Mountains, giving access to the western slopes and plains, was discovered by a small party seeking pastures for their growing herds. It was immediately exploited; and the next quarter-century saw what can only be called an explosive spread of pastoral settlement, in which exploration and the taking up of land proceeded side by side.

The spread of settlers inland from the other ports took place without the initial delay suffered by the Sydney colony. There was a rapid

20

increase in Australia's population as a result of the gold rushes that followed the discoveries of rich deposits, starting in the 1850's. This increased the demand for rural products, and so acted as an additional stimulus to settlement. The widespread prospecting that took place also speeded up the acquisition of detailed information about the inland.

The pioneer settlers of Australia were remarkable in their enterprise and dauntlessness. In no more than three-quarters of a century after the First Fleet landed its human cargo, they had penetrated into almost every habitable corner of the continent—a land mass with an area almost as great as that of the United States of America. By the 1860's or 1870's places which today are still regarded as the back of beyond had mining settlements or cattle station homesteads. Maps of Australia in atlases published at that time differ surprisingly little from those of today.

While the place names on the century-old maps might suggest an almost modern Australia, for the most part the landscape would have been very different. In the better rainfall regions now given over to intensive agriculture, much if not most of the original forest would still have been standing. Inland of the Divide, where (in response to decreasing rainfall) the forest gives way to savannah-woodland, and then by stages to open plain, the vegetation would have been significantly more dense 100 years ago. Records show that the grazing country which today is an open eucalypt parkland once had many more bushes and shrubs, thickening here and there to form extensive areas of scrub. The suppression of seedling regeneration by stock was probably the main reason for its disappearance; but old-time diaries show that inland holdings often employed big gangs of scrub-cutters for years on end. (These gangs were sometimes made up of Chinese, who had come into Australia in considerable numbers during the gold rushes.) We know too that the native pasture must originally have been richer and more close than it is now. Some of the dominant species (e.g. kangaroo grass, *Themeda australis*) failed to maintain their status under continuous stocking.

Thus we find that when the rabbit was spreading and establishing itself in Australia, settlement was widespread though naturally 'thin' by present-day standards. Pastoral holdings were large; transport was difficult and primitive; and landholders were so fully occupied in

21

coping with a strange and difficult environment that they were virtually helpless when faced with the problem presented by the invading pest. Moreover, the changes in the environment that were taking place steadily as a result of their labours and the activity of their stock—the clearing of forest and scrub, the thinning of the pastures and the provision of permanent water—tended to favour the rabbit, although they mostly did not favour the survival of the native fauna.

INTRODUCTION AND COLONIZING SPREAD
OF THE RABBIT

Oryctolagus cuniculus accompanied the earliest settlers to Australia, the records showing that five animals arrived with the First Fleet in 1788. During the next 50 or 60 years further batches of domestic rabbits must often have been imported. By the middle of the nineteenth century rabbits seem to have been kept in every city and town, and they are known to have escaped or been liberated in many places. Although they had established themselves in the neighbourhood of Sydney and several other centres, these feral populations remained localized and apparently gave little or no trouble.

The first records of rabbits occurring in numbers in the wild state, and causing concern, came not from the Australian mainland but Tasmania. The Hobart *Colonial Times* of 11 May 1827 reported that the 'common rabbit' was 'running about some large estates in thousands'. These were probably the descendants of 'wild grey' rabbits imported by a Dr William Crowther a few years before: whether this introduction consisted of wild-caught animals or merely agouti-coated hutch rabbits we do not know.

There is no doubt that the rabbit hordes which ultimately overran the southern half of continental Australia were descendants of the 1859 *Lightning* shipment mentioned in the preceding chapter, because almost every important stage in the colonizing spread from the liberation site of the animals was observed and has been recorded. The extraordinary difference between the history of the descendants of the 1859 introduction and that of the many other 'wild' populations which developed from liberations and escapes must have been due, in the main at any rate, to the fact that the *Lightning* shipment included genuinely (i.e. genetically) wild rabbits.

The rabbits imported by Thomas Austin were liberated on his property, Barwon Park, some twenty miles west of Geelong and 150 miles south of the Murray River and the Victoria–New South Wales border. It did not take long for their progeny to reveal what might be termed their pest potential. By 1865, 20,000 rabbits had been killed on the estate, and they abounded on neighbouring properties. When the population increase had gained momentum, the rabbits proceeded to spread northward and westward with a speed which we believe to be without parallel in the whole history of animal invasions. By 1880 the advance guard had reached and succeeded in crossing the Murray River; and rabbits were first reported on the Queensland border in 1886, having traversed the State of New South Wales from south to north at a rate of about 70 miles a year. The spread from South Australia into and across Western Australia (which was channelled mainly along an inland route) was achieved at a comparable speed. Rabbits appeared on the shores of the Indian Ocean at a point north of Geraldton, W.A., sixteen years after their first sighting at Fowler's Bay, S.A., about 1100 miles to the east.

People were very anxious about rabbits and were keeping a continuous watch for them during the period of their colonizing spread, so there is unlikely to be any serious error in the dates of first sightings. The speed of these early long-distance movements must therefore be accepted as fact, even though it might seem almost incredible when one realizes that those long 'jumps' could not have been completed within one rabbit generation, and that the migrating animals must have paused at times to drop and rear litters.

THE BARRIER FENCES

While the rabbit advanced, it was preceded by reports (growing ever more alarming) of its destructiveness in those areas where it had become established and had built up in numbers. With no experience and little biological understanding to guide them, the authorities tried to cope with the situation in various ways which proved ineffectual and sometimes very costly. The building of barrier fences deserves special mention, if only because of the vast scale of the effort involved. Many thousands of miles of rabbit-netted fencing were erected with the idea of protecting regions that were still rabbit-free, the best known fences being those which followed State boundaries, for example between Victoria and South Australia, and along the Queensland–New South Wales border. Perhaps the most famous of all, however, was the 'No. 1 rabbit fence' in Western Australia, the story of which exemplifies the cost in time, money and human endurance of constructing these barriers across long stretches of dry and nearly, or quite, uninhabited country.

The 'No. 1 fence' ran for over 1100 miles northward from a point on the south coast near Hopetoun to the southern end of the Eighty Mile Beach, east of Port Hedland. It took five years to build, and was completed in 1907. The northern section of the fence, which passed through virtually unexplored country on the edge of the Great Sandy Desert, was naturally the most difficult and costly to erect. Water had first to be provided for the large numbers of men and draught animals that were to be employed (this, of course, was before the days of motor transport) and a series of bores were sunk at intervals of fifteen to twenty miles along the line of the fence. The average haulage distance from the railhead was 200 miles, with a maximum of 450 miles. Much of the fencing material had to be packed on camels for the final stages of the journey.

Although they may, in places, have brought temporary benefit, the barrier fences failed in their object of stemming the rabbit's spread, not only because they were often built too late, but because it proved impracticable to maintain them at 100% efficiency, which was essential if they were to fulfil their expected role. Looking back, it is easy to condemn the effort as biologically unrealistic, and a waste of public funds; but at the time the building of the barrier fences seemed to offer a solution, in fact the only solution, of a desperately serious problem.

ESTABLISHMENT AND CONSOLIDATION

The primary dispersal (which we have called the colonizing spread) took the rabbit into all those parts of the Australian continent in which it was able to maintain itself permanently. There then had to follow a process of local spread and ecological consolidation, which of course occurred at different times in different parts of Australia: for example, in the Western Division of New South Wales it took place in the years between 1885 and 1892, whereas in Western Australia it may be said to have occupied the second and third decades of this century.

The rabbit succeeded in establishing itself, admittedly sometimes precariously, over more than one half of the continent's approximately 3,000,000 square miles, and in country ranging from desert to high alps and coastal heaths. Climate sets a northern limit to its distribution, and it is an unimportant constituent of the fauna anywhere north of the tropic of Capricorn. Presumably because of its Mediterranean ancestry, the rabbit is better adapted to a dry than to a humid heat. It is only to be expected, therefore, that it does not extend as far north along the sub-tropical eastern seaboard as it does in the centre and the west of the continent.

For the most part the northern limit of the rabbit lies in country which, if not virtually uninhabited, is at best but sparsely settled. This is not the case in the east, however, and here (in Queensland) some barrier fences are still being maintained. One follows the Queensland–New South Wales border from the coast inland for 100 miles or more, and links up with the southern end of the Darling Downs rabbit fence. This in turn, after running north for about 100 miles, joins a third fence that follows the line of the railway westward from Miles to Mitchell, and then veers away to the north-west. There are rabbits on both sides of the last-mentioned, more inland fence; but the colonies to the north of it are generally static and of little economic consequence. In fact there is good reason to believe that these three fences approximate fairly closely to the rabbit's natural northern limit in eastern Australia.

The rabbit is well entrenched throughout the temperate zone, which supports the great majority of Australia's rural population. Within those regions where extremes of climate are not the determinants of the survival and success of the species, its abundance and economic importance depend on a variety of factors. Thus there are certain soils, notably the heavy self-mulching black or grey earths, in which the rabbit finds it impossible to dig warrens. Where these soils occur extensively, and to the exclusion of other types, the rabbit is not normally a serious problem. *Oryctolagus cuniculus* is not by nature a forest-dweller; and in the very extensive areas of forest that still remain on the ranges of south-eastern Australia the rabbit is relatively uncommon, living mostly as a transient or in clearings. When the forest is cleared for settlement, however, it is quick to take advantage of the altered conditions.

ECONOMIC IMPORTANCE

Where areas of open level country with fertile soils are exploited for intensive agriculture, ordinarily good farm management usually keeps the rabbit in adequate check. As a general rule, the 'small' landholder is only extended in his efforts to cope with the rabbit if he farms in an area where there is an admixture of uncleared wasteland, or intrusions of bracken and bramble. These conditions can be regarded as normal, however, in most hill settlements.

While the rabbit will damage cereal and other crops to an extent that can be locally serious, its most important impact on the rural economy, by far, is made as a competitor of grazing stock and destroyer of pasture. Australia is still very largely a pastoral country; and most of the rabbit-infested land is held in the form of grazing properties, ranging in size from 500–1000 acres (on the average) in better rainfall areas to tens of thousands, or even hundreds of thousands of acres further inland. The hard core of the rabbit problem may be said to stem from the fact that intensive effort is essential for effective control, and intensive effort becomes uneconomic where land is of such low intrinsic value that it has to be exploited on an extensive basis.

The rabbit-infested portion of the continent is so vast in area, and the conditions (both for rabbits and for primary production) vary so widely between regions, and at different times in any one region, that it is impossible to give a meaningful expression of the rabbit's economic importance in Australia within the compass of a few paragraphs. The following points and examples should, however, help towards an understanding of the nature and magnitude of the problem.

In the third season of its spread in south-eastern Australia (1952–53) after a useful but localized reduction in the rabbit population in the preceding years, myxomatosis produced a widespread general mortality of a very high order. Australia's wool clip reached a record figure in 1953; and, after due allowance had been made for other contributing factors, it was estimated that the reduced competition from rabbits had resulted in the growth of 70,000,000 lb. of additional wool, valued on current prices at £24,000,000, to which should be added some £10,000,000 representing the value of the additional sheep and lambs slaughtered and on hand (Reid, 1953). No attempt was made to assess the benefit of the reduction in the rabbit population on beef, dairy and crop production.

Rabbits had been at high density in south-eastern Australia for at least half a dozen years before 1953; and the effect of the catastrophic mortality was often immediately obvious to a traveller who had a detailed familiarity with the landscape. Ranges of hills that had been kept in a semi-denuded state by the pressure of rabbit grazing became grass-covered and green almost overnight. This broad picture relates to the sudden reduction of an exceptionally high rabbit population. Perhaps a more significant illustration, because it stems from experience over an extended period, is provided by the record of a 10,000 acre grazing property in eastern central New South Wales. The owner took the extreme step of eradicating rabbits from his land, and keeping it rabbit-free. He was able to compare his production before and after the clearance, against that of his neighbours who took what might be regarded as normal measures to reduce rabbit numbers. He found that complete freedom from rabbits doubled the number of sheep that he could safely carry.

Figures such as those quoted above tell but half of the story. A full appreciation of the significance of the rabbit in the rural economy can only be achieved through an intimate acquaintance with some particular district, and a piecing together of the personal histories of the various local landholders through the vicissitudes of changing commodity prices and seasons. Some of us acquired knowledge of this kind in the course of a decade's field work on myxomatosis, and our conclusions can be summed up as follows. The efficient manager with adequate financial backing will cope with rabbits effectively; but control is achieved at the cost of considerable time and energy, and too much of what should be profit is spent in the process. The not quite so efficient and fortunate is likely to be reduced to marginal prosperity, and will be much more sensitive, financially, to the effects of periodic droughts or falls in prices. The type of landholder most seriously affected is the one who, even without the rabbit, would only just be successful. The extra demand on his energy, organizing ability and pocket will almost inevitably tip the scales against him, and he is likely to find himself continually in debt. In this connexion it might be mentioned that one of the most important by-products of the success of myxomatosis in the years immediately following its liberation in Australia was a substantial overall reduction in rural debt.

CONTROL—POLICY AND TECHNIQUES

In the hundred years since the rabbit first raised its head as a serious pest in Australia there has naturally been what might be termed evolutionary change in both the policy and the techniques of dealing with the problem. The Australian Commonwealth is a Federation of six States which retain very substantial sovereign rights. Except on Commonwealth land, such as the Australian Capital Territory, the federal authorities have no legal rights or responsibilities in connexion with the rabbit and its control (though the Commonwealth Government has responded, through C.S.I.R.O., to the need for research in this field). The shaping and administration of regulations dealing with the rabbit remain the concern of State Governments. While the relevant legislation and the administrative machinery that derives from it vary from State to State in several minor and a few major respects, there is a tendency as one might expect for all to share certain essential features. Thus in every State rabbits are 'declared vermin' and the owner or occupier of the land has the responsibility, under law, to deal with them effectively—'exterminate' is the term that usually appears in the acts.

The only really important differences between the policies of the several States relate to what might be called the localization of the authority enforcing the vermin regulations, and the degree to which the administrative authority is backed, politically and financially. In some States (e.g. Tasmania and Victoria) the authority is centralized, and the vermin regulations are administered by a branch of either the Department of Agriculture or of Lands. In one or two States the authority is vested in district or regional bodies (e.g. the Pastures Protection Boards in New South Wales) which have rabbit control as one of many and diverse responsibilities. On the whole, and in practice, a centralized form of authority seems more likely to be effective. (Those States which have switched to it, after experience with local-body administration, insist that the change was not only desirable but necessary.) For one thing, it makes it easier to build up a team of technically trained men to advise and plan. It was those States which were best organized in this respect that contributed most to the effort in exploiting myxomatosis and the acquisition of information on its performance in the field.

The marked difference from State to State in the degree to which the attack on the rabbit has received governmental support can be attributed to a number of factors—the relative extent of low-value land on which rabbit control appeared virtually hopeless, and the strength of the vested interests in the commercial exploitation of the rabbit are two that might be mentioned. The most important factor of all, however, seems to have been the chance occurrence in key positions of individuals who after gaining a first-hand appreciation of the cost of the rabbit to their State developed an unshakable conviction that something should and could be done about it. These men have broken through the understandable reluctance in political circles to become involved in a problem that was ill defined and seemed to call for action that would be unpopular as well as costly. Although there have been numerous examples of local inspectors who have managed to achieve a significant improvement in the rabbit position throughout a whole district by their enthusiasm and their understanding approach to landholders, by and large officialdom has had to be content, until recently, with a policeman's role—prosecuting or threatening to prosecute property owners with rabbit-infested land, and too often seeing court actions fail or result in little better than a token effort.

The individual most affected by the rabbit, of course, is the landholder himself; and a good manager rarely needs any encouragement from the local inspector to take effective action against the pest. As the rabbit is mobile in addition to being a rapid reproducer, the landholder who takes his rabbit problem seriously will usually go to the trouble and expense of netting his boundary, to prevent his efforts being neutralized by the infiltration of rabbits from his neighbours' land (see Pl. X, fig. 1). When the boundary of a property has been rabbit-netted, if the owner still has money and energy to spare he will consider the next logical step—that of ridding his block of the last

29

rabbit and maintaining the boundary as a complete barrier to re-infestation.

In this way it came about that much of the grazing country, at any rate in south-eastern Australia, developed a reticulation of rabbit-netted fences that were maintained with varying degrees of efficiency. Much of this fencing was erected in the period between two world wars; and where a serious attempt was made to eradicate rabbits from the blocks so protected the job was often done by simple and primitive methods, that is the organized use of dog packs backed by the destruction of cover and the digging out of warrens. It is difficult to estimate what proportion of properties that had gone to the expense of boundary netting followed this up with a complete clearance of the rabbit; but it is possible that in some districts up to one-quarter of the netted blocks were freed of rabbits during certain periods. When rabbits increased so startlingly in the years immediately after the Second World War, they succeeded in infesting some properties that had been rabbit free for nearly a quarter of a century.

Although more than mere killing is involved in effective rabbit control, killing is an essential part of the process. Various methods are employed, in addition to the simple ones just mentioned, the most important being poisoning, fumigation and trapping. Australians have always used poisons with a wholesale abandon that horrifies visitors from other countries; and with the very effective poison, sodium fluoracetate ('1080'), coming into use since the war, and improved methods of application (including the laying of poison baits from aircraft) poisoning has come to be the main technique of direct control employed. Fumigants remain a valuable part of the armoury of weapons used against the rabbit in Australia, but they are normally effective only when employed in conjunction with other methods. Trapping can be said to have fallen into disrepute; for although it can sometimes be exploited as part of a genuine effort in control, the main use of traps is to obtain skins and carcasses, and thus they have become associated with commercial exploitation, with its recognized tendency to conserve rather than control the rabbit.

One other technique for dealing with rabbit infestations deserves mention. The burrow ripper was developed in the years immediately following the last war: it consists of one or more tines, usually linked to the hydraulic lift of a tractor, and it enables warrens to be torn up to a depth of about two feet. This piece of equipment has rendered the old laborious 'digging out' virtually obsolete; and it is very effective. Thousands of rippers were involved in the drive against rabbits in the late 1940's and early 1950's; and there is no longer any valid excuse for tolerating active warren colonies on open land.

CONTROL—PLANNING AND POLICY CHANGES

A number of weapons are thus available to the landholder faced with a rabbit problem; but on any but a small property their effective use, in various combinations, calls for biologically intelligent planning. When control on a district or regional basis is to be undertaken, fore-sighted planning becomes even more important. It was primarily the recognition of this fact that led to the change in official policy which took place about a dozen years ago. The new approach was first adopted in Tasmania (where myxomatosis took several years to show any significant effect), with Western Australia and Victoria almost immediately following suit.

The new official approach is activated by the conviction that, under most conditions, the rabbit can be brought under effective control with the techniques available; but it recognizes the difficulties and the economic strain involved in the process, and thus the need for technical advice and assistance to landholders. Where the 'new deal' is in operation, the local inspectors have assumed a broader and more positive role than that of mere policemen: they help landholders to plan their control efforts, arrange the loan or rental of equipment, and can sometimes carry out a job, for example of poisoning, on a contract basis. (Even though it may be played down, the coercive element in official policy has to remain, of course, to deal with unco-operative landholders whose attitude can seriously disrupt the progress of a regional control campaign.) The planning and co-ordinating role of the official machine is best exemplified in the exploitation of aircraft for poisoning, which has been a feature of the control programme in the State of Victoria. This has involved the zoning of aerial contractors, the mass purchase of bait material, the organizing of landholders into groups to ensure the most economic use of flying time, and a check on the efficiency of the operations.

In combination with myxomatosis (the effects of which will be discussed in later chapters) the more positive and helpful official policies have made a very significant impact on the rabbit problem, brushing aside, as it were, the power of the vested interests in the commercialization of the animal where these had obstructed the development of effective control, as they did in, say, Tasmania. It would be wrong to suggest, however, that a final solution of the problem has been found, or is even in sight. The rabbit is a most formidable enemy, because no victory over it, short of total eradication, can ever be permanent. (Even the property owner who completely clears his land of rabbits has to spend time and trouble patrolling and repairing his boundary fence, and dealing with infiltrations.) Action against rabbits involves the outlay of hard cash; and landholders in difficult country and in difficult times cannot always produce it, at any rate in sufficient quantity to ensure a worthwhile achievement. It would also be wrong to give the impression that the new official approach has been adopted by all the Australian States; and even in those which have adopted it, the vermin-control authorities sometimes have to struggle with inadequate funds and staff. It is the constructive and stimulating side of the work that tends to suffer when there are not enough men to go round, for inspection and enforcement have to continue if a serious deterioration in the situation is to be prevented.

Before passing on from relations between the rabbit and man to those aspects of the animal's ecology and behaviour that affect its relations with the myxoma virus, the question of 'decommercialization' should perhaps be mentioned. The authorities in New Zealand decided that the prospect of dealing successfully with the rabbit as a pest would always be endangered while there was a vested interest in the sale of its meat and fur. By a series of planned steps the value of the rabbit as an article of trade was first reduced and ultimately obliterated: it is now illegal to sell rabbits or rabbit products, or to offer them for sale. New Zealand used to have a lucrative export trade in rabbit skins and carcasses, and this was deliberately sacrificed in furtherance of an objective which people were convinced was of greater value to the country as a whole.

Decommercialization never emerged from the discussion stage in Australia. The federal structure would make the adoption of the policy much more difficult there than it was in New Zealand; moreover it is doubtful whether any great benefit would result from decommercialization until rabbit control throughout the country was better organized and better backed than it is at present. It is not inconceivable, however, that something comparable with the New Zealand policy might have to be seriously considered at some time in the future.

CHAPTER 4

RABBIT BIOLOGY AND BEHAVIOUR

Although it might be said that every aspect of the rabbit's ecology had some bearing, albeit slight and indirect, on the relations between the animal and the myxoma virus, it is quite obvious that there are certain features of its biology and behaviour that are directly relevant to the likelihood of infection, to the recovery of the population from outbreaks of myxomatosis, and to the performance and prospects of the virus as an agent of control. Among these are (a) the daily time-table of activity; (b) the dispersion of the rabbit population in relation to surface water; (c) the timing and length of the breeding season, and its relation to outbreaks of the disease; (d) the reproductive potential and rate of population turnover under different climatic conditions; (e) the frequency and extent of movements; and (f) social behaviour, which to some extent channels and limits contact between individuals.

Precise and adequate ecological information on the rabbit has been very slow to accumulate, and even today it is far from complete. For a number of reasons direct observation of wild populations has proved of limited value. No practicable way has been discovered of catching rabbits in large numbers, alive and undamaged, without destroying their burrow system or otherwise disrupting their normal routine.[1] (Live-catching is an essential prerequisite to individual marking and recognition upon which, in turn, many crucial ecological data depend.) Most of the worthwhile ecological information obtained in recent years has been derived from the analysis of large, random dead samples taken periodically from wild populations in different parts of the country, and from the observation of colonies maintained in a semi-natural state in large enclosures, where the animals have

---

[1] Knotted snares and hand-netting by spotlight are now being used successfully in Britain and Australia respectively. Limitations to both techniques are set, however, by topography and the nature of the local vegetation, particularly the ground cover.

become accustomed to carry on their activities under artificial illumination. Enclosure studies were carried out by C.S.I.R.O. concurrently at Canberra and about 150 miles south-west, near Albury.

Although individual investigators, such as H. N. Southern and R. M. Lockley in Britain, have made important contributions to our knowledge of the wild rabbit, the major advances can be said to date from the time when the wildlife workers in the C.S.I.R.O. (and of the sister organization in New Zealand, the D.S.I.R.) started field investigations backed by adequate facilities and personnel. The interpretation of data obtained from large dead samples was pioneered by Brambell (1944), working in north Wales on material supplied at monthly intervals by commercial trappers. His techniques, improved by the development of more precise methods of age determination, were adopted in New Zealand and are still being employed by C.S.I.R.O. workers in Australia.

BURROWING AND ABOVE-GROUND ACTIVITY

It is rather surprising that no leporid other than *Oryctolagus* has developed the burrowing habit, for it has such obvious survival value under extreme conditions. Over much of southern Australia the middle and late summer tends to be hot and dry, and this applies for most of the year in the desert and semi-desert regions. A hot dry climate is normally characterized by wide daily fluctuations in temperature and relative humidity which are, of course, negatively correlated. It is self-evident that the rabbit, by going underground when the sun is up, manages to avoid most of the direct stress that such a climate imposes; and it is noticeable that in the hotter and drier areas rabbits postpone their emergence above ground until just before darkness sets in.

Hayward (1961) has assessed with some precision the physiological advantages of a burrow harbour. His work was carried out near Albury in the eastern Riverina, during a summer that was hotter and drier than the average. He found that, allowing for a lag of a few days in following long-term changes, the temperature within rabbit burrows remained remarkably steady, fluctuating over only 2 or 3° about the ruling mean, and that the relative humidity of the burrow atmosphere never dropped below virtual saturation. When the atmospheric maximum reached 104° F., with a night minimum of about 52° F., the temperature within a burrow remained at about 77° F. Thus warren-dwelling rabbits need never be exposed to temperatures in the upper half of the diurnal range, or to excessively low atmospheric humidities. Hayward calculated that, under the conditions current at the time of his experiments, a rabbit underground would lose 40% less water in respiration than if it were breathing the outside air.

It is only when and where conditions are severe that the rabbit is, one might say, forced to find harbour in burrows. Although there is not much precise information on this point, it is well known that in almost any infested area some rabbits will be found above ground during the daytime, lying up in sheltered 'squats' or in hollow logs. In parts of the Tasmanian highlands, where the soil overlying basic rock is very shallow, most of the rabbits appear to be permanently surface living, even the litters sometimes being produced in depressions under low bushes. H. V. Thompson (personal communication) has reported that surface living was noticeably more prevalent among rabbits in Britain after the population was reduced to low numbers by myxomatosis. Such a change in habits does not seem to have taken place in Australia; indeed in Victoria, according to Douglas, the proportion of surface squatters seems to have declined noticeably. All in all it would probably be safe to say that periods of surface living, mostly of short duration, occur in the lives of a high percentage of wild rabbits, and that they are most likely to be indulged in at change-over times between the breeding and non-breeding season, or when new foraging areas are being explored, and probably mostly by individuals that are insecure in the social structure. A surface-dwelling rabbit seems to have one or two chosen squats to which it returns with some regularity.

Under temperate conditions an established rabbit population usually becomes active above ground in the late afternoon, an hour or so before dusk; and most activity ceases an hour or so after sunrise. Although to the casual observer it may appear that almost every rabbit in the local population is moving and feeding by the time darkness descends—and sometimes this may be the case (Myers, 1957)—observations on colonies comprising known numbers of individuals, such as those in the C.S.I.R.O. enclosures, indicate that there is a good deal of coming and going between the surface and the warrens even during the night, and that the main afternoon or evening emergence may involve little more than half the total population. Young rabbits, in particular, are more erratic in their routine than are the adults.

Oryctolagus cuniculus is both crepuscular and nocturnal in its habits; although, as we have seen, hot dry conditions may impose a strictly nocturnal rhythm upon it. The extension of its activities into the hours of daylight seems to be encouraged by a feeling of security. Populations that have remained unmolested for some time tend to begin their activities well before sundown; and conversely one that has recently been trapped or dogged[1] will be very hard to observe, because the survivors seem reluctant to show themselves while there is light.

The effect of the absence of disturbance on the activity timetable was clearly demonstrated by the behaviour of an experimental colony of wild rabbits established by the C.S.I.R.O. Division of Animal Health in the early years of the work on myxomatosis. The colony was maintained for over eighteen months in the grounds of the Division's headquarters in Melbourne; and it remained undisturbed, except for the introduction of food and the search for corpses (which was carried out when the rabbits were in their burrows) and the twice-weekly introduction of a few freshly caught animals to balance the disease mortality. It was noted (M. W. Mules, personal communication) that the emergence above ground became progressively earlier, advancing by about two hours during the period of the colony's existence. The initial rhythm was presumably that ruling in the well-trapped Bacchus Marsh area, from which all the rabbits were obtained; and the fresh introductions must have adapted themselves almost at once to the current convention within the enclosure.

It became obvious during the course of field observations on myxomatosis that the survivors of a rabbit population which had passed

---

[1] The dog packs which used to be maintained on all big properties in the rabbit-infested zone deserve special mention. With the subdivision of large holdings, and the recent increase in the popularity of poisoning with sodium fluoracetate, the rabbiting packs have almost disappeared from the land. Their value was, and in many places still could be, immense. The efficacy of fumigation and warren 'ripping' is greatly increased if the area to be treated is first dogged to drive all the rabbits underground. More important, when a landholder decides to eradicate rabbits completely from an area that has been boundary-netted, there is really no satisfactory substitute for dogs when it comes to 'getting the last rabbit'.

through a sharp epizootic tended for a time to be more cryptic in their habits, presumably postponing their emergence above ground. This was brought to our notice when observing the effect of the first outbreak of myxomatosis in southern Queensland, in 1951, when the estimate of the local mortality, arrived at just after the cessation of disease activity, had to be reduced from over 90 % to nearer 50 % as a result of a check made a month or two later (Ratcliffe, Myers, Fennessy & Calaby, 1952). Mykytowycz (1961) noted a marked change in the emergence behaviour of the rabbits in his enclosure after an outbreak of myxomatosis. How strong or general is this reaction to the disturbing effect of an intense epizootic it is impossible to say, however, as repeated observations and counts have been recorded but rarely. An estimate of mortality, exaggerated because it was made too soon after an outbreak, will of course lead in turn to an exaggerated estimate of the subsequent recovery in numbers by breeding.

It is clear that the periods of the rabbits' above-ground activity coincide with, or at any rate substantially overlap, those of the great majority of myxomatosis vectors and predators, which are predominantly crepuscular and nocturnal. Even the so-called day-biting species of mosquitoes usually tend to be most aggressive towards dusk. The rabbit's timetable affords it some protection from diurnal birds of prey; but these concentrate on the young, which tend to emerge earlier and return to their burrows later than the adults.

RABBITS AND WATER

The water relations and requirements of *O. cuniculus* are probably of greater ecological importance in Australia than in any other country colonized by the species, for aridity is a seasonal or permanent feature of so much of its Australian habitat. In the winter-rainfall areas, and of course throughout the low-rainfall regions of the inland, the pasture plants dry off soon after the beginning of a hot dry spell or season. As we have seen, the burrowing habit protects the rabbit from the direct effects of high day temperatures and atmospheric aridity; but it still requires water for the effective utilization of dry feed and for temperature regulation. Rabbits start drinking regularly, if they can, when the pastures dry off in hot weather; and they continue to do so until there is enough rain to stimulate the production of succulent new growth or the germination of seedlings. The fact that rabbits seek out and regularly visit water when it is available during the summer months tends to increase their accessibility to water-breeding blood-sucking insects.

The European rabbit lacks the special physiological adaptations possessed by some desert-living mammals; but even without them it fares remarkably well under the stress of water restriction. Hayward (1961) studied the performance of rabbits held in enclosures without water at the same time as he was investigating burrow microclimates, that is in an unusually hot and dry summer. The major mortality in his experimental colony did not occur until about two months after the onset of hot weather and the complete drying up of the pastures. The large adult rabbits were much more successful in withstanding the testing conditions than smaller and younger individuals, surviving a weight loss that ranged up to 50 %, much of which was rapidly regained when drinking water was provided. There was a very wide variation in the performance of individuals within the experimental group, one male rabbit surviving the regime, and being in good condition nearly three months after the last effective fall of rain. This suggests that rigorous selection might produce a local population that would surpass Hayward's experimental colony in average performance, which was impressive enough.

Despite the resilience revealed by Hayward's experiment (in rabbits, incidentally, drawn from a district where climatic conditions are not extreme) shortage of water undoubtedly constitutes the most important mortality factor over most of inland Australia; and generally speaking the main natural pruning-down of the rabbit population occurs during the summer—remembering, however, the high mortality of kittens during the breeding season itself.

BREEDING SEASON

What might be termed the ancestral reproductive timetable of *O. cuniculus* was presumably evolved as an adaptation to the seasonal availability of suitable food in a Mediterranean climate. There is a strong tendency for animals to produce their young when food of the right kind is most readily available; and although it has not yet been experimentally proved, there is good evidence for believing that some constituent of freshly growing plants provides a necessary stimulus for reproduction in the wild rabbit. In countries where the summers are normally dry and hot and the rainfall has a predominantly winter distribution, as in southern Australia, the main flushes of plant growth occur in the autumn and in the spring. The summer is typically the non-breeding season for the rabbit, with the does in anoestrus and the testes of the bucks small and withdrawn. While breeding tends to tail off rather unevenly as summer advances, it usually starts fairly abruptly in the autumn or early winter, in response to the generally rather sudden appearance of new growth.

The percentage of does that become pregnant in the autumn, when breeding starts in Australia, is usually less than at the spring peak; and mid-winter conditions (at any rate in the cooler parts of the continent) depress reproductive activity to some extent. Thus the graph of breeding tends to be bimodal, with a small peak in the autumn and a large one in the spring.

The commencement of breeding, and its cessation, will depend very much on seasonal conditions. Any spell of very hot weather that results in a drying-off of the pasture will bring breeding to a stop: lactating does will cease producing milk, and suckling young will die. Females of the wild European rabbit apparently cannot lactate on dry food; and thus it is normal in Australia for the last-born litters of the season, irrespective of the level of population density and the apparent health of the mother, to lose weight and starve, owing to a cessation of the milk (Myers, 1958; Mykytowycz, 1960). This physiological reaction on the part of the female rabbit contrasts with that of, for example, the cow, which will utilize its own tissues to provide milk for its young (Lines, 1952). It possibly has an adaptive value in that it would tend to preserve the breeding stock at the expense of young animals which would almost certainly succumb anyway, for the ability of kittens to survive on dry feed is markedly less than that of adult rabbits.

The timetable of reproduction outlined above applies in particular to south-eastern Australia, where the C.S.I.R.O. studies have been undertaken. In Britain and New Zealand, where parallel investigations have been carried out, for example by Brambell (1944) and Watson (1957), the seasonal timetable is a little different, almost

40

certainly as a result of the less well-marked seasonal changes in the nature of the food supply. The main difference, as compared with the Australian picture, is the absence of autumn activity, the more gradual decrease in the pregnancy rate as summer advances (with the late litters successfully weaned), and the tendency of at any rate a few females to become pregnant at every season of the year.

Within the Australian continent the rabbit has become well established in regions that do not have a rainfall distribution of the Mediterranean type: thus it has penetrated into southern Queensland, which has a predominantly summer rainfall, and into desert and semi-desert areas where precipitation is quite erratic. The sampling of populations in these regions is being undertaken at the present time; and the findings to date (K. Myers, personal communication) indicate that breeding is governed by the incidence of growth-producing rains, the reproductive response to such rains being depressed if they occur during the summer. The incomplete response on the part of local rabbit populations to summer rain, even when the vegetational response is adequate, is possibly linked with the cycle of male fertility (Myers & Poole, 1962). Outside the regions having a reliable rainfall with a Mediterranean incidence the breeding season is likely to be short, permitting fewer litters to be dropped than in the naturally more favourable zone. In the arid areas even the weaning of one litter per female may be difficult in many years.

REPRODUCTION: PRE-NATAL MORTALITY

Ovulation in the rabbit is induced, normally by coition. During the breeding season does exhibit oestrous behaviour at intervals of 7 days, or multiples thereof (Myers & Poole, 1962). Of predominant importance is the post-partum oestrus, at which the majority of conceptions occur. Bucks wait for the does at the mouth of the nesting burrows and mate with them immediately they emerge after dropping their litters. It is thus quite normal for a doe to be both pregnant and lactating throughout most of the breeding season.

The period of gestation ranges from 28 to 33 days, mostly falling between 29 and 32 days, with a mean of just over 30 days (Myers & Poole, 1962). The nesting burrow is either a blind diverticulum of an established warren, or a separate, shallow 'stop' (to use the British term) which the doe can, and sometimes does, excavate in the course of a single night. The litter is dropped in a nest of dry grass, lined with fur plucked by the mother from her abdomen. The doe does not remain with her young, but visits the nestlings during the night to feed them—often only once, and for a surprisingly short time—generally blocking the entrance of the nesting burrow with tamped-down soil when she leaves. The young normally make their first appearance above ground at the beginning of the fourth week of their existence; and during that week they have to learn to feed themselves if their dam conceived after their birth.

Under reasonably favourable conditions, all young does will be ready to breed at six to ten months, when they will have attained a body weight of about 1200 g.[1]: occasional individuals breed at four months, or even younger. Thus females from the earliest litters may produce litters of their own before the breeding season closes. Males take somewhat longer to attain sexual maturity—about nine months.

The number of young per litter increases during the course of the breeding season, and also with the age of the doe. The great majority of litters contain three to seven young. The mean litter of a large sample (or population over a whole season) will usually fall between 3·75 and 5·25. Local populations sometimes show differences in mean litter size that appear to be significant, although the ecological explanation of these differences is rarely apparent.

Brambell (1942, 1944) was the first to draw attention to the importance of pre-natal mortality in the wild rabbit, and to attempt an assessment of its incidence. In addition to the death and resorption of one or two embryos—that is partial litter loss—he showed that there was a tendency for the resorption of litters *in toto*, and that this seemed to occur mainly at or just before mid-term. Because of the nature of his material, Brambell could only assess the incidence of total litter loss indirectly; but he was able to show this type of pre-natal mortality to be highly significant, calculating that more than half the litters conceived were resorbed *in utero*.

The loss of litters as a result of intra-uterine mortality was investigated in detail in the course of the C.S.I.R.O. enclosure studies

---

[1] The normal range of (unpaunched) body weights of adult *Oryctolagus* in good condition is 1500–2200 g. During the dry season food shortage substantial loss in body weight can be expected.

(Mykytowycz, 1960; Myers & Poole, 1962). As their experimental animals were alive, with pregnancy in the does revealed and its approximate stage determined by palpation at the monthly censuses, the C.S.I.R.O. workers were in a position to get much more precise information on this phenomenon. They confirmed the fact that the incidence of litter resorption could be very high at times, and showed that it tended to reach a peak during mid-winter—providing a partial explanation for the decreased drop of young at this season. Attempts to correlate the incidence with the social status of the does, and with environmental stress, were on the whole inconclusive. All that it is safe to say is that pre-natal loss of litters is one of the factors (among several, mostly as yet not clearly defined) which combine to depress breeding success as population density, especially within warrens, builds up.

The effect of population density on reproductive success is very marked. The Canberra enclosure was provided with two burrow systems. During the first season of the study, one of these (North Warren) was maintained as the preserve of the dominant doe, which permitted no other females (except her own offspring) to use it. In 1957, during the seven-months breeding season, the dominant doe reared seven litters, and one of her daughters one litter, in the North Warren. In 1958, when the breeding season extended over nine months and ten does permanently inhabited the North Warren, a total of only ten litters were successfully reared (Mykytowycz, 1960).

RATE OF POPULATION INCREASE

It is very difficult to suggest a figure for the maximum increase, over one year or breeding season, likely to occur in a rabbit population after numbers had been reduced (e.g. by an epizootic or control action) to a level at which intra-uterine and pre-adult mortality attributable to high-density stress was unimportant. The performance of the most successful breeders in experimental colonies under favourable conditions does not provide a safe basis for calculation: thus a doe in one of the Canberra enclosures produced seven litters between June and the end of December, 1957, with an average litter size of 6·5, giving a total of 45 young (Mykytowycz, 1959a).

The data from the enclosure studies carried out at Albury (Myers

43

& Poole, 1962) probably provide the most useful guide, and indicate that in a favourable breeding season, with sexual activity starting in March or April and the last litters dropped at the end of November or early in December, an average doe might produce and wean just over twenty-four kittens. This agrees closely with Watson's (1957) figure derived from his analysis of samples from Hawke's Bay in New Zealand, and is equivalent to a 12-fold population increase—assuming a 50:50 sex ratio and disregarding adult mortality.

This figure, as it stands, is virtually meaningless. In the first place, as has already been pointed out, in many parts of Australia at any rate, the breeding season would be significantly shorter than the one postulated. Then there is the question of mortality: the mortality among kittens will be more important in the calculation than that of adults. Kitten mortality is usually very high, particularly in late litters; and the mortality attributable to predators will be proportionally higher in low-density populations (see below). Taking these two points into consideration, it would seem safe to say that an eight-fold population increase was all that one could reasonably expect in a year, and that most populations would be doing very well to achieve a five-fold increase. Even the latter, however, would permit the rabbits in an area to recover from an 80 % mortality within twelve months.

POPULATION TURNOVER: MORTALITY FACTORS

The potential life span of the rabbit is fully ten years; and there are one or two known instances of wild rabbits in Australia having survived as long as that. Trappers are apt to talk of four-, five- and six-year old rabbits as if they were commonplace, without of course being able to advance good evidence to support their estimates. Southern (1940), whose warren colony study in England was based on observation of individually marked animals, was the first to suggest that 'the life span of the rabbit in nature is probably quite short, the majority of them living only about a year.'

Unpublished observations made by Myers (before myxomatosis had become widely established) confirmed Southern's conclusions. In the Murray Valley and eastern Riverina he found that individuals more than one year old formed an insignificant proportion of the local population. More recently, he informed us that population

44

sampling further afield—currently in progress, as has been mentioned earlier—has revealed that in regions less favourable for rabbit breeding, such as southern Queensland and far-western New South Wales, individuals of up to and over two years old constitute quite an important fraction of the population.

Comparable findings have been reported (Bull, 1956) from New Zealand, old individuals being much more numerous in the rabbit populations of the sub-antarctic Auckland Islands than in mainland samples. Bull attributes this to the absence of regular control on the islands. However, nothing more than a difference in the reproductive rate (caused in these cases by differences in the effective length of the breeding season) need be postulated to explain the difference in age structure, if there is no suggestion of differences in the incidence of age-specific mortality factors.

Mortality from myxomatosis, is, of course, highly selective, but in respect of recovered and susceptible individuals rather than age groups. Thus frequent epizootics, particularly if associated with attenuated virus strains and a high infection rate, will result in an increase in the number of immunes in the local population. Should conditions (perhaps assisted by a decreased population pressure) permit the survivors to live longer than normal, the percentage of immunes in the population could well become very substantial.

Mass mortalities of rabbits occur from time to time as a result of what might be termed environmental catastrophes, such as droughts and flooding; rabbits, especially kittens and nestlings, are particularly susceptible to waterlogging of the soil. Drought, of course, is a normal and recurrent feature of much of the rabbit's habitat in Australia, and as has already been mentioned shortage of water is the most important mortality factor in the arid zone, not only being the cause of the low population density (in all but exceptional years), but also the main determinant of the pattern of the animals' distribution.

Although it has been the least successfully studied, and is the most difficult to discuss with confidence and precision, undoubtedly the mortality factor of the greatest importance ecologically is predation. Australia may be popularly regarded as a land poorly supplied with predatory animals; but from the rabbit's point of view it is very well stocked with efficient enemies. In approximate order of importance these comprise the introduced European fox, feral cats, diurnal and nocturnal birds of prey, with the domestic dog, and of course man himself, entering the picture. The present-day restricted range of the dingo does not overlap that of the rabbit sufficiently (at any rate in country where the latter is economically important) for this species to call for comment.

It would be true to say that the Australian ecologists' assessment of the significance of predation has undergone a radical change since they began making observations in the pre-myxomatosis period of 1949–50. At that time heavy concentrations of predators occurred in the areas of high population density, where hunting was easy, but they failed to effect a significant reduction in the numbers of young. The rabbit population indeed was only prevented from eating itself out by the application of intensive control measures.

When myxomatosis, backed by organized control, had achieved a great and general reduction in rabbit numbers, a different picture began to emerge. Rabbits in the C.S.I.R.O. enclosures at Canberra and Albury (which received no supplementary feed) invariably bred up rapidly to densities that would be abnormally high in nature, while the population of the surrounding grazing country remained low, hardly ever attaining a level that necessitated or stimulated control action. The fences did not protect the experimental colonies from myxomatosis or flooding; and mortality from both causes occurred within the enclosures. The rabbits inside them were, however, protected from predation—completely from mammalian predators, and substantially from avian ones. The conclusion seemed inescapable that the contrasting performance of the populations inside and outside the enclosure fences was due, in the main, to predation.

During this period of generally low to very low rabbit numbers, in addition to the enclosure studies, natural populations were under observation at times in various parts of south-eastern Australia, for example in the Northern Tableland of New South Wales and in valleys of the southern coastal ranges. The field observations failed to reveal any regular substantial recruitment of young animals into the population. On one occasion (A. L. Dyce, personal communication) when a few groups of weanlings appeared above ground, their numbers were actually observed to fall away virtually to zero within a week or two. On another occasion, sampling provided evidence that successful conceptions were occurring although results in the form of weaned young were not seen at all. Once again it seemed that predation must have been an important factor in the observed state of affairs, if not the most important.

The situation is probably straightforward enough—in fact, the logical outcome of the interrelationship between the rabbit and a group of predators all of which are (a) non-specific, (b) dependent to a limited extent only on rabbit density for the maintenance of their own status, (c) largely uninterested in adult rabbits, but (d) prone to seek out kittens when they are available. A predator-prey relationship of this kind can be expected to result in an intensity of predation that varies inversely with prey density; that is to say, within limits and reason the percentage kill of kittens will increase as rabbit density decreases.

Predation on the rabbit in Australia may be likened to a poor handbrake on a car, which will hold the vehicle on a gentle slope, but becomes less and less effective after the car starts to move and as it gathers momentum. Predation will have the effect of slowing down the rate of population increase at low-density levels, perhaps even preventing any significant increase for a year or two. However, once the rabbits achieve a break-through, so to speak, predation is likely to become unimportant.

Both foxes and feral cats, the most important and efficient rabbit predators, are much more abundant in Australia than most people realize. The fox's interest in young rabbits starts before they emerge above ground, for it is adept at digging out breeding burrows, after apparently pin-pointing the position of the nest by ear. Unfortunately perhaps, foxes have earned the enmity of sheep-men because of their occasional attacks on lambs. Even when no specific action is taken against them, poisoning with sodium fluoracetate (which has become the most widely used method of rabbit control) almost automatically results in a substantial reduction in local fox numbers. In common with other canids, foxes are very susceptible to '1080' and are invariably killed when they come to feed on poisoned rabbit carcasses.

47

SOCIAL BEHAVIOUR

Passing reference has already been made to social status in the rabbit; and it is desirable, particularly before discussing movements, to outline current knowledge of social behaviour and organization among wild *Oryctolagus*. In the course of his observations on a warren colony, mentioned earlier in this chapter, Southern (1948) threw useful light on social behaviour, and pointed the way to the intensive enclosure studies initiated by C.S.I.R.O. workers in 1957 (see Myers & Mykytowycz, 1958; Mykytowycz, 1958a, 1959a, 1960; Myers & Poole, 1959, 1961, 1962).

In a way, the acquisition of information on the rabbit's social organization has been rather unbalanced. The behaviour of experimental colonies living and breeding in enclosures has been recorded in great detail; but very few observations have so far been made on natural populations to determine the extent to which the phenomena revealed in the enclosures are manifest in populations whose movements are not physically restricted. While one must expect that some relationships are intensified by confinement (even though the study enclosures were up to two acres in area) there can be little doubt that the pattern of social behaviour seen in the experimental colonies was a reflection of inborn traits which would also govern the actions of individuals in a normal wild population. In this connexion, it may be recalled that Southern's (1948) observations were made on a natural warren, the inhabitants of which were free to move in and out through gaps in the encircling fence; and his findings were fully compatible with those of the C.S.I.R.O. workers. Furthermore Myers has informed us (personal communication) that the inhabitants of three small warrens which he kept under observation in South Australia, and which were not subject to human manipulation of any kind, were socially organized in exactly the same way as the rabbits he had studied in enclosures.

On the approach of the breeding season, rabbits (practically all which would normally be adult, or nearly so) form themselves into social groups. The groups comprise any number of animals up to a maximum of about seven, that is three or four does and two or three bucks. Each group establishes a territory; trespassing on this by members of other groups is resented and normally provokes immediate and successful attack on the trespasser. Within the groups' territory each member comes to have a 'home range', a not very clearly defined area to which the individual's main daily activities are restricted. A rabbit's home range overlaps those of other members of its group, and is not defended.

There is no doubt that every rabbit in a social group recognizes the other members individually, coming to accept them and their proximity with somewhat conditioned tolerance. A striking feature of the group structure is the linear hierarchy of dominance developing among the males, by combat which initially at any rate can be serious and even fatal. Once the 'peck order' has been established it is adhered to strictly, and fights between males become rare. The home range of the dominant buck will be virtually co-extensive with the group territory; and he is usually the first to notice and deal with trespass across the territory boundary.

In addition to and supplementing aggressive action, the seal of ownership is placed on the group territory in various ways. Thus urination and defaecation, and particularly rubbing with the chin, are used to mark points on the territory boundary and important features within it. Mykytowycz (1962 a) has shown that the secretion of the chin gland (which is bigger in males than in females, and biggest of all in dominant bucks) has an important territorial function. 'Features such as the edges of posts, tips of grass blades and branches, edges of burrow entrances...and even kittens and does during amatory behaviour, are subjected to "chinning" during which they are smeared with small amounts of the secretion.'

Aggressiveness in females varies greatly between individuals. There is usually no doubt whether a particular doe is the dominant or a subordinate animal; but the females do not seem to establish a linear hierarchy among themselves as do the bucks. They may even form what might be termed tolerant attachments to other does; and their aggressiveness is most often exhibited in the immediate neighbourhood of their breeding burrows. In males as well as females both age and heredity obviously contribute to dominance, but 'local knowledge' is also important. Whatever its sex an alien rabbit which gains acceptance into an established group achieves only subordinate status, unless it comes in as a kitten.

Myers & Poole have produced good evidence that group formation is based on female behaviour: does apparently like to associate fairly closely with others while breeding. A remarkable synchronization of conception and littering was recorded among the does of a group. This is probably attributable to the stimulating effect of the post-partum oestrus of the dominant—almost invariably the oldest—doe which is normally the first to litter at the beginning of a breeding season. The bucks merely distribute themselves between the groups that the does have formed. Their self-imposed hierarchy, in addition to reducing the overall amount of conflict, has a marked effect in limiting paternity. The dominant buck will sire most of the litters conceived during the season, though the fact that the does tend to come into oestrus together will give the second ranking male some opportunity to mate with females of the group. Lower-ranking bucks rarely have a chance of mating, and generally live a rather harassed existence.

One thing emerges clearly from the C.S.I.R.O. observations: the European rabbit is what might be called a positively gregarious species. Every animal attempts, persistently and sometimes desperately, to become accepted into a social group if it is not already a member of one. There is a general tendency for group structure to persist from one breeding season to the next, although changes in group composition are continually taking place. While they are still sexually immature young rabbits are treated in the main with general tolerance, and are even permitted to move between group territories with relative impunity. As they attain sexual maturity the young of the early litters are regarded by the adults with hostility, and a proportion of them are driven away. If these fail to establish themselves in other groups, they will try and establish new groups in untenanted territory. When a group acquires more than six or seven adult members it will usually subdivide into two groups which soon attain independent status. The upper limit to the size of a social group is probably determined by the rabbit's inability to recognize and remember, as individuals, more than about half a dozen of its fellows.

Social grouping is essentially a development of the rabbit's reproductive behaviour. It is the approach of the breeding season, recognized in some way by the animals, that stimulates group formation; and when breeding ceases—in Australia usually as a result of the intervention of dry conditions—the rigidity of the group structure and territorial aggression immediately weaken. It is noteworthy that while competition for receptive females invariably leads to strife, as does the proprietary feeling of a female for its breeding burrow, competition for a dwindling and inadequate food supply does not.

In the enclosure populations each group's territory included both harbour and feeding ground. In the case of a natural unconfined population it is probable that social behaviour and territorial consciousness will develop mainly in relation to the burrow systems and their immediate environs, and that there would normally be a mingling between groups on the feeding grounds, which might be some distance away. This is a point on which field observations are needed.

The significance of social grouping in relation to myxomatosis is probably not great, as the activity of the vector insects would be unaffected by the rabbits' social organization and territory boundaries. However, there do appear to be certain minor ways in which the group structure might affect disease performance. Close physical contact, both sexual and when lying-up during the day, is virtually confined to members of a group. During those periods when contact transmission may become epidemiologically important (e.g. in winter, between epizootics) the pattern of disease spread could be largely determined by the social structure. K. Myers (personal communication) encountered what appeared to be a clear instance of this. It was also strongly indicated during the course of an outbreak of myxomatosis within the Canberra enclosure. At this time Mykytowycz (1960) also found evidence that rabbits infected by myxomatosis straying beyond the boundaries of their own group territories tended to succumb more quickly than those which remained within them, presumably as a result of attacks made upon them as trespassers. In this way a proportion of individuals which might normally have recovered could become casualties.

It should perhaps be mentioned that the recently acquired information on the rabbit's social behaviour has confirmed certain popular beliefs and claims which were formerly regarded almost as delusions. Thus trappers long ago recognized the existence of the dominant buck, which they were likely to refer to as 'king of the warren'. A Rabbit Board official in New Zealand some years ago told one of the authors that he regarded signs of 'colonization' among survivors as evidence of a really effective kill by poisoning. It is now clear that

4-2

colonization, as he termed it, was the deliberate aggregation of survivors, which had been left as well-separated individuals, obeying their instinct to reconstitute social groups.

One aspect of the rabbit's gregariousness deserves special mention, as it happens to have considerable practical significance. It is commonly believed that rabbits on one side of a netted barrier will always try to get through or under it if the food supply on the other side is obviously better than on their own. Observations by C.S.I.R.O. field workers, confirmed by the experience of more than one landholder, have shown that the 'pressure on a fence', to use the term in common parlance, will be reduced to a vanishing point if there are no rabbits at all on the other side. (From the landholder's point of view the 'other side' would of course normally be the inside.) If there are a few rabbits on one side of the fence, and more on the other, the insecure and hungry members of the dense population seem to be attracted to the barrier and will do their best to find a way through it.

MOVEMENTS

Precise information on rabbit movements is virtually unavailable; but from a miscellany of observations and records it is possible to synthesize a picture which is unlikely to be radically modified if and when a comprehensive investigation, based on marked animals, is carried out. For half a century or so after the beginning of its colonizing spread Australians had reason to be impressed by the mobility of the rabbit. Spectacular mass movements were frequently recorded, but these seemed to have become less obvious and frequent since the first decade or so of this century. Recent studies, already discussed, have tended more to emphasize the basic stability of a rabbit population, at any rate in a reasonably favourable habitat.

The bigger and more obvious mass movements are essentially phenomena of the inland, or other regions where rainfall is not only unreliable but tends to be patchy. All the reports known to us relate to areas that have received the moving mobs: no observations seem to have been recorded in country where and at the time when 'migrations' originated. Thus it is impossible to indicate the distances actually covered by these mass movements, although it is popularly believed that they may approach or exceed 100 miles.

We have received one or two first-hand accounts of large scale rabbit movements that have taken place within recent, or fairly recent years. All were migrations out of drought-stricken areas in a direction that would normally lead the animals to better conditions, for example movements eastward across the boundary of the Western Division in central New South Wales, and movements into and across properties lying to the south of the Trans-Continental Railway and the Nullarbor Plain. In general, one can say with confidence that movements of this kind are stimulated in rabbit populations that have built up a high density in an unusually favourable run of seasons by the onset of hard, and more normal, conditions. The ecologist will be curious to know what particular circumstances lead to a mass exodus from an area that has become unfavourable, rather than to a local dying-off of the animals without any concerted attempt to escape. As far as we know the latter happens just as, if not more, frequently. It seems likely that the speed at which the local conditions deteriorate, and whether or not drinking water is available to the rabbits, might be crucial factors: they would certainly affect the physical fitness of the population at the time when the situation became critical. However, the possibility that the chance emergence of 'leaders' endowed with exceptional initiative (such as are now known to occur in several animal species) might be of importance. In this connexion, the observations of our informant on the central New South Wales movements may be significant. On at least one occasion he had the opportunity of watching a moving mob of rabbits that had been held up on a river bank: it was the lead given by one or two individuals which plunged unhesitatingly into the water and swam across to the other side that solved the problem for the migrants.

Related to the mass movements just discussed are the seasonal shifts of the rabbit population between coastal or river-frontage zones and what might be termed hinterland areas, which have become part of the traditional rabbit lore in South Australia and the lower Murray Valley. Knowing that immigrations of rabbits are often postulated by landholders when alternative explanations for the rather sudden appearance of the animals in numbers were much more probable, at one time we tended to discredit this belief. A few observations of directional night movements seem to support it; but perhaps the most telling evidence in its favour is the fact that in the regions concerned trappers, and carcass-collecting trucks that operate in association with them, concentrate their activities in different areas in the summer and winter seasons. The confirmation and elucidation of this local migration in regions where summer drought tends to be acute would be of considerable biological interest.

We come now to the minor, more individual, movements on which the adaptive dispersion of the rabbit population must largely depend. These tend to fall into two categories: (a) movements stimulated by food shortage, and (b) the wanderings of what might be termed social outcasts.

There can be no doubt that a rabbit prefers to carry out its activities in an area which it knows intimately. When it moves away from its normal home range, for the first three weeks or so—the time a rabbit seems to take to become confidently familiar with a new territory— it will almost invariably head for its old haunts, if they are nearby, when disturbed, for example by a dog pack. Rabbits normally forage no further from their daytime harbour than they need to. When shortage of food drives them to search further afield, and they locate an exploitable supply at a distance of more than quarter or half a mile from their burrows, our observations indicate that after a short period during which they return every morning to their old harbours they tend to establish new ones in the area where they have found food.

Observations in the C.S.I.R.O. enclosures (see in particular Myers & Poole, 1961) showed clearly that as soon as the pasture dried off, leading to a cessation of reproductive activity and the disappearance of territorial aggression, rabbits began to move beyond their group territories and to explore the possibility of escape through the fences. Thus one may conclude that the period of summer stress is the one in which most of the movements in search of food will be concentrated. It must be remembered, however, that this is also a period of considerable biological risk, when the availability of drinking water becomes important and the advantages of an underground harbour increase. Altogether, it is very easy to understand why hot dry spells must produce a high incidence of mortality in adults, as well as in the much more susceptible young.

The movement of that fraction of the population which finds itself surplus to the local 'establishment' will occur, in the main, at two times of the year, that is in the pre-breeding period, when the social groups are being re-formed after the reproductive inactivity of the middle and late summer, and again in the late spring and early summer when the maturing survivors of the autumn litters are forced out of the breeding groups by their parent adults. It should be noted that these movements occur at times when seasonal conditions are normally favourable. The wanderings of any particular individual will continue, as long as it escapes fatal accident, until it finds a place to settle down in company with one or more other rabbits, and the total distances covered could be quite considerable. One may assume that a substantial number of rabbits are involved in the twice-yearly search for social security; and the efficiency of the species in discovering and exploiting new territory, such as land freshly cleared and developed for agriculture, and the recolonization of areas in which the rabbit may have been exterminated, must depend largely on the activities of such purposeful wanderers.

The statement made some pages back regarding the lack of precise data on movements has to be qualified in the light of some recently acquired information. For over a decade, the Vermin Branch and the Victorian Lands Department has ear-tagged wild rabbits as opportunity offered. The records of recoveries made over the past half dozen years (G. W. Douglas, *in litt.*) are of considerable interest. Twenty-five individually marked animals have been recaptured since the end of 1956. Of these, sixteen were recovered within a mile of their release points, while two had moved a couple of miles. The remaining seven animals were all marked and released by Douglas in the neighbourhood of Goroke, near the South Australian border just south of the so-called Little Desert. The distances of the recoveries from the release points, in miles, and the time interval between release and recovery (in brackets) were as follows: 5 (2 years), 9 (1 year 9 months), 13 (3 years 10 months), 14 (6 months), 30 (2 years), 60 (2 years) and 85 (2 years).

It is hard to say whether these Victorian records do much more than confirm the correctness of our conclusion that the wanderings of a rabbit which leaves its home ground for some reason or other must often result in displacement over a considerable distance. The fact that the far-travellers were from a dry-country population, with the three really long movements all eastward and into higher-rainfall areas, may perhaps be significant. A point worth mentioning is that ear-tagged rabbits had all been handled, and undoubtedly scared in the process. It is not inconceivable that on their release some of them inadvertently moved beyond the limits of the terrain with which they were familiar, and thus became launched on a spell of footloose wandering. Men who have had long experience in rabbit control are emphatic that human interference, for example by trapping or dogging, usually causes some degree of dispersal from the area of disturbance. Thus the classification of types of rabbit movements which we have attempted should include dispersal stimulated by man's activities, which of course may occur at any season of the year.

To sum up, there are three main types of rabbit movements. The first are local 'migrations', stimulated by desperation, out of areas that have (probably rather suddenly) become unfavourable. These mass movements are confined to the more drought-risky regions of the continent; they probably represent a dead loss to the rabbit population as a whole, as the areas invaded would normally have their full complement of the animals. The second is the seasonal shift that probably occurs into and away from well-watered country in those parts of the continent which have a reliable winter rainfall and a summer that is, equally reliably, a period of drought. If confirmed, such movements—which must cover scores of miles—would obviously be of very great ecological significance and value to the species. Finally, there is the aggregate of individual movements, for the most part socially motivated, which provide the basis for the continual exploration and exploitation of all favourable habitats within the range of the species.

MISCELLANEOUS

There remain a few points, not already touched upon, which should be mentioned to complete this summary of rabbit ecology.

From his base at Albury, Myers (1962) organized an annual survey of rabbit distribution and population trends, starting in 1951 and covering a decade during which outbreaks of myxomatosis occurred almost annually. He selected three blocks of country, each approximately 100 square miles in area, which included representative samples of the major habitat types of the region. The greater proportion of each block consisted of 'open flat or undulating country with light timber'—the most valuable land from the grazier's point of view. The other habitats represented were 'creek frontages and swamps', 'rocky hills', 'heavily timbered hills', and 'stands of pine and box on sandy soils'.

During the period of the survey the regional rabbit population first declined, reaching a minimum in 1956, and then staged a slight recovery. The observed nature of the recovery is of peculiar interest. It did not involve the 'open' country, which remained free of rabbits from 1954 until the survey was discontinued—this applied to all three blocks. The rabbit populations on the creek frontages and swamps fared almost as badly; and it was only in the rocky and heavily timbered hills, and in the stands of pine and box on the patches of light soil, that infestations persisted and built up in numbers. Myers concluded that 'there are certain habitats within the region that remained perpetual refuge areas for rabbits in times of adversity. During good years the populations increase and spill out over the open and more valuable country.'

It is difficult to say how widely applicable are the implications of Myers' findings. The Riverina is well known to be a region presenting a mosaic of habitats which differ markedly in their favourableness for the rabbit. There can be no doubt, however, that the same sort of thing is to be found, albeit not so obviously, in many parts of the continent. It is possible that the temporary shrinkage of the rabbit population into refuge areas, so to speak, followed by an outward diffusion and the recolonization of the deserted habitats (which under certain conditions may become more favourable than the refuges) may occur quite widely. Where it is found to occur in marked degree, it could obviously be exploited in the strategy of control.

New Zealand workers seem to have been the first to point out (Bull, 1956) that a dense lush pasture did not provide a favourable habitat for $O.$ $cuniculus$, which prefers either a more open growth or a short more lawn-like sward. In regions where the adequacy and regularity of rainfall leads to the development of dense and lush ground cover it can be converted into a more favourable state by the feeding activity of the rabbits themselves, or by overgrazing by stock, or burning. Induced changes of this kind in the pasture have apparently occurred widely in New Zealand. In Australia, a country much less generously endowed with effective precipitation, a ground cover of plants unfavourably dense from the rabbit's point of view was, and is, uncommon. Myers has suggested (personal communication) that the alpine grassland of the Kosciusko region may have been sufficiently unfavourable to rabbits, in its virgin state, to have required a degree of overgrazing by stock before rabbit numbers could build up to a significant level.

In the more arid parts of Australia, the normal state of the rabbit population is one of patchiness and low density. The occasional wet periods of adequate length, on which the regeneration of long-lived perennial plants largely depends, can also be expected to result in a marked build up of rabbit numbers. The desperate foraging of this inflated population, when hot dry conditions return, usually results in the destruction of virtually all seedlings. Thus we are faced with the anomaly that the most serious vegetational changes are likely to develop in those parts of the rabbit's range where it is on the average least abundant. The best example of the threat to the survival of a locally dominant and economically important component of the dry-country flora is provided by the mulga (*Acacia aneura*) in the semi-desert pastoral country of South Australia.

Except where soil erosion intervenes, giving the stamp of permanence to vegetational damage and degradation, changes in the composition and density of the pasture due to rabbit activity are unlikely to prove irreversible. Indeed in all but low-rainfall areas, the recovery of the vegetation is usually fairly rapid once rabbits have been brought under effective control.

Digestion in the rabbit and its relatives has an interesting complication in the form of coprophagy ('refection'), which insures that material which has undergone bacterial fermentation in the caecum normally passes through the gut a second time. During the inactive period (i.e. the day in the case of the wild rabbit, the night for domestic ones) faeces of a special type are produced, consisting of soft, moist, mutually-adhering pellets. These are swallowed by the animal as they emerge from the anus. Physiologically, refection in the leporids can be regarded as a clumsy, and rather less efficient, substitute for the ungulate process of rumination.

Oryctolagus cuniculus is infested, both in domestication and in the wild, by a normal spectrum of internal and external parasites, and is also subject to various microbial and protozoan infections. None of the rabbit's diseases, however, is of sufficient importance ecologically —as a factor competing, one might say, with myxomatosis in determining population levels—to merit specific mention here. Coccidiosis provides a possible exception (for discussion and references, see Mykytowycz, 1962 b).

In the years immediately following the Second World War, some 100,000,000 rabbits passed through human hands annually, in Australia, in the form of carcasses or skins for export alone, with unknown but certainly considerable numbers finding their way to local markets or being killed but not recovered; and this toll had no marked effect on overall rabbit abundance. These figures will give some idea of the great size of Australia's rabbit population before the advent of myxomatosis and the improved organization of control.

During the past half century and more, with the rabbit established in numbers over most of its present range in Australia, there has been a tremendous reduction (amounting to extinction or virtual extinction in the case of many species) in the small to medium-sized herbivorous ground-dwelling marsupials, other than those species inhabiting the dense forest. To what extent the rabbit, rather than sheep and cattle, has been responsible for this eclipse of the competing native fauna it is impossible to say. In some pastoral areas outside the range of the rabbit the disappearance of small wallabies and rat kangaroos seems to have been almost as complete as it has been in rabbit-infested country. There can be no doubt, however, that the combined activities of rabbits and stock have contributed very importantly to the impoverishment of the indigenous mammal fauna of the savannah-woodland and open plain, which together constitute Australia's pastoral zone and the region in which the rabbit is most solidly entrenched.

CHAPTER 5

MYXOMA VIRUS:
ITS CLASSIFICATION, STRUCTURE
AND PROPERTIES

A substantial amount of experimentation has been carried out on the lesions produced in laboratory rabbits by myxoma virus and the closely related virus of Shope's fibroma. Much of the early work was motivated by the idea that since these viruses caused tumours when inoculated into domestic rabbits, their study might illuminate problems of carcinogenesis. No attempt will be made to survey this work here, but some aspects of it will be discussed in chapter 8.

Basic virological studies with myxoma virus are meagre, largely because it has until very recently been much more difficult to manipulate in the laboratory than the prototype of the poxvirus group, vaccinia virus, which has been a popular agent for such work. However, enough information has been accumulated to enable myxoma virus to be classified with confidence as a member of the poxvirus group. In this chapter we will review the facts upon which this classification is based, and also discuss what is known of the physical, chemical, and antigenic properties of myxoma and fibroma viruses.

THE CLASSIFICATION OF MYXOMA VIRUS

We will adopt Lwoff's (1957) definition of a virus, which excludes agents of the psittacosis group which possess RNA and DNA, have cells walls containing muramic acid, and probably multiply by binary fission (Moulder, 1962). Investigations of the composition of vaccinia virus and its mode of replication (see chapter 6) leave no doubt that the poxviruses are true viruses. They are the largest and most complex agents infecting vertebrates which can be classified thus. Some of the important properties of the poxvirus group, and the relationships between its different members, are shown in Table 1.

In 1950 the Fifth International Congress of Microbiology in Rio

60

de Janeiro laid down eight criteria which it was suggested should be applied to the classification of animal viruses (Andrewes, 1951). Fenner (1953a) discussed the classification of myxoma and fibroma viruses according to these criteria, and concluded that the resemblances between these two viruses and the prototype poxvirus, vaccinia, were so close that they should be included within the poxvirus group. This proposal was subsequently adopted by the appropriate subcommittee of the International Congress of Microbiology (Fenner

TABLE I. *The poxvirus group*

Shape and size of virion	Ovoid; approximately 300–350 by 200–250 by 100 mμ
Nucleic acid	DNA
Antigens	Several, but common internal antigen
Site of multiplication	Cytoplasm
Non-genetic reactivation	Confined to members of group; active within and between subgroups
Genetic recombination	Only within subgroups

Subgroups			Ungrouped
Vaccinia	Fowl-pox	South American myxoma	Molluscum contagiosum
Variola	Canary-pox	Californian myxoma	Monkey-tumour poxvirus
Ectromelia	Pigeon-pox, etc.	Shope's fibroma	Swine-pox
Rabbit-pox		Squirrel fibroma	Bovine papular stomatitis
Cow-pox		Hare fibroma	Sheep-pox
			Contagious pustular dermatitis

& Burnet, 1957), thus confirming the correctness of the judgement made by Aragão (1927) long before, for he had called attention to the close resemblance between the causative agents of myxoma and those of variola, molluscum contagiosum, and bird epithelioma (fowlpox).

Recently, with increased knowledge of the properties of animal viruses, more rational schemes of virus classification have been proposed, and the poxviruses form one of the best 'natural' groups. The homogeneity of the group gains further support from the demonstration by Woodroofe & Fenner (1962) that all the poxviruses they were able to test shared a common internal antigen. The evidence for this was as follows. Partially purified suspensions of vaccinia and myxoma viruses, prepared by scarification of the rabbit's back, were extracted with NaOH as described by Smadel, Rivers & Hoagland (1942), and antibodies to these 'NP' antigens were produced in rabbits.

These antibodies showed some homologous neutralizing capacity, which was removed by adsorption with suspensions of homologous virus particles. However, such adsorption did not remove the capacity of the sera to precipitate, and fix complement, when tested with either 'NP' antigen. In addition, convalescent serum produced after the infection of experimental animals with a wide variety of poxviruses (but not with other viruses) precipitated both the vaccinia 'NP' and myxoma 'NP' antigens (Table 2).

TABLE 2. *Ring-precipitin tests with the poxvirus group antigen, a component of 'NP' antigens (data from Woodroofe & Fenner, 1962)*

						Antiserum to				
Antigen	Vaccinia	Ectro-melia	Myxoma	Fibroma	Fowl-pox	Vaccinia NP	Myxoma NP	Herpes	Influenza A
Vaccinia NP	++++*	+++	+++	++	+++	+++	++	−	−
Myxoma NP	+++	++	++++	++++	++	++	+++	−	−

* −, No precipitin ring, + + to + + + +, moderate to very heavy precipitin ring.

Within the poxvirus group there are several well-defined subgroups, the members of which exhibit serological cross-reactivity when tested by neutralization tests (Table 3), and complement-fixation or precipitation tests with 'soluble' antigens. Members of the subgroups also show cross-immunity in experimental animals. Such tests show no cross-reactivity between members of the different subgroups, or between them and the 'ungrouped' poxviruses (Woodroofe & Fenner, 1962).

Experiments on non-genetic reactivation and recombination (review, Fenner, 1962) are in accord with the classification shown in Table 1. Two independent groups of investigators (Hanafusa, Hanafusa & Kamahora, 1959; Fenner & Woodroofe, 1960) showed that all poxviruses, and only poxviruses, were able to reactivate suitably heat-inactivated preparations of another poxvirus by a non-genetic mechanism. By this criterion also myxoma and fibroma viruses must be classified as poxviruses. Genetic recombination was found to occur only between members of the subgroups previously determined on serological grounds. No viable hybrids were obtained between myxoma or fibroma virus and viruses of the vaccinia-variola subgroup, but hybrids (recombinants) were demonstrated between several members of the vaccinia-variola subgroup, and between myxoma and fibroma viruses (Woodroofe & Fenner, 1960).

TABLE 3. *Plaque neutralization tests with poxviruses belonging to different subgroups, showing cross-neutralization within subgroups but not between members of different subgroups (data from Woodroofe & Fenner, 1962)*

Virus	Antiserum to		
	Vaccinia	Myxoma	Fowl-pox
Vaccinia	−1·6*	−0·1	−0·1
Cow-pox	−2·4	−0·3	−0·1
Ectromelia	−2·2	−0·4	−0·1
Myxoma	+0·3	−1·8	−0·3
Fowl-pox	0	+0·1	−1·4

* Figures show the ratios of plaque counts obtained with normal sera and the antisera indicated, in $\log_{10}$ units. Values of −1·0 or less (shown in heavy type) indicate neutralization.

Myxoma virus is thus a poxvirus which infects certain mammals, but which is antigenically related to the so-called 'mammalian poxviruses' (vaccinia, variola, cowpox, etc.) only in the possession of a common internal antigen. Although in this and the succeeding chapter we usually speak of 'myxoma virus' as a single entity, recent work (see p. 67 and chapter 15) has shown that there are two distinct varieties of myxoma virus, the Californian type and the South American type. In addition, two other viruses have been described which belong to the same subgroup as myxoma virus, namely rabbit fibroma virus and squirrel fibroma virus (see chapter 7). Quite recently a virus has been recovered from fibromas of hares in France and Italy. This has now been shown to be a member of the myxoma-fibroma subgroup. It is more closely related to rabbit fibroma virus than to the myxoma virus but shows some cross-protection with both (Woodroofe & Fenner, 1965; Fenner, 1965).

Reciprocal cross-immunity has been demonstrated between myxoma and rabbit fibroma viruses both in *Oryctolagus* and in *Sylvilagus* (Shope, 1932, 1936a). A variety of other serological tests; agglutination (Ledingham, 1937), complement-fixation (Shaffer, 1941) and gel-diffusion precipitation (Fenner, 1965) have confirmed the close relationship between them. Squirrel fibroma virus is passed with difficulty in *Oryctolagus*, but Kilham, Herman & Fisher (1953) were able to demonstrate neutralizing antibodies active against rabbit fibroma virus in the serum of squirrels (*Sciurus carolinensis*) which had recovered from infection with squirrel fibroma virus, and in rabbits which had received two inoculations of the latter virus.

Apart from deer fibroma, which is unrelated to myxoma virus (Shope, Mangold, Macnamara & Dumbell, 1958), the occurrence of multiple skin fibromas has been recorded by Herman & Reilly (1955) in the porcupine (*Erethizon dorsatum*) and the fox squirrel (*Sciurus niger*), but no etiological studies have been made. In view of the limited investigations carried out on viral infections of wild North American mammals it would be wise to bear in mind the possibilities that myxoma and rabbit or squirrel fibroma viruses may cause natural infections in wild animals other than those in which they have so far been recognized, and that other viruses possibly related to them may cause lesions in other North American mammals.

A report by Guo (1937) that cross-immunity could be demonstrated in the rabbit between vaccinia (neurolapine) and fibroma virus, but not between myxoma virus and either neurolapine or fibroma, is contrary to a great deal of other evidence (e.g. Andrewes, 1936; Parker, 1940; Fenner, 1958), and can be ignored. The claim by Takahashi, Kameyama, Kato & Kamahora (1959) that serological crossing could be demonstrated between vaccinia, myxoma and fowlpox viruses by complement-fixation tests using crude suspensions of virus also lacks confirmation (Woodroofe & Fenner, 1962; Harada & Matumoto, 1962).

In summary, myxoma virus is a member of the poxvirus group, and together with rabbit fibroma virus, squirrel fibroma virus and hare fibroma virus, it forms a subgroup within that group.

THE SIZE AND SHAPE OF MYXOMA VIRIONS

Lwoff, Anderson & Jacob (1959) coined the term virion to describe the mature viral particle, the form described by earlier writers on the poxviruses as the 'elementary body'.

The poxviruses are large enough to be visualized with the light microscope, after appropriate staining, and Buist (1886) probably saw the actual virions of vaccinia virus in calf-lymph. Lipschütz (1927), Aragão (1927) and Lewis & Gardner (1932) all described the presence within myxoma-infected cells of minute granules which they regarded as the infectious agent. Van Rooyen (1937) found large numbers of minute particles (which were certainly myxoma virions) in the conjunctival exudate of infected rabbits. Herzberg & Thelen (1938) used Victoria blue staining to demonstrate intracellular and extracellular virions of fibroma virus, and in our laboratory the Morosow stain (Gispen's modification, Gispen, 1952) has proved a useful method for following the purification of poxviruses (Joklik, 1962 a).

TABLE 4. *The dimensions of virions of myxoma, fibroma, and vaccinia viruses, determined on specimens fixed with osmium tetroxide*

Virus	No. of particles measured	Mean length ($m\mu$)	Mean width ($m\mu$)	Mean thickness ($m\mu$)	Reference
Vaccinia	25	305 ± 15	227 ± 20	80 ± 10	Farrant & Fenner
Myxoma	24	286 ± 15	230 ± 20	75 ± 10	(1953)
Fibroma	40	283	244	100	Lloyd & Kahler (1955)

Electron micrographs of unshadowed (Ruska & Kausche, 1943) and shadowed (Farrant & Fenner, 1953) virions of myxoma and vaccinia viruses showed that they were indistinguishable in size and shape (Table 4 and Pl. II) a conclusion supported by recent studies of the same viruses examined by the negative staining technique (Nagington & Horne, 1962; Chapple & Westwood, 1963). The complex structure of the myxoma virions is well depicted in the thin sections illustrated in Plate VII. The existence of a central body resistant to peptic digestion has been demonstrated in the virions of vaccinia (Dawson & McFarlane, 1948), fowlpox (Bang, Levy & Gey, 1951), and myxoma (Farrant & Fenner, 1953); and is probably a general feature of the poxviruses (Peters, 1960). Lloyd & Kahler (1955) found that in electron micrographs of shadowed preparations rabbit fibroma virus was morphologically indistinguishable from myxoma virus.

Further details of the structure of poxviruses have been sought by three methods; electron microscopy and enzymatic digestion (Peters, 1960), electron micrographs of thin sections of pellets of virions deposited by centrifugation (Epstein, 1958), and negative staining with phosphotungstic acid (Nagington & Horne, 1962; Padgett, Wright, Jayne & Walker, 1964). From all these studies it seems that morphologically myxoma and vaccinia viruses are very similar, and differ from certain other poxviruses, notably contagious pustular dermatitis virus (Nagington & Horne, 1962) and bovine papular stomatitis virus (Nagington, Plowright & Horne, 1962).

THE ANTIGENS OF MYXOMA AND FIBROMA VIRUSES

As might be expected from their large size and complicated structure, the poxviruses contain several antigens. Detailed investigations have been made of the soluble antigens produced in poxvirus infected cells by workers at the Rockefeller Institute in the period 1933–42 (review, Smadel & Hoagland, 1942), and more recently by two groups in England (Appleyard, Westwood & Zwartouw, 1962; Rondle & Dumbell, 1962). Smadel & Hoagland described four antigens, L, S, NP, and the X agglutinogen, which they thought were represented on the surface of vaccinia virions, although antibodies to none of these neutralized infectivity.

More recent work has made use of gel-diffusion precipitation. In experiments with cultured cells infected with rabbitpox virus, Appleyard et al. (1962) were able to distinguish no fewer than fifteen lines. Some of these are due to antigens of the virion which are presumably produced in excess, others may be associated with new enzymes concerned with viral synthesis, and the functions of others are unknown. In experiments with cowpox and vaccinia virus, Rondle & Dumbell (1962) distinguished nine precipitable substances and made the interesting observation that one component which was produced by wild-type cowpox virus was lacking in its white variants.

The Rockefeller Institute group carried out some preliminary studies with soluble antigens of myxoma virus (Rivers & Ward, 1937; Rivers, Ward & Smadel, 1939; Smadel, Ward & Rivers, 1940; Teixeira & Smadel, 1941). They described the recovery and partial purification of two proteins of different specificity, and found these

66

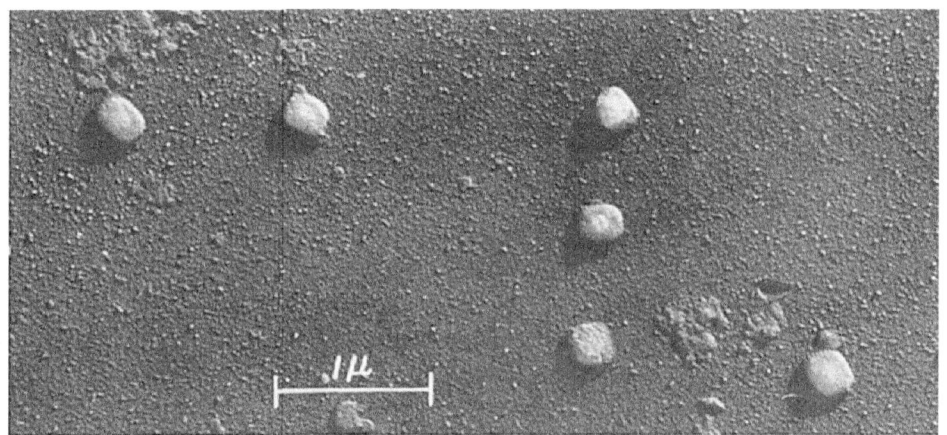

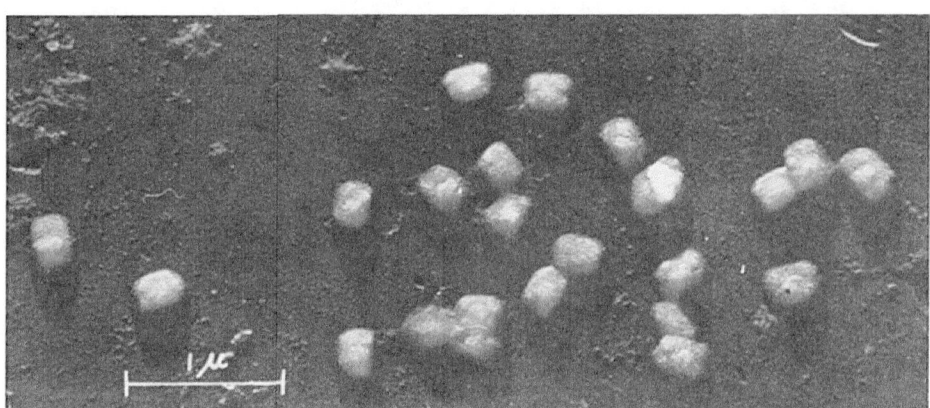

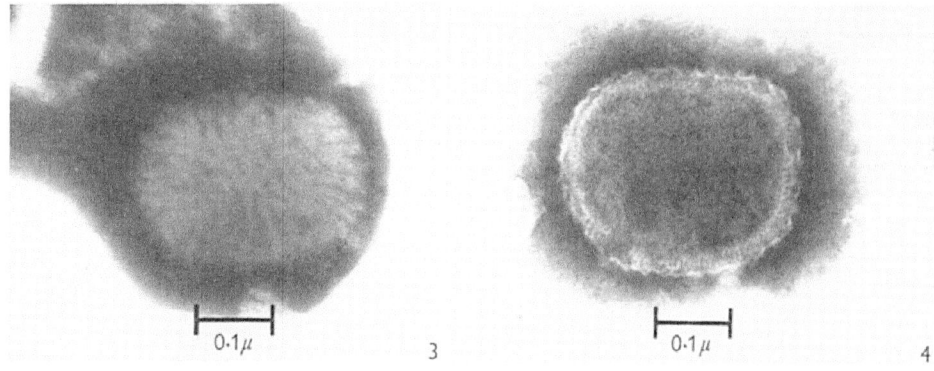

PLATE II. Electron micrographs of the virions of myxoma virus. Fig. 1, myxoma virus, and Fig. 2, vaccinia virus, fixed with osmium tetroxide and shadowed with uranium (from Farrant & Fenner, 1953). Figures 3 and 4, negatively stained with phosphotungstic acid (from Padgett, Wright, Jayne & Walker, 1964). Both types of particle are seen in preparations of both vaccinia and myxoma virions, that shown in Fig. 3 being more common.

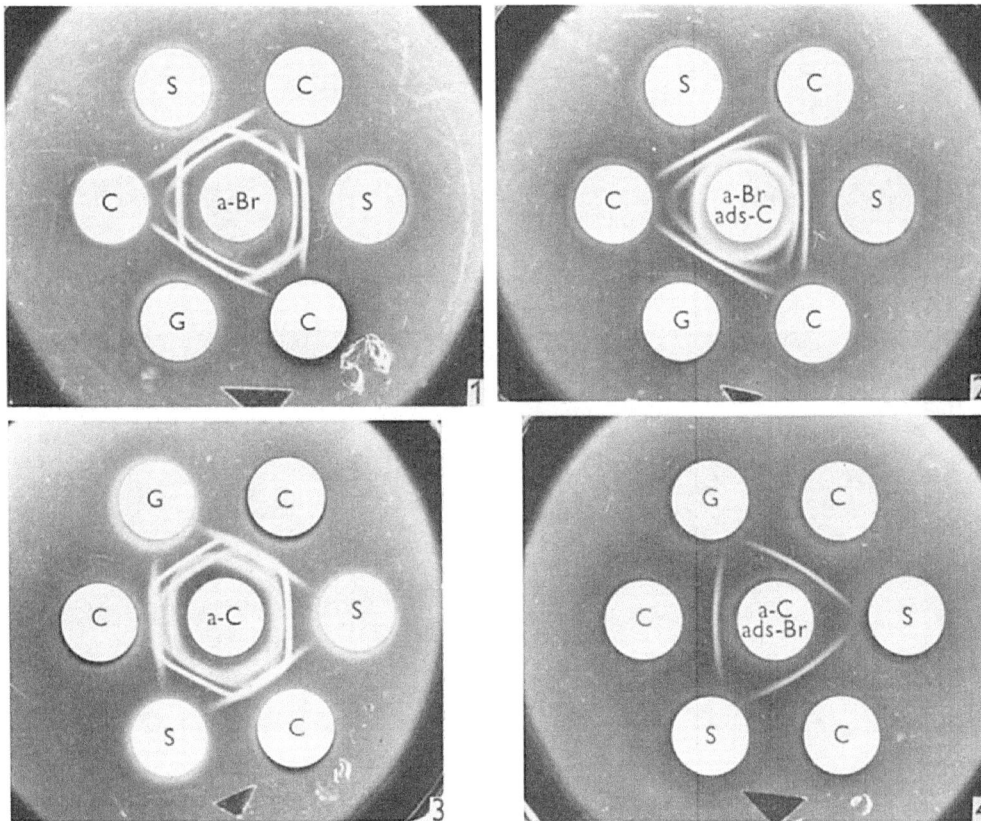

PLATE III. The results of gel-diffusion tests with a Californian strain of myxoma virus and its antiserum (C, a-C), a Brazilian strain and its antiserum (Br, a-Br), and two Australian strains of virus derived originally from Brazil (S, G). Fig. 1 and 3 show direct reactions of a-Br and a-C respectively with both types of virus. Figs. 2 and 4 show the results after prior treatment of the central cup with the heterologous antigen preparation (ads-C and ads-Br respectively) (modified from Reisner, Sobey & Conolly, 1963).

in rabbit lesion material, in the serum of acutely ill rabbits, and in extracts of infected chorioallantoic membrane. No further studies were made until 1957, when Mansi (1957b) demonstrated three precipitin lines in gel-diffusion tests between myxoma antiserum and myxoma infected rabbit tissue. He found that different treatments of the tissue extract (varying pH, heating, etc.) acted differentially on the capacity of the extract to produce the different lines but he did not pursue the characterization of the antigens any further (Mansi & Thomas, 1958). Like all other workers he found that there were cross-reactions between the soluble antigens of myxoma and fibroma viruses.

Reisner, Sobey & Conolly (1963) extended this approach and showed that there were at least five soluble antigens in myxoma-infected tissues, and Walker (personal communication, 1963) has distinguished no fewer than eight antigens by immuno-electrophoresis. Reisner et al. also found that there were substantial differences between the soluble antigens of myxoma virus obtained from California, and strains derived originally from Brazil. These differences are illustrated in Pl. III. The Californian and Brazilian strains have common antigens, and each type has antigens which are apparently peculiar to it. The latter form 'spurs' in the unabsorbed plates (Pl. III, figs. 1, 3), and these are the only lines left after adsorption in the central cup with the heterologous antigen preparations (Pl. III, figs. 2, 4).

Using a more extensive suite of virus strains derived from several outbreaks in California, and various parts of Central and South America (see chapter 15) Fenner (1965) found that several Californian strains were antigenically similar, and differed from all the South American strains. There were some differences among the latter, however. Two strains from Colombia, and two from Panama, were antigenically closer to the Californian type virus than the Brazilian. No antigenic differences have yet been found amongst strains obtained from field cases in Australia and Europe.

CHEMICAL COMPOSITION OF MYXOMA VIRUS

Vaccinia and cowpox viruses are the only poxviruses which have been obtained in large enough quantities, in a purified form, for

5-2

chemical analysis. The Rockefeller Institute group (Smadel & Hoagland, 1942) reported that vaccinia virus contained carbon 33·7 %, nitrogen 15·3 %, phosphorus 0·57 %, copper 0·05 %, cholesterol 1·4 %, phospholipid 2·2 %, neutral fat 2·2 %, reducing sugars 2·8 %, and deoxyribonucleic acid 5·8 %. More recently Joklik (1962 a) showed that the DNA's of four different viruses of the vaccinia-variola subgroup have molar ratios of adenine to thymine and guanine to cytosine very close to 1; and molar dissymmetry ratios for adenine + thymine to guanine + cytosine of 1·70. He studied the DNA of cowpox virus in detail and showed that it was two-stranded,

TABLE 5. *Inhibition by 5 bromodeoxyuridine (BDU) of plaque production by DNA viruses, but not by an RNA virus*

Virus	Nucleic acid	Concentration of BDU				
		$10^{-4·3}$M	$10^{-4·9}$M	$10^{-5·5}$M	$10^{-6·1}$M	Nil
Semliki Forest	RNA	61, 69	66, 87	—	—	62, 79
Vaccinia	DNA	—	0, 0	0, 0	66, 100	70, 80
Myxoma	DNA	0, 0	0, 4 (small)	17, 17	84, 78	79, 68

Plates of rabbit kidney cells were infected with 50–100 pfu of the viruses indicated, and various concentrations of BDU added to the overlay medium.

and could be extracted in molecules with a molecular weight of 80 million (Joklik, 1962 b). Joklik found only one-quarter as much copper in his preparations of purified virus as that reported by the Rockefeller Institute workers, and he suggested that copper was probably an adsorbed impurity rather than an integral part of the virus.

No chemical examination has yet been made of myxoma virus. Kato, Takahashi, Miyamoto & Kamahora (1963) found that cells infected with myxoma or rabbit fibroma virus, and exposed to a pulse of tritiated thymidine, show cytoplasmic DNA factories closely resembling those demonstrated by Cairns (1960) for vaccinia virus. This fact, and the inhibition of myxoma virus multiplication in cultured cells by 5 bromodeoxyuridine (Table 5) provide presumptive evidence that myxoma virus has DNA as its genetic material, as would be expected if virus classification has any biological meaning.

SUSCEPTIBILITY TO CHEMICAL AND PHYSICAL AGENTS

Almost all investigations of viral inactivation have been concerned with the destruction of viral infectivity, that is the destruction of the capacity of the virions to produce new infectious virus when they enter susceptible cells. It has long been known, especially from work with influenza virus, that other biological effects of the virus may be inactivated at rates quite different from infectivity (Henle & Henle, 1947) and recent work shows that the situation is rather similar with the poxviruses also. Indeed the response of viruses to different modes of inactivation can be used to analyse the functional role of viral components (Fenner, 1962; Hanafusa, 1962).

Depending upon the type of damage inflicted upon vaccinia virus, particles may lose infectivity but retain other biological properties. Damage to the viral coat may so alter the surface properties of the virion that it will not adsorb to cells, and so cannot be phagocytosed by them. Such particles could exhibit no other biological activities, since they could not enter cells. By other treatments (e.g. heat, urea) the protein which is thought to be responsible for inducing the enzyme which uncoats the nucleic acid is damaged (Joklik, 1962c), and the infectivity of the particle is destroyed without loss of its capacity to adsorb, and without damage to its genetic material. Such virus can be reactivated by a non-genetic mechanism, and it retains several other biological activities such as a cell-killing effect and the capacity to interfere with homologous virus.

Treatment of virus with radiation, or with certain chemicals, may so damage the genetic material of the virus that single virions are unable to replicate to the stage of producing complete virus. However, such virions may retain many biological activities, such as the capacity to interfere (Galasso & Sharp, 1963), to reactivate heat-inactivated virus (Joklik, Abel & Holmes, 1960), to produce toxic effects (Brown, Mayyasi & Officer, 1959), and to be reactivated by genetic mechanisms (Abel, 1962).

Most of the effects just described have been investigated with vaccinia virus, but exploratory experiments have shown that many apply also to myxoma and fibroma viruses. Indeed the very interesting phenomenon of non-genetic reactivation was first demonstrated with these viruses (Berry & Dedrick, 1936). In the sections which follow a brief account will be given of the inactivation of myxoma virus by various physical and chemical treatments, but no account can be given of the relative damage to all the biological activities of the virion produced by each type of treatment, since few of the required experiments have been carried out.

Inactivation by heat

Bronson & Parker (1943) carried out a careful study of the loss of infectivity of suspensions of myxoma virions suspended in normal rabbit serum. For the first time in experiments upon heat-inactivation of viruses, they emphazised the importance of the inactivation rate (rather than the 'thermal death point'), and showed that inactivation followed first-order kinetics at all temperatures tested.

TABLE 6. *Observed and calculated half-life times of myxoma virus suspended in normal rabbit serum, and held at temperature indicated*

Half-life times	Temperature				
	$60°$	$50°$	$30°$	$20°$	$4°$
Observed	0·3 min.*	22·3 min.*	1·3 days†	4·8 days†	10 days†
Calculated	—	—	50 days	5×10^3 days	5×10^6 days

* From Bronson & Parker (1943).
† Data from Day, Fenner, Woodroofe & McIntyre (1956).

Suitable treatment of their data show that there is a linear relation between the logarithm of the rate constant and the reciprocal of the absolute temperature. Extrapolation allows the calculation of a half-life at various temperatures, assuming that inactivation is due to heat alone. In Table 6 we have compared some observed and calculated half-life times.

It is apparent that at the lower temperatures infectivity is lost very much more rapidly than would be expected on the basis of heat-denaturation of an essential protein. The more rapid inactivation is almost certainly due to a variety of chemical effects, including the activity of enzymes in the suspending medium. It is impossible to guess at the nature and concentration of such factors on the mouthparts of an infected mosquito, which is the situation in which the length of survival of infective virus is of considerable epidemiological importance.

Inactivation by ether

Andrewes & Horstmann (1949) first drew attention to the value of testing the susceptibility of viruses to inactivation by ether as an aid in their classification. They reported that myxoma and fibroma viruses were relatively susceptible, their infectivity dropping over a thousandfold after treatment with 20 % ethyl ether overnight at 4° C., whereas other poxviruses like vaccinia showed no loss in titre. This test presumably reflects the importance for the structural and functional integrity of viruses of lipids and lipoproteins in the viral coat.

Kilham, Lerner, Hiatt & Shack (1958) found that myxoma virus was completely inactivated by ether treatment for $\frac{1}{2}$–2 hours at room temperature. Virus so treated was reactivable, which indicates that its genetic material was not damaged by this treatment.

SUMMARY

Myxoma virus is a large and chemically complex virus belonging to the poxvirus group. A variety of different antigens have been demonstrated in the virion and in soluble form in myxoma-infected cells. The virion contains an internal antigen which shows cross-reactivity with all other poxviruses, but serological crossing with the surface antigens, and with other soluble antigens, has been demonstrated only between myxoma, and rabbit, hare and squirrel fibroma viruses. These four viruses, which show varying degrees of cross-protection in experimental animals, form a subgroup of the poxvirus group.

CHAPTER 6

INTERACTIONS BETWEEN MYXOMA VIRUS AND THE HOST CELL

In the previous chapter we described some of the characters of the myxoma virion, the mature form in which the virus is transferred from one host cell to another, and from one host to another. The essence of a virus, however, is the way it is altered in, and then alters, the susceptible host cell in which it sets up infection and undergoes replication. This is not the place to describe in any detail the nature of viral replication, and indeed the information available on myxoma virus multiplication is incomplete and fragmentary. Nevertheless, it is desirable that an account should be given of the general process of replication of the poxviruses, which we shall do by using vaccinia virus as the model. Available information on the growth of myxoma and fibroma viruses in cultured cells will be viewed against this background. The developing egg and the new-born mouse resemble cultured cells rather than mature animals in their susceptibility to viral infection, and their role as hosts of myxoma and fibroma viruses will also be described in this chapter.

THE MULTIPLICATION OF VACCINIA VIRUS IN CULTURED CELLS

The multiplication of the prototype poxvirus, vaccinia, has been extensively studied, and this brief account of the growth cycle of vaccinia in cultured cells is taken mainly from review articles by Fenner (1962) and Joklik (1962c).

Adsorption of the virion to the cell membrane occurs without any particular orientation, at a slower rate than would be predicted for a simple particle-surface interaction. Penetration is effected by the phagocytic ingestion of the virion. As soon as the virion is ingested the phospholipid and some of the protein of the viral coat are degraded. This process, which does not proceed to the exposure of

72

the viral DNA, is effected by enzymes already active in the cell. With infective virus it is immediately followed by a more extensive degradation of the viral coat, a process which depends upon the induction of a new 'uncoating enzyme' and is followed by release of the viral DNA. It is the failure of heated virus to induce this enzyme which is responsible for its lack of infectivity, and the susceptibility of the inner protein of the heated particle to enzyme induced by other active poxvirus virions that results in non-genetic reactivation. This stage begins the eclipse phase, which ends with the production of new virions.

Exposure of the viral DNA is promptly followed by the synthesis of new DNA and new proteins, some of which are the new enzymes required for the synthesis of viral components. Viral antigen and DNA are produced at the same cytoplasmic site and each infective particle in a cell produces a separate factory for their synthesis (Cairns, 1960).

Within the matrix which constitutes the factory area viral membranes enclose some of the ground substance of the matrix, producing the so-called 'immature forms' of the electron microscopist. These become more compact and more complex in structure until they look like mature virions. Assays show that new infectious virus is first produced at about the fourth hour after infection, and the titre increases rapidly between the sixth and the tenth hours. Few of the mature virions are released from the cell, most remaining cell-associated until the cell is finally destroyed.

THE MULTIPLICATION OF MYXOMA AND FIBROMA VIRUSES IN CULTURED CELLS

Reports on the multiplication of myxoma and fibroma viruses in cultured cells date back to Benjamin & Rivers (1931), and Plotz (1932), but quantitative studies were not carried out until very recently. Using plasma-clot cultures, Chaproniere (1956) found that myxoma virus multiplied well in both rabbit kidney epithelial cells and in rabbit heart fibroblasts, but it produced strikingly different changes in these two types of cell. Infected cells underwent profound morphological changes but appeared to survive for many days. In the rabbit kidney cells there developed eosinophilic cytoplasmic inclusions and vacuolation of the nucleus, lesions similar to those reported by Rivers (1930) in the epithelium over myxoma nodules in the skin of the rabbit. Rabbit heart fibroblasts were changed by infection with myxoma virus into large stellate cells with basophilic granules in the nucleus and very little 'inclusion material' in the cytoplasm. Their appearance closely resembled that of the 'myxoma cell' which most histopathologists regard as the pathognomonic feature of myxoma infection of the domestic rabbit (Hurst, 1937a; Ahlström, 1940a).

In 1962 three groups of American workers independently described methods of plaque assay of myxoma or fibroma virus. Schwerdt & Schwerdt (1962) showed that myxoma virus (South American strain) produced plaques in primary rabbit kidney cells monolayers and in an established line of suckling rabbit kidney cells. They used dilute agar as an overlay, which was slipped off the cells before they were stained with carbol fuchsin. Counts were made at $\times 15$ magnification. The assay was as sensitive as, and more accurate than, intradermal titrations in rabbits.

Verna & Eylar (1962) reported the production of plaques by fibroma virus in monolayers of rabbit kidney cells, using a liquid overlay medium. They had no success with an agar overlay, but Padgett, Moore & Walker (1962) showed that both myxoma and fibroma viruses would produce macroscopic plaques in rabbit kidney cells monolayers overlaid with agar. If the number of plaques per bottle was not too high their distribution was Poissonian. Myxoma virus produced a plaque of damaged cells, which reached a diameter of up to 3 mm. after six days and did not take up neutral red. Fibroma virus, on the other hand, produced foci up to 1·5 mm. in diameter, which were composed of small piles of cells, which took up and retained neutral red as well as the normal cells. In rabbit embryo fibroblasts, however, fibroma virus produces a small, clearly defined plaque of damaged cells similar to those produced by the Californian type of myxoma virus (see Pl. IV). It would be of considerable interest to see whether the myxoma viruses would produce proliferative rather than destructive lesions in cells derived from their natural hosts, *Sylvilagus bachmani*, and *S. brasiliensis*, in which they produce benign fibromas. However, no such studies have yet been reported.

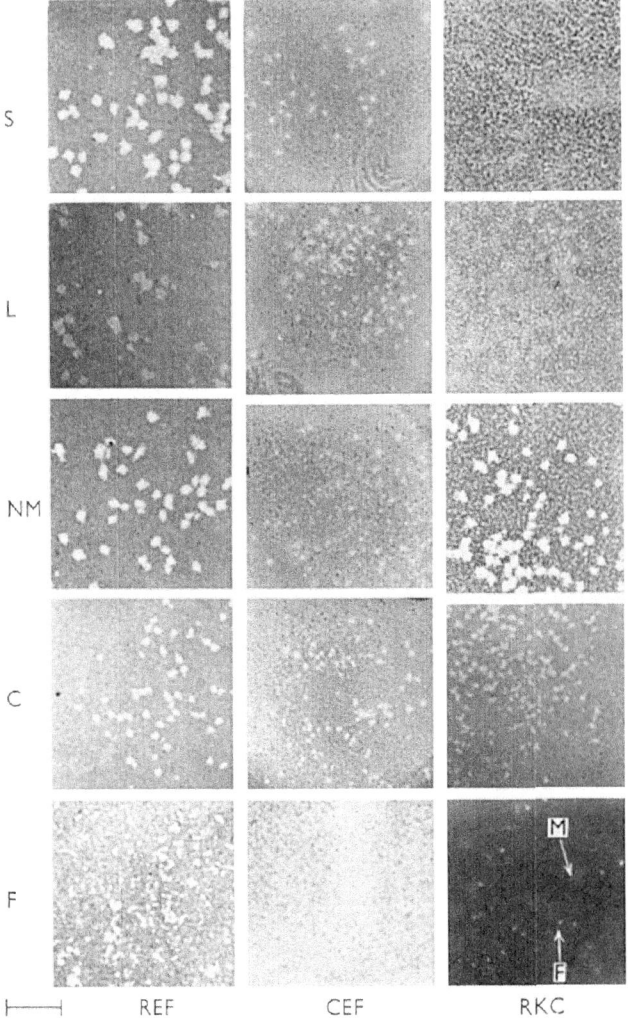

PLATE IV. Plaques produced by myxoma and fibroma viruses. Myxoma viruses: S, standard laboratory strain; L, Lausanne; NM, neuromyxoma; C, Californian; F, rabbit fibroma virus. Scale = 1 cm. Left: rabbit embryo fibroblasts, stained with neutral red six days after inoculation (seven days for rabbit fibroma virus). Centre: chick embryo fibroblasts with DEAE-dextran in overlay medium. Stained with neutral red six days after inoculation. Right: suckling rabbit kidney cells, stained with neutral red six days after inoculation. Lower right-hand corner: rabbit fibroma and South American myxoma viruses in rabbit kidney cells, with 5 % anti-fibroma rabbit serum and no neutral red in overlay medium (F, fibroma; M, myxoma) from Moore & Walker (1963). All other preparations from Woodroofe & Fenner (1965).

Woodroofe & Fenner (1965) have examined the plaques produced on rabbit kidney cell and rabbit embryo fibroblast monolayers by several prototype strains of myxoma virus (see chapters 8 and 15). Three different types of plaque were seen. In rabbit kidney cell monolayers all the prototype Australian and European field strains (deriving originally from Brazil) produced rather hazy plaques 2–3 mm. in diameter on the sixth day, which could be clearly seen only with indirect lighting. Neuromyxoma produced clear plaques about the same size, which could be readily seen with transmitted light. Californian strains, on the other hand, produced minute hazy plaques, which attained a diameter of only 1 mm. by the sixth day. The position of all of these plaques could be more readily visualized by the inclusion of myxoma antiserum in the overlay medium (Moore & Walker, 1962). Under these conditions neuromyxoma produced an area of precipitation with well-defined edges, about the same size as the plaque, other derivatives of Brazilian myxoma virus produced larger areas with irregular edges, and Californian type virus produced small irregular areas of precipitation which were slightly larger than the small well-defined foci found with fibroma virus. All strains of myxoma virus produced well-defined plaques on rabbit embryo fibroblast monolayers. As with rabbit kidney cells, the Californian strains were characterized by smaller plaques than the Brazilian strains and their derivatives (Pl. IV), and differences were seen amongst the latter.

Plaque assay promises to be a valuable method for the analysis of the genetics of myxoma viruses, as well as for their accurate assay. Padgett, Goodson & Walker (personal communication 1963) have used plaque assays to carry out one-step growth curves of myxoma and fibroma viruses in rabbit kidney cells. Their results are summarized in Fig. 2, which compares one-step growth curves obtained with vaccinia virus (from Easterbrook, 1961), myxoma virus (Padgett & Walker, 1962, personal communication 1963) and fibroma virus (Goodson, 1963). Despite the complicating effect of the differing multiplicities of infection, different cells and different modes of dispersion of the cells, the growth curves are basically similar. In all cases over 90 % of the virus remained cell-associated throughout the growth cycle. As in vaccinia-infected cells (Cairns, 1960) specific immune fluorescence was detected before new infective virus, and

Padgett & Walker (1962) detected inclusion bodies by Giemsa and acridine orange staining as early as the third hour after inoculation.

French workers have studied the multiplication cycle of fibroma virus *in vivo*, and in rabbit spleen fibroblasts. By assaying the 'soluble' and sedimentable antigen in the nictitating membrane of the newborn rabbit infected with a large dose of fibroma virus, Edlinger & Harel (1956) showed that new soluble antigen could be

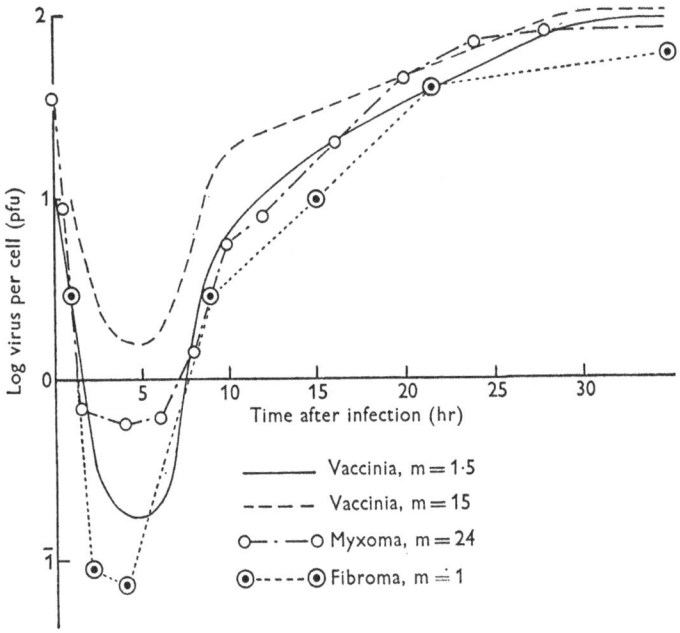

Fig. 2. One step growth curves of vaccinia virus in suspended KB cells (from Easterbrook, 1961) and myxoma and fibroma viruses in monolayers of rabbit kidney cells (modified from Padgett & Walker, 1963; Goodson, 1963). Curves show titre of cell-associated virus expressed as pfu per cell. *m*, Multiplicity of infection.

detected by the fourth hour. This was confirmed by Constantin, Febvre & Harel (1956), who examined the multiplication of fibroma virus in rabbit spleen fibroblasts.

An interesting feature of this work was the persistence of virus in high titre for a prolonged period, without apparent cellular destruction (Constantin, Febvre & Harel, 1956). Kato, Takahashi, Kameyama, Morita & Kamahora (1959) found a similar effect in cultures of FL cells infected with myxoma or fibroma virus, but a more critical analysis showed that in this case there was apparently a balance between the multiplication of uninfected cells and destruction of infected cells. However, it has been the experience of many investigators that fibroma virus in particular is not a cytocidal virus, and infected rabbit kidney cells certainly continue to multiply. Hinze & Walker (1964) for example, found that serially cultured rabbit kidney cells, if maintained under carefully controlled conditions of temperature and nutrition, continued to multiply even faster after infection with fibroma virus than did the control uninfected cells, and demonstrated a cell morphology and growth pattern distinctly different from that of the normal controls.

ATTENUATION OF MYXOMA VIRUS BY PASSAGE IN CULTURED CELLS

McKercher & Saito (1964) have successfully used serial passage of Californian myxoma virus (MSD strain) in cultured rabbit kidney cells to produce an attenuated vaccine. The virulence of the Californian virus for *Oryctolagus* was not altered by 110 serial passages on the chorioallantoic membrane, but when this material was passed 32 times in cultured *Oryctolagus* kidney cells incubated at $33.6°$ it diminished greatly in virulence, although still producing generalized symptoms. After a further eight passages the virulence had diminished still more. Large doses inoculated intradermally then produced only a small localized nodule which underwent rapid regression, and with small doses the only signs of infection were a rise in temperature of the inoculated rabbit, the production of antibody to myxoma virus, and immunity to challenge infection with virulent myxoma virus.

GROWTH OF MYXOMA AND FIBROMA VIRUSES IN THE DEVELOPING CHICK EMBRYO

After a period of nearly 40 years during which it was thought that myxoma virus multiplied and produced lesions only in *Oryctolagus*, three groups of workers showed independently that serial passage could be carried out on the chorioallantoic membrane of developing hen's eggs (Lush, 1937; Haagen & Du, 1938; Hoffstadt & Pilcher, 1938). Lush's paper was the most informative, for she showed that with dilute suspensions discrete pocks were produced. There was a roughly linear relationship between dilution and pock count, so that growth on the chorioallantoic membrane could be used both for infectivity titrations and for the assay of neutralizing antibodies. Lush's results were confirmed and refined by Fenner & McIntyre (1956). Twelve-day-old eggs inoculated on the chorioallantoic membrane with suitably diluted suspensions of virus showed discrete foci 0·5–1 mm. in diameter three days later. No adaptation was required, fully developed foci appearing after the inoculation of the original rabbit material. Lush has described and illustrated the histology of the lesions.

Serial passage of highly virulent South American type virus on the chorioallantoic membrane for 75 passages altered neither the pathogenicity of the virus for rabbit or the egg (Fenner & Marshall, 1957), nor the relative titres of virus assayed in rabbit or egg (Fenner & McIntyre, 1956). Haagen & Du (1938), on the other hand, reported that after twenty passages inoculated embryos died on the second day, while egg-passaged virus produced local nodules only, with minimal signs of generalization, after intradermal inoculation into rabbits. Inoculated rabbits usually recovered and were immune to reinfection with virulent myxoma virus.

We have recently investigated the effect of the age of the inoculated embryo on the number of pocks produced on the chorioallantoic membrane, and on the viral yield per membrane, using the highly virulent standard laboratory strain and the attenuated neuromyxoma strain. Eggs from a single batch were inoculated with a standard dose of each of the two strains after they had been pre-incubated for periods ranging from nine to sixteen days. They were incubated at 35° and 39° for a further three days. After counting the specific pocks on the membranes they were homogenized and their content of infectious virus determined (Table 7). Eggs pre-incubated for twelve days gave the highest pock counts, and the most clearly defined pocks; but nine-day-old eggs yielded about twenty times more infectious virus per membrane. In the same experiment eggs incubated at 39° after inoculation failed to produce either pocks or infectious virus with either strain of myxoma virus.

Using twelve-day embryos incubated for three days at 35° C., Fenner & Marshall (1957) were able to distinguish between the pocks

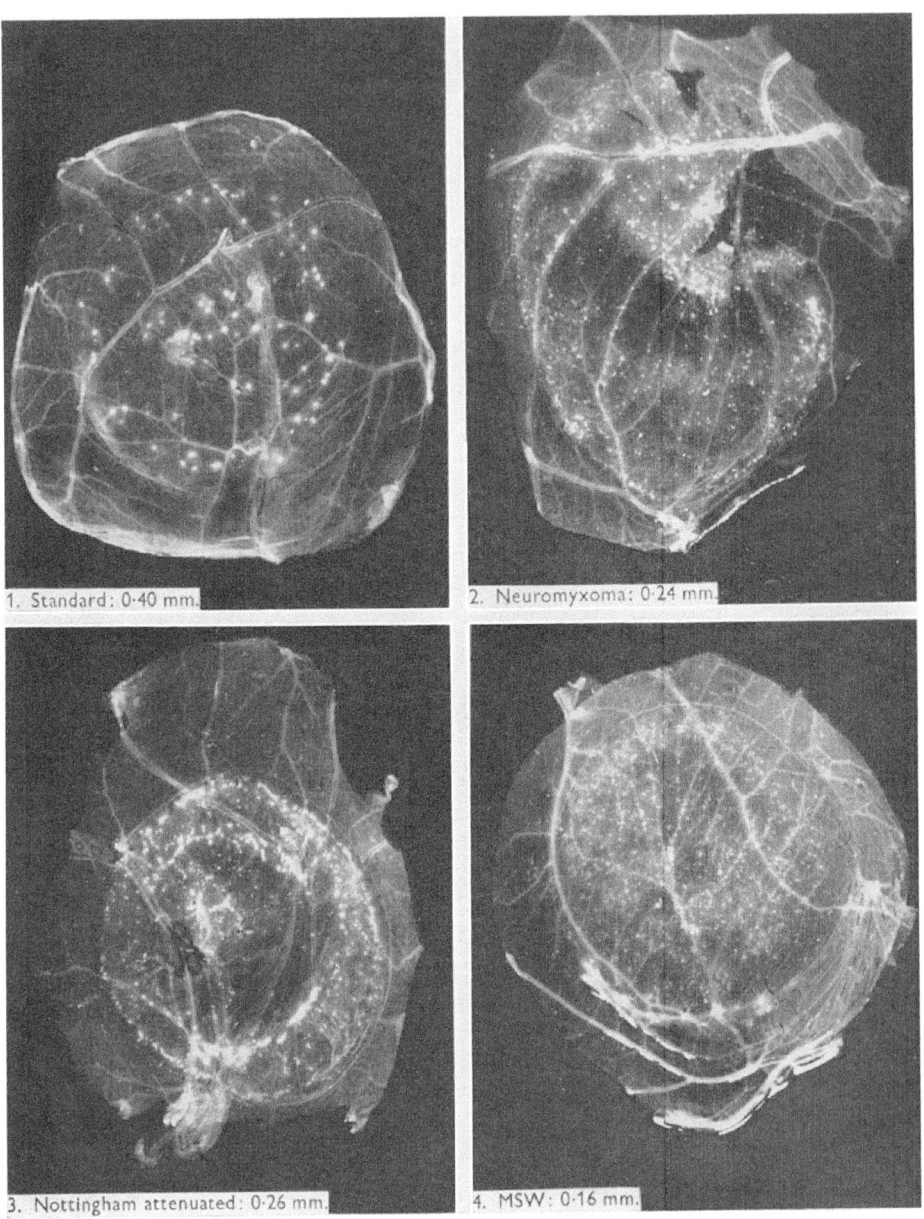

PLATE V. Appearance of pocks produced by growth of different strains of myxoma virus on the chorioallantoic membrane. Twelve-day-old eggs incubated at 35° C. for three days after inoculation. The density of pocks on all the membranes illustrated (except standard laboratory strain) is considerably greater than usually used for pock-counting, or for the measurement of pock size.

Fig. 1. Standard laboratory strain, mean diameter 0·40 mm.
Fig. 2. Neuromyxoma strain, mean diameter 0·24 mm.
Fig. 3. England/Nottingham/4-55/1, attenuated, mean diameter 0·26 mm.
Fig. 4. U.S.A./San Francisco/1950/1 (MSW), mean diameter 0·16 mm.

produced in the chorioallantoic membrane by certain strains of myxoma virus (Pl. V). All strains derived from infections with Californian myxoma virus produced minute pocks (average diameter 0·16 mm.), whereas the South American and Panama strains produce much larger lesions (average diameter 0·40 mm.). Hurst's neuro-myxoma produced lesions of intermediate size (average diameter

TABLE 7. *The relationship between the pock count and the virus yield, and the age of the embryo at the time of inoculation (F. Fenner, unpublished observations)*

	Pock count (means of 9 eggs)			Virus yield*	
Age of embryo at inoculation (days)	Standard laboratory strain	Neuro-myxoma	Non-specific lesions	Standard laboratory strain	Neuro-myxoma
9	53	51	+ + †	5·9	3·9
10	56	84	+ +	4·7	3·8
11	140	87	+	4·9	3·6
12	205	90	−	4·6	2·7
13	116	25 irr.‡	−	4·3	2·8
14	6 irr.	16 irr.	−	3·6	2·0
15	32 irr.	−	−	3·5	2·6
16	−	−	+	1·0	1·0

* In pfu per membrane in $\log_{10}$ units (from the membranes used for pock counts).
† −, +, + +, No, slight, and severe non-specific lesions.
‡ irr., Irregular; some CAM show pocks, others negative.

0·24 mm.). In a suspension of myxoma virus recovered from a naturally infected rabbit from Nottinghamshire, England, Fenner & Marshall (1957) were able to distinguish two viral populations by the pock characters. Single pock isolations showed that the one associated with pocks of the usual size (0·40 mm. average diameter) was highly virulent for rabbits, whereas the virus producing small pocks (0·26 mm. average diameter) was greatly attenuated. No significant differences in pock character have been detected amongst the hundreds of samples recovered from naturally infected rabbits in Australia over the last decade, all of which produce lesions like those of the South American strains.

Fibroma virus also grows on the chorioallantoic membrane, but produces very minute lesions which cannot be counted. Smith (1948) carried the OA strain of fibroma virus through eighteen serial passages on the chorioallantoic membrane. It did not invade the

79

embryo and she could recognize it only by back-passage of membrane suspensions in rabbits. Egg passage produced no change in pathogenicity for the rabbit. Indirect confirmation of Smith's findings

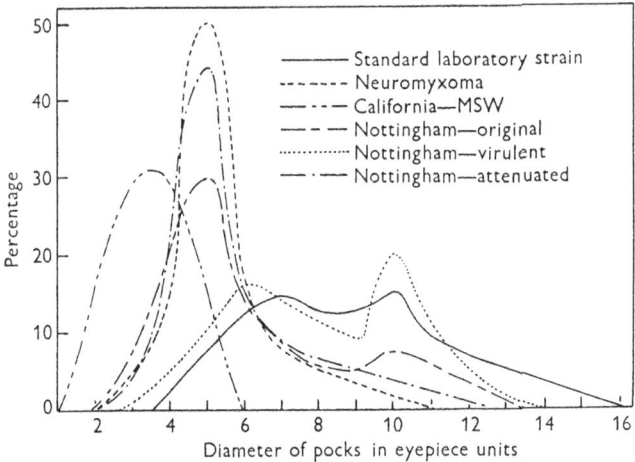

Fig. 3. Distributions of the diameters of pocks caused by the growth of a variety of strains of myxoma virus on the chorioallantoic membrane. All strains inoculated on twelve-day-old eggs which were then incubated for three days at 35° C. (from Fenner & Marshall, 1957).

were provided by the demonstration that OA and Boerlage strains of fibroma virus could reactivate heat-inactivated vaccinia virus on the chorioallantoic membrane (Fenner & Woodroofe, 1960).

GROWTH OF MYXOMA AND FIBROMA VIRUSES IN NEWBORN MICE

Like the developing chick embryo, the newborn mouse is susceptible to many viruses which fail to multiply in the adult animal. Myxoma and fibroma viruses are no exceptions. Andrewes & Harisijades (1955) demonstrated an increase in titre (maximal at three to five days) in the brains of mice inoculated when one day old, and with some difficulty they carried out serial passage of myxoma virus for thirty passages in them. The passaged virus had not altered in its virulence for rabbits. Claringbold & Sobey (1955) have confirmed these results.

Examination by fluorescent antibody staining of the brains of mice infected as newborns three days earlier showed scattered infected cells in the meninges, ependyma, and chorioid plexuses (Fenner, unpublished observations), just as Mims (1960) found with other poxviruses in mature mice. Dalmat (1958c) showed that fibroma virus would persist for periods of up to a month in the brains of mice inoculated when they were one day old, and he carried out three serial passages of fibroma virus in baby mouse brain. The tumours produced in adult *Oryctolagus* by virus which had been passaged in mouse brain did not differ in time of appearance or in rate of growth from those produced by rabbit-passaged virus, but in contrast to these they were infectious for mosquitoes, when tested fifteen days after inoculation of the rabbit.

INFECTIVITY ASSAYS OF MYXOMA VIRUS

Most infectivity assays of myxoma virus have been carried out by the intradermal inoculation of rabbits with serial dilutions of virus, the appearance of a skin lesion being taken as indicating the presence of virus (Parker, 1940), or by pock counting on the chorioallantoic membrane of the developing egg (Lush, 1937). A detailed comparison of these methods was reported by Fenner & McIntyre (1956), using several strains of myxoma virus of differing virulence for rabbits and with different passage histories in eggs. All strains behaved in the same way, titration on the rabbit back being on the average $2\frac{1}{2}$ times more sensitive than assay on the chorioallantoic membrane. The evidence was consistent with the hypothesis that a pock or a skin lesion was initiated by one infective unit of virus, and that the infective unit was stable with dilution. There was some individual variability in susceptibility to infection in rabbits, and a great deal of inter-egg variation in the pock count assays. The maximum value for the average probability that an infective unit would produce a lesion on the rabbit skin was 0·6, and on the chorioallantoic membrane 0·25.

Claringbold & Sobey (1955) have adapted a principle commonly used in the assay of pharmacological preparations to the biological assay of myxoma virus. They showed that with doses of more than 100 ID 50 there is a linear relationship between lesion diameter and log dose. Using a known strain of virus which produces a well-defined skin lesion the method provides a rapid and economical assay for relatively large doses of virus, in terms of some standard. However,

6   81   F & R M

since different strains of myxoma virus may produce quite dissimilar skin lesions (see chapters 13 and 15) this type of assay could not be used except for the comparison of various preparations of a known strain.

Myxoma virus produces minute plaques on chick embryo fibroblast monolayers, the clarity of which is greatly improved by the addition of DEAE dextran to the overlay medium (Woodroofe & Fenner, 1965). These can be used for infectivity assays and neutralization tests (Woodroofe & Fenner, 1962). Tests with several different strains of myxoma virus (including the Californian type) failed to reveal any differences in their plaque morphology on chick embryo fibroblasts.

TABLE 8. *Comparison of several assay methods for myxoma and fibroma viruses (modified from Padgett, Moore & Walker, 1962)*

	Virus titre* by various assay methods			
Virus	CAM pock count	Rabbit skin lesions	Rabbit kidney cell tube culture	Plaque count on rabbit kidney cells
South American myxoma	$3 \cdot 0 \times 10^5$	$1 \cdot 3 \times 10^6$	$7 \cdot 6 \times 10^5$	$3 \cdot 3 \times 10^6$
Patuxent fibroma	—	$1 \cdot 7 \times 10^6$	$1 \cdot 3 \times 10^6$	$1 \cdot 6 \times 10^6$

* Titre calculated as: pock-forming units per ml. on CAM; ID 50 per ml. in rabbit skin; TCID 50 per ml. in tube cultures; plaque-forming units per ml. in plaque assay.

However, the most reproducible method of assay of both myxoma and fibroma virus is by plaque counts on rabbit kidney cell or rabbit embryo fibroblast monolayers, as described earlier in this chapter. We have confirmed the experience of Padgett *et al.* (1962) that the sensitivity of such assays is as great as that of intradermal titration in rabbits, and greater than that of pock counts of myxoma virus on the chorioallantoic membrane (Table 8).

ELECTRON MICROSCOPE STUDIES OF MYXOMA AND FIBROMA VIRUS MULTIPLICATION

French workers have conducted detailed electron microscopic studies of fibroma virus multiplication, which have been summarized by Febvre (1962). The sequence of intracellular events is very similar to that found with other poxviruses (e.g. Ozaki & Higashi (1959) with ectromelia virus, Dales & Siminovitch (1961) with vaccinia virus). After about three hours an area of 'viroplasm' develops in the cytoplasm of infected cells. This appears as a diffuse, very finely granular material in electron micrographs, and contains both DNA (demonstrable by Feulgen staining) and viral antigen. Between the eighth and the tenth hour after infection small areas of condensation of the fine granular component of the viroplasm appear, and these are then progressively surrounded by membranes and take on the characteristic appearance of mature virions.

Cells infected with fibroma virus grow more luxuriantly than the controls, and after the first twenty-four hours the viroplasm increases in size and many more virions appear within it. After the third day more and more of the particles lose their inner core of DNA and are seen as hollow membranes, although because of the production of new virions the infectious titre remains unchanged up to the seventh day. At this late stage the cytoplasm is full of ring-shaped empty forms of the virus, with double membranes.

Time sequence studies of myxoma virus in cultured cells have yet to be carried out, but Epstein, Reissig & De Robertis (1952), and Pötz (1957) have described the electron microscopic appearances of infected rabbit skin. Because of the failure of myxoma virus to produce a rapid cytocidal effect, many of the cells of the skin contained old viral forms of the type described for fibroma by Febvre. Interpretation of the sequence of intracellular events is impossible with this material. However, recent electron micrographs by Dr E. H. Mercer of thin sections of rabbit skin infected with myxoma virus show very clearly the complex structure of the myxoma virions (see Pl. VII).

SUMMARY

Experiments with cultured cells show that the virions of the poxviruses are taken up by phagocytosis, and in susceptible cells they then lose infectivity (undergo eclipse) prior to initiating the production of new viral DNA and protein. Each infective particle sets up a separate factory in the cytoplasm. New virions are produced from about the fifth hour onwards, and production continues until about the thirtieth hour. Most of them remain cell-associated. Cells infected with myxoma virus may be killed, or may survive for a long period and even undergo multiplication. The latter situation is still more common with fibroma virus.

Under suitable conditions plaques are produced in monolayers of rabbit kidney cells, rabbit embryo fibroblasts, and chick embryo fibroblasts; and plaque counts are a sensitive and accurate method of assay of both myxoma and fibroma viruses. Certain strains of myxoma virus produce distinctive plaques, especially in rabbit kidney cell monolayers, and plaque morphology may be used in genetic studies of the virus.

The chorioallantoic membrane of the developing egg can be readily infected with a myxoma virus, with the production of small pocks. Pock counting is a useful method of assay, although not as reproducible as plaque counting. Pock size can be used to characterize certain strains of myxoma virus. Rabbit fibroma virus multiplies on the chorioallantoic membrane but does not produce countable pocks. The newborn mouse can be infected by the intracerebral route with both myxoma and fibroma viruses, without symptoms. Weaned mice are insusceptible.

CHAPTER 7

THE HOST RANGE OF MYXOMA AND FIBROMA VIRUSES

In textbooks of virology it is usual to include, in the descriptions of an animal virus and the disease associated with it, an account of its host range. It is not often that the reader pauses to consider how inadequate the investigations on this subject usually are. In all probability, the virus in question will have been tested in a few laboratory animals, often by one route of inoculation only, and these animals will then have been observed for a few days or weeks to see whether they respond with recognizable signs of infection. Sometimes a more searching analysis of the host response will include assay of the organs for evidence of viral multiplication, or tests for an antibody response. But almost always the range of species inoculated is extremely limited, and the number of individuals tested (when the usual response is negative) is very small. One implication of the first factor is exemplified by the successful use of ferrets for the recovery of human influenza virus (Smith, Andrewes & Laidlaw, 1933), a by-product of investigations of canine distemper. On the other hand, although very large numbers of European hares have been exposed to natural infection with myxoma virus, only a few cases of myxomatosis have been recognized in them. This emphasizes the necessity of testing a very large number of individual animals before pronouncing with certainty that a particular virus does not multiply, or does not produce disease, in a particular host animal.

With these qualifications, myxoma virus has been tested as extensively as any other animal virus, for one of the most important prerequisites for the deliberate use of myxoma virus for rabbit control in Australia was an assurance that it would not adversely affect other animals.

85

THE RESPONSE OF VERTEBRATES TO MYXOMA VIRUS

Over the last decade experience in the field on a vast scale in Europe and Australia has confirmed the early laboratory impression (Aragão, 1927; Hyde & Gardner, 1933; Martin, 1936), and the later more extensive trials by Bull & Dickinson (1937), that under natural conditions myxoma virus will produce recognizable lesions only in some species of *Sylvilagus*, *Oryctolagus cuniculus*, and two species of *Lepus*.

Tables 9 and 10 summarize available information on the production of signs of disease in animals by the inoculation of myxoma virus (usually by the intradermal or subcutaneous route). Three points should be emphasized, (*a*) the criterion of infection has usually been merely the production of signs of disease, (*b*) only small numbers of each species of animal specified have been tested, and (*c*) very few of the large number of potential reservoir hosts of myxomatosis in the Americas have been tested. It is apparent from the tables that apart from the Leporidae, myxoma virus has been found to produce lesions only in embryonic or newborn animals, or in animals infected by highly artificial routes.

The remainder of this chapter will consist of a description of the response of leporids other than *O. cuniculus* to infection with the South American and Californian strains of myxoma virus, where information on this is available. A comparison will also be made of the susceptibility of cultured cells, and the animals from which they are derived, to infection with myxoma virus. The response of both *Oryctolagus* and *Sylvilagus* to fibroma virus closely resembles the response of *Sylvilagus* to myxoma; and the features of fibroma virus infection in different leporids will be described briefly. Succeeding chapters are devoted to descriptions of the pathogenesis and immunology of myxomatosis in *O. cuniculus*.

SOUTH AMERICAN MYXOMA VIRUS

The South American myxoma virus, or derivatives of it, are now enzootically established in *Sylvilagus brasiliensis* or *Oryctolagus cuniculus* in many parts of South and Central America, in Australia, and in Europe. Most published laboratory work has been carried out with material originally derived from South America, principally with the strain recovered by Moses (1911).

TABLE 9. *The response of animals other than leporids to infection with myxoma virus (South American strain)*

Both the common and scientific names of the animals listed in this table have been corrected to conform with the most widely accepted current usage.

	Reference
Domesticated animals	
No signs or symptoms of disease	
Sheep, goats, horses, pigs, cattle, dogs, cats, fowls, pigeons, duck	Data summarized by Bull & Dickinson (1937)
Man	Sanarelli (1898); F. Fenner (unpublished observations 1951).
Laboratory animals	
No signs or symptoms of disease	
Monkey, guinea-pig, mice, rats, ferrets, hamsters	Data summarized in Bull & Dickinson (1937)
Viral multiplication demonstrated, but no disease	
Newborn mice	Andrewes & Harisijades (1955)
Adult mice inoculated via the bile duct	Mims (1964)
Viral multiplication, and lesions	
Chick embryo inoculated on chorioallantoic membrane	Lush (1937)
Wild animals	
No signs or symptoms of disease	
Brush possum, *Trichosurus vulpecula*	
Tasmanian wallaby, *Thylogale billardieri*	
Black-tailed (swamp) wallaby, *Wallabia bicolor*	
Bennett's wallaby, *W. rufogrisea frutica*	
Rufous rat-kangaroo, *Aepyprymnus rufescens*	
Common wombat, *Vombatus hirsutus*	
Fruit bat (flying fox), *Pteropus poliocephalus*	
Water rat, *Hydromys chrysogaster*	Bull & Dickinson (1937)
Southern blue-tongue lizard, *Tilinqua nigrolutea*	
Bearded dragon (jew lizard), *Amphibolorus barbatus*	
Goanna, *Varanus varius*	
White-backed magpie, *Gymnorhina hypoleuca*	
Sulphur-crested cockatoo, *Cacatua galerita*	
Galah, *Cacatua roseicapilla*	
Wedge-tailed eagle, *Aquila audax*	
Grey teal, *Anas gibberifrons*	
No signs of disease, and no production of antibody	
Grey teal, *Anas gibberifrons*	I. D. Marshall (personal communication, 1961)
Black duck, *A. superciliosa* (Gmelin)	I. D. Marshall (personal communication, 1961)
Hedgehog, *Erinaceus algirus algirus*	Placidi (1957)

87

TABLE 10. *The susceptibility of leporids to infection with myxoma and fibroma virus*

| Species | Virus* | Natural infection | Local lesions produced | | | Infectivity of skin lesions for mosquitoes | Reference |
			After inoculation	After bite of infective mosquito	Generalization		
Sylvilagus brasiliensis	South American	+, in Brazil	+†	+	– in adult + in juvenile	+	Aragão (1943) Aragão (1953)‡
S. bachmani	South American	–	+, moderate	...	–	+	Marshall & Regnery (1963)
	Californian	+, in California	+, moderate	+, moderate	–	+	Marshall & Regnery (1963)
	Shope's fibroma	–	–	–	–	...	Regnery & Marshall (1962)§
S. floridanus	South American	–	+, small	...	–	–	Regnery & Marshall (1962)
	Californian	–	...	+, small	–	–	Regnery & Marshall (1962)
	Shope's fibroma	+, in eastern U.S.A.	+, large	+, moderate	– in adult + in newborn	+	See text, this chapter and chapter 11.
S. audubonii	South American	–	+, large	+, moderate	–	+	Regnery & Marshall (1962)
	Californian	–	+, very small	+, very small	–	–	
	Shope's fibroma	–	–	–	–	...	
S. nuttallii	South American	–	+, large	...	+	+	Regnery & Marshall (1962)
	Californian	–	+, small	+, small	–	–	
	Shope's fibroma	–	–	–	–	...	
S. idahoensis	Californian	–	...	+, small	–	–	Regnery & Marshall (1962)
S. transitionalis	South American	–	–	...	–	...	Hobbs (1931)

Oryctolagus cuniculus	South American	+, in Australia, Europe, and South America	+, large prominent	+	+ extensive and severe	+	See text, chapter 7, 8 and 11.
	Californian	+, in California	+, small inconspicuous	+	+ extensive	+	
	Shope's fibroma	−	+, moderate	+, moderate	− in adult (rarely +) / + in newborn	− in adult / + in newborn	
Lepus europaeus	South American	+, in Europe	+, rare	...	+, very rare	...	See text, chapter 7
	Californian	−	+, moderate	...	−	...	F. Fenner (unpublished observations)
	Shope's fibroma	−	−	...	−	...	
L. timidus	South American	+, in Ireland	...	...	...	...	Anonymous (1955)
L. americanus	South American	−	...	...	−	...	Hobbs (1931)
L. californicus	South American	−	...	...	−	...	Hyde & Gardner (1933)
	Californian	−	...	...	−	...	Regnery & Marshall (1962)

* South American, South American type of myxoma virus; Californian, Californian type of myxoma virus; Shope's fibroma, Shope's rabbit fibroma virus.

† −, Negative result; +, positive result; ..., tests not made.
‡ H. de B. Aragão (personal communication, 1953).
‡ D. C. Regnery & I. D. Marshall (personal communication, 1962).

Natural infections with the South American virus or derivatives of it have been described only in *S. brasiliensis, O. cuniculus, Lepus europaeus* and *L. timidus*. Experimentally, lesions have been produced in other North American *Sylvilagus* spp. (Regnery & Marshall, personal communication, 1962), and cultured cells derived for a variety of animals have been shown to support viral multiplication (Chaproniere & Andrewes, 1957, 1958).

Sylvilagus brasiliensis

Most of our information about myxomatosis in *S. brasiliensis* is derived from Aragão (1943), who found that 40 % of wild *S. brasiliensis* bought in a local market in Rio de Janeiro were susceptible to myxomatosis, and developed a localized tumour at the site of subcutaneous inoculation, or where an infected mosquito probed through their skin. His contention that those not susceptible had been immunized by prior infection was supported by the discovery of the naturally infected wild tapeti captured in the State of Rio.

Tumours appeared in the skin of susceptible *S. brasiliensis* five to seven days after they had been bitten by an infected mosquito. They developed slowly, and remained localized at the inoculation site, disappearing gradually after a period of 10–40 days. Tapetis could be infected by mosquito bite, by scarification of the conjunctiva or the skin, or by contact with a diseased domestic rabbit. In a juvenile animal inoculated intradermally, Aragão (personal communication, 1953) found that generalized lesions including mild blepharoconjunctivitis were produced, and virus was isolated from heart blood taken when the animal died seventeen days after the inoculation.

There appears to be substantial immunity to re-infection, but its duration is not known. Under natural conditions in Brazil myxomatosis in *S. brasiliensis* appears to be a summer, mosquito-transmitted disease. The mechanism by which it overwinters has not been studied.

North American Sylvilagus *and* Lepus

Hyde & Gardner (1933) inoculated myxoma virus into twenty western jack-rabbits (*Lepus californicus melanotis* Mearne), six Maryland cottontails (*S. floridanus mallanus* Thomas) and nine Kansas cottontails (*S. floridanus* Allen), with negative results. Most other workers, however, have reported that myxoma virus does multiply in *S. floridanus*, usually producing a trivial lesion but conferring immunity to subsequent challenge with fibroma virus (Shope, 1936*a,b*; K. E. Hyde, 1936; L. Kilham, personal communication).

D. C. Regnery & I. D. Marshall (personal communication, 1962) tested South American myxoma virus in several Californian leporids. In *S. bachmani*, the natural host of the Californian strain of virus, intradermal inoculation with the South American virus produced a localized tumour, and comparative titrations of the virus in *Oryctolagus* and in *S. bachmani* gave identical endpoints. In this host the tumours produced by South American virus were more protuberant than those produced by the Californian virus, and resembled, on a smaller scale, those produced in *Oryctolagus*. The skin titre in *S. bachmani* lesions never rose to a high level, and all attempts at mosquito transfer of the virus from these lesions were negative (Marshall & Regnery, 1963).

The response of *S. nuttallii* and *S. audubonii* to the intradermal inoculation of relatively large doses of the South American virus was strikingly more severe than that found in those animals after infection with the Californian virus. With *S. nuttallii* a very large local lesion was produced, and obvious generalization occurred with ocular and anogenital lesions not unlike those produced in *O. cuniculus* by moderately attenuated strains of the South American virus. Mosquito transmission to *Oryctolagus* was achieved from the lesions produced by the South American virus in *S. nuttallii*, *S. audubonii* and *S. floridanus*, and from the lesion in *S. nuttallii* to *S. audubonii* also.

Lepus europaeus—*the European hare*

Bull & Dickinson (1937) were the first investigators to inoculate the European hare with myxoma virus. Nine animals caught in the field in Australia were injected with a large dose of highly virulent virus, with no effect. Hares are found in many parts of Europe and Australia where *Oryctolagus* occurs, usually in very much smaller numbers; and in both continents the hare population has remained static or has slightly increased during the periods of the great epizootics in the rabbits. Nevertheless, several cases of infection of hares with myxomatosis have now been reported, mainly from France (Magallon, Bazin & Bazin, 1953; Lucas, Bouley, Quinchon & Toucas, 1953; Jacotot, Vallée & Virat, 1954*a*), and also from Great Britain

(Whitty, 1955; Collins, 1955), Poland (Kejdana, 1955), and Australia. Confirmed infection of a mountain hare (*L. timidus*) in Northern Ireland has also been reported (Anonymous, 1955).

Photographs (see Pl. VI, fig. 1) and the recovery of myxoma virus from some of the affected hares, leave no doubt that these hares were infected with myxoma virus. The general consensus is that exceptional animals may show symptoms, rather than that there is any essential alteration in the virus by which it is adapted to multiply and produce lesions in the hare. Jacotot, Vallée & Virat (1955 a) inoculated thirteen hares captured in various parts of Europe with large doses of myxoma virus. They used two strains, one obtained from an acutely fatal case of myxomatosis in a rabbit, and the other from a fatal case in a hare (Jacotot *et al.* 1954 a). In one of the inoculated hares a small lump developed at one inoculation site on the fifth day and persisted until it was removed on the fifteenth day. It contained myxoma virus, and this hare developed specific complement-fixing antibodies. Virus was recovered from the testes of three hares nine, twelve and fifteen days after they had received an intratesticular inoculation. Thus occasional hares appear to be slightly susceptible to infection with myxoma virus, and the cases recovered from the field indicate that very rarely an animal may suffer from severe generalized myxomatosis. Virus obtained from such an animal was no more virulent for other hares than was the virus obtained from the rabbits. The so-called hare fibroma virus of southern France and northern Italy is discussed on page 99.

The relative susceptibilities of intact animals and cultured cells

Chaproniere & Andrewes (1957) compared the relative susceptibility of cultured cells derived from several different species of animals with the response of the intact animals to infection with South American myxoma virus. Their general finding (Table 11) was that cells or organs cultured from a variety of animals supported multiplication of myxoma virus, whereas there was no evidence that multiplication occurred in the corresponding intact animals. In some cases (Chaproniere & Andrewes, 1958) they found that multiplication also occurred in homografts of guinea pig and rat kidney, both when these were infected *in vitro* before grafting or infected by inoculation at the time of grafting. Animals in which such virus growth

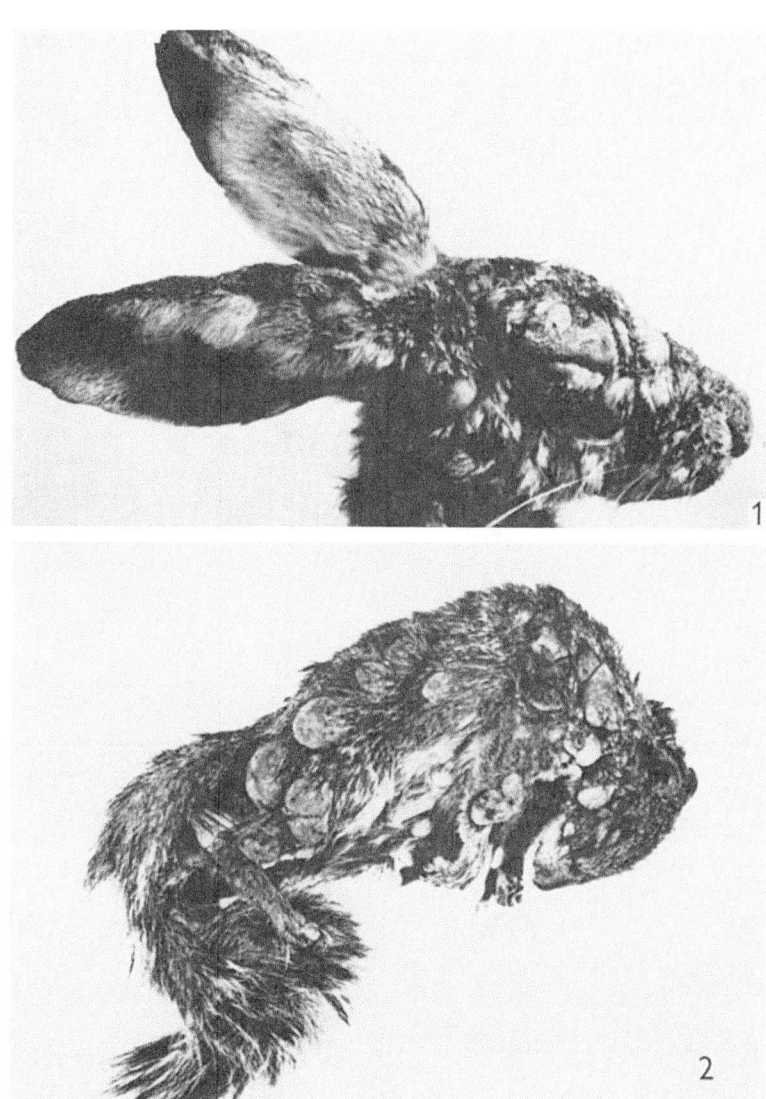

PLATE VI. Fig. 1. Severe generalized myxomatosis in a European hare (*Lepus europaeus*) (from Jacotot, Vallée & Virat, 1954a).
Fig. 2. Naturally occurring severe generalized fibromatosis due to squirrel fibroma virus, in the grey squirrel (*Sciurus carolinensis*) (from Kilham, Herman & Fisher, 1953).

occurred developed neutralizing antibody. They were unable to determine the reason for the insusceptibility of the intact animal.

Some recent experiments by Mims (1964) provide evidence bearing on this problem. Using fluorescent antibody staining to detect viral multiplication, he found no evidence that either myxoma or influenza virus would multiply in the livers of mice injected intravenously with large doses. However, when he injected either of these viruses up to the bile duct they multiplied readily in the periportal

TABLE 11. *The host range of myxoma virus in vitro and in vivo (data from Chaproniere & Andrewes, 1957; Kilham, 1958; F. Fenner, unpublished observations)*

| | | Intact animal | |
Species	Cultured cells or organs	Suckling	Adult
Oryctolagus cuniculus	+*	+	+
Sylvilagus floridanus	+	...	+
Sciurus carolinensis	+	...	−
Man	+	...	−
Ferret	−	...	−
Rat	+	−	−
Monkey	+	...	...
Mouse	+	+	−
Chick embryo	+	+	

* +, Growth; −, no growth; ..., not tested.

hepatic cells. The macrophages of the liver, which phagocytosed large amounts of virus after the intravenous injections, destroyed these viruses and thus protected the susceptible hepatic cells. Inoculation via the bile duct bypassed this protective barrier.

CALIFORNIAN MYXOMA VIRUS

Fenner & Marshall (1957) observed that two strains of myxoma virus recovered from infected rabbits in California (U.S.A./San Francisco/ 1950/1; and U.S.A./San Diego/1949/1) were highly lethal for laboratory rabbits, but produced 'typical' myxomatosis only in rare rabbits which survived for more than twelve days. Most animals died before generalized symptoms had become severe (see Pl. XIII). More recently Marshall *et al.* (1963) obtained eight strains of myxoma virus from naturally infected domestic rabbits and one from *Sylvilagus bachmani* in California. These behaved in laboratory rabbits and on the chorioallantoic membrane in the same way as had the earlier recoveries from California. These differences, the different type of plaque produced on infected rabbit kidney cell monolayers, and the consistent differences between the soluble antigens of the South American and Californian strains, justify treatment of the Californian myxoma virus as the product of a different evolutionary history from South American myxoma virus (see also chapter 15).

TABLE 12. *The susceptibility of several Californian leporids to mosquito bite infection with Californian myxoma virus, and the infectivity of the lesions produced in them for mosquitoes (data from D. C. Regnery & I. D. Marshall, personal communication, 1962)*

First host	Mosquito	Second host	Mosquito	Third host
S. bachmani	26/54*	*S. bachmani*	153/253†‡	*O. cuniculus*
S. bachmani	14/16	*S. audubonii*	0/32	*O. cuniculus*
S. bachmani	8/11	*S. nuttallii*	0/20	*O. cuniculus*
S. bachmani	3/3	*S. idahoensis*	0/15	*O. cuniculus*
S. bachmani	4/7	*S. floridanus*	0/21	*O. cuniculus*

* Numerator = positive results on attempted transfers from *S. bachmani* to recipient host indicated; denominator = number of transfers attempted.

† Numerator = positive results on attempted transfer from recipient host to *O. cuniculus*; denominator = number of transfers attempted.

‡ The rather low efficiency of transfers from *S. bachmani* (first line) is due to the inclusion of many attempts at unfavourable times, for example with poorly developed or very old tumours. All other transfer attempts (lines 2–5) were made at the optimum times.

D. C. Regnery & I. D. Marshall (personal communication, 1962) examined the response of several American leporids to mosquito-bite infection with Californian myxoma virus, with the results summarized in Table 12. Localized tumours were produced in all five species of *Sylvilagus*, but mosquito transfer could be effected only from lesions in *S. bachmani*, which had earlier been identified as the natural host of myxomatosis in California (Marshall & Regnery, 1960). Thus the response of North American *Sylvilagus* to the Californian strain differed considerably from that observed with the South American virus, for the latter produced relatively severe lesions, which were infectious for mosquitoes, in *S. audubonii, S. nuttallii* and *S. floridanus*; whereas the Californian strain produced much larger lesions in *S.*

bachmani than in the other species, and only *S. bachmani* furnished tumours which were infectious for mosquitoes.

The features of the infection of *S. bachmani* with Californian myxoma virus have been described by Grodhaus, Regnery & Marshall (1963). Animals infected by mosquito bite developed small localized tumours between the seventh and the ninth day after infection. Generalization was never observed. The tumour at the site of the infective bite gradually increased in size, and after the fourth or fifth week it usually scabbed and regressed. Mosquito transmission, effected by bites through the tumour, was achieved as early as the seventh day after mosquito bite infection, and the tumours were usually infectious for mosquitoes up to the end of the fifth week. Occasionally a tumour persisted, and was a source of virus for mosquitoes, for a much longer period; in one instance as long as 86 days after infection.

Complement-fixing antibody had usually declined to very low levels by the third month after infection, but animals were immune to re-infection at this time. After intervals as long as eighteen months, however, about half the rabbits tested by mosquito bite infection produced tumours at the site of the infective bite. These usually regressed more rapidly than in normal rabbits, but occasionally they were found to be infective for mosquitoes (D. C. Regnery, personal communication, 1962).

FIBROMA VIRUS

Shope (1932) first recovered fibroma virus from a localized tumour on the leg of an Eastern cottontail shot near Princeton, New Jersey, and since then it has been frequently recovered from similar lesions on wild cottontails from many of the States east of the Mississippi River. Herman, Kilham & Warbach (1956) reported that over half the wild *S. floridanus* taken at Patuxent Research Refuge, Laurel, Md., between July and December 1951, bore a fibroma or gave serological evidence of past infection with fibroma virus. The incidence was much lower in other years. Similar fluctuations have been reported by the Rose Lake Wild-life Experiment Station in Michigan (cited by Herman *et al.* 1956), and New Jersey (R. E. Shope, personal communication, 1962). Natural fibromas in *S. floridanus* almost invariably occur on the legs or feet, occasionally on the nose. Sometimes there are two or three tumours of comparable size on a single animal (R. E. Shope, personal communication, 1962).

Kilham & Fisher (1954) investigated the pathogenesis of fibroma in *S. floridanus*. After the inoculation of large doses of fibroma virus, skin thickenings first appeared within seven or eight days and progressed slowly to produce large fibromas. They persisted for as long as five months, and then disappeared rather rapidly. Virus was found at higher concentrations in the epidermis than in the body of the fibroma, and the maximum titre was not reached until the end of the second month, in spite of the coincident presence of high titre neutralizing antibodies in the serum. In a subsequent communication, Kilham & Dalmat (1955) reported that fibromas which developed in a baby cottontail rabbit did not regress for almost a year, and served as effective sources of virus for mosquito transmission for at least ten months. A wild caught cottontail rabbit which had fibromas of natural origin remained infective for mosquitoes from December through May.

T. M. Yuill (personal communication, 1962), who recovered fibroma virus from a naturally infected *Sylvilagus floridanus* in Wisconsin (Yuill & Hanson, 1964), has carried out a few experiments on the susceptibility of young *S. floridanus*. In very young animals, enormous tumours developed locally after intradermal inoculation, and small secondary lesions were found in other parts of the skin and in the internal organs. Such infections were fatal, rabbits inoculated when one to four days old usually dying by the twenty-fifth day. Slightly older animals (seven to nine days old) developed very large tumours but no apparent secondary lesions, and they usually recovered. In one experiment in which two out of four one-day-old cottontails were inoculated with fibroma virus, tumours appeared in the inoculated rabbits on the fourth day. Eight days after the appearance of these tumours, their two littermates each had small fibromas on the feet. These did not become very large. The mother rabbit was infested with fleas (*Cediopsylla simplex* and *Odontopsylla multispinosus*) and other ectoparasites, and it is probable that transmission was effected by an arthropod vector.

Fibroma virus infection of other American leporids

K. E. Hyde (1936) reported that fibroma virus grew well in jackrabbits and cottontails. In a more systematic survey, D. C. Regnery &

I. D. Marshall (personal communication, 1962) found that fibroma virus failed to produce lesions in *S. audubonii*, *S. nuttallii* and *S. bachmani* (see Table 10).

Infection of Oryctolagus cuniculus *with rabbit fibroma virus*

The South American and Californian types of myxoma virus cause lesions in their natural hosts (*S. brasiliensis* and *S. bachmani*, respectively) very similar to those produced by fibroma virus in *S. floridanus*. However, whereas the myxoma viruses cause a highly lethal generalized infection when they are transferred from their native hosts to *Oryctolagus cuniculus*, Shope's fibroma virus produces only a benign and localized tumour. More careful study (Hurst, 1937c) showed that some degree of generalization often occurs, and it may well be that careful study of fibroma in *S. floridanus* would reveal a similar degree of generalization. But in general fibroma virus produces a localized tumour in both *Sylvilagus* and *Oryctolagus*.

TABLE 13. *The results of intradermal inoculation of a young hare (Lepus europaeus), and domestic rabbits, with serial dilutions of rabbit fibroma virus, hare fibroma virus, and Californian myxoma virus (data from F. Fenner, unpublished observations)*

	Titre of virus	
Virus	In hare	In rabbit
Hare fibroma virus	10^5	10^5
Rabbit fibroma virus	No lesions	10^5
Californian myxoma virus	10^3	10^7

In newborn *O. cuniculus* fibroma virus causes a generalized and often lethal infection (Duran-Reynals, 1945; Harel & Constantin, 1954), and the severity of the disease in adult domestic rabbits can be greatly enhanced by treatment with cortisone or tar (Harel & Constantin, 1954; Ahlström & Andrewes, 1938). Jacotot, Levaditi, Vallée & Virat (1954) described a case of generalized fibromas in an untreated adult rabbit (1 out of 200 under test)—possibly a case of exceptional susceptibility comparable to the occasional hare that exhibits generalized myxomatosis.

Infection of Lepus europaeus with rabbit fibroma virus

The recent descriptions of epizootics of fibroma in hares in southern France and northern Italy (see below) has led to consideration of the susceptibility of the European hare to rabbit fibroma virus. Table 13 sets out the results of parallel titrations of hare fibroma virus, rabbit fibroma virus and Californian myxoma virus in a young hare and in domestic rabbits. P. J. Chapple (personal communication, 1963) was also unable to produce lesions in hares with rabbit fibroma virus.

Infection of squirrels and other North American animals with rabbit fibroma virus

The discovery of squirrel fibroma virus (see next section) led to tests of the susceptibility of squirrels (*Sciurus carolinensis*) to rabbit fibroma virus and myxoma virus. Such tests as were carried out were negative (Kilham *et al.* 1953; Herman & Reilly, 1955; Chaproniere & Andrewes, 1957). A variety of other North American animals which are encountered in the same habitat as cottontails were tested for their susceptibility to fibroma virus, with negative results (Table 14).

TABLE 14. *The response of animals other than leporids to infection with fibroma virus*

No signs of symptoms of disease	References
Weasel, *Mustela frenata noveboracensis*	
Opossum, *D. marsupialis virginiana*	
Woodchuck, *Marmota m. monax*	
Flying squirrel, *Glaucomys v. volans*	Herman, Kilham & Warbach
White-footed mouse, *Peromyscus leucopus noveboracensis*	(1956)
Meadow vole, *Microtus p. pennsylvanicus*	
Chipmunk, *Tamias striatus fisheri*	
Bobwhite quail, *Colinus virginianus*	
Box turtle, *Terrapene carolina*	
Grey squirrel, *Sciurus carolinensis*	Kilham, Herman & Fisher (1953); Chaproniere & Andrewes (1957)

SQUIRREL FIBROMA VIRUS

The fourth member of the myxoma-fibroma subgroup was recovered by Kilham *et al.* (1953) from naturally infected grey squirrels

(*Sciurus carolinensis*) captured in Maryland. Over a period of a year, six animals were captured which exhibited multiple fibromas. Plate VI, fig. 2, illustrates a severe case. Subsequently Herman & Reilly (1955) found references to squirrels bearing multiple fibromas having been shot or collected between 1936 and 1950 from the States of Virginia, North Carolina and New York, as well as from Maryland.

The virus was passed in series in squirrels, both by intracutaneous inoculation and by interrupted and delayed feeding of mosquitoes, the infective donor tissues being 21- and 29-day-old tumours in suckling squirrels (Kilham, 1955). It was also passed to woodchucks (*Marmota monax*) in which localized lesions were produced at the site of inoculation. Material from woodchucks, but not that taken directly from a squirrel, produced small tumours in the skin of the domestic rabbit. Squirrel fibroma virus could be passed in rabbit or squirrel kidney cells, but tests in newborn rabbits, mice and hamsters were completely negative (Kirschstein, Rabson & Kilham, 1958). Neutralization tests confirmed the close relationship of the causative virus to Shope's rabbit fibroma virus; a relationship originally suspected on histological grounds. Generalized myxomatosis was produced in domestic rabbits challenged with virulent or attenuated myxoma virus four weeks after inoculation with squirrel fibroma virus, but it was much less severe than in control animals inoculated with vaccinia virus (Woodroofe and Fenner, 1965).

HARE FIBROMA VIRUS

In 1959 epizootics of cutaneous tumours in hares were reported from the Mediterranean coast of France between Montpellier and La Camargue (Lafenètre, Cortez, Rioux, Pages, Vollhardt & Quatrefages, 1960) and in the Po Valley (Leinati, Mandelli & Carrara, 1959). Similar epizootics had been reported in hares in Germany as long ago as 1908–9 (Dungern & Coca, 1909) and in the Po Valley in 1929 (Mello, 1929). Inoculation of laboratory rabbits with cell suspensions (Dungern & Coca, 1909) or cell-free filtrates (Lafenètre *et al.* 1960; Leinati, Mandelli, Carrara, Cilli, Castrucci & Scatozza, 1961) produced fibromas.

Leinati *et al.* (1961) and Woodroofe & Fenner (1965) showed that there was a substantial level of cross protection, in both directions, in tests carried out with domestic rabbits. In limited experiments Lafenètre et al. (1960) found no cross protection between myxoma virus and hare fibroma virus in rabbits. However Woodroofe and Fenner (1965) showed that rabbits which had recovered from myxomatosis were completely immune to challenge inoculation with hare fibroma virus, and rabbits which had been inoculated four weeks earlier with hare fibroma virus were protected against the lethal effects of inoculation with highly virulent myxoma virus, although they suffered from generalized myxomatosis.

Hare fibroma virus produces very small skin lesions in adult domestic rabbits, and they do not appear until about the twelfth day, whereas comparable concentrations of Boerlage rabbit fibroma virus produce tumours within six days, and grow to a much larger size. Hare fibroma virus produced large tumours in newborn rabbits, which were indistinguishable histologically from those induced by Shope's fibroma virus (Leinati et al. 1961). A. Lucas (personal communication, 1962) found that hare fibroma virus failed to produce a lesion in the single specimen of Sylvilagus he was able to test.

In view of the fact that no other member of the myxoma-fibroma subgroup of the poxviruses had been found to occur naturally elsewhere than in the Americas we at first thought that the hare fibroma virus might have arisen by adaptation of rabbit fibroma virus (or, less likely, myxoma virus) to the European hare since 1952. However, the evidence adduced by Leinati et al. (1959) is highly suggestive of the identity of recent outbreaks with the disease which occurred in Germany in 1909 and in Italy in 1929, long before Shope's rabbit fibroma had been discovered. In addition, workers in England and Australia failed to produce lesions in hares after the inoculation of rabbit fibroma virus (Table 13), and hare fibroma virus fails to produce plaques in cultured cells which support the growth of rabbit fibroma virus (Woodroofe & Fenner, 1965).

Our preliminary serological tests (by gel-diffusion precipitation) confirm the cross-protection tests and show that there are close relationships between hare fibroma virus and other members of the myxoma-fibroma subgroup. It seems certain, therefore, that hare fibroma virus is not an altered form of myxoma or Shope's fibroma virus, but represents a disease sui generis.

SUMMARY

Bearing in mind the inadequacies of laboratory testing, and the large numbers of possibly susceptible wild animals which have not been carefully observed, myxoma and rabbit fibroma appear to have as restricted a host range as any animal virus. Lesions have been produced by them only in members of the Leporidae.

In almost all leporids tested, both the South American and Californian types of myxoma virus and Shope's fibroma virus produce lesions which can best be described as fibromas. Only in *O. cuniculus* is a generalized and usually lethal disease produced by myxoma virus. Shope's fibroma virus produces a fibroma in this host also. Two other related viruses, squirrel fibroma and hare fibroma virus, produce small tumours in *Oryctolagus*, no lesions in *Sylvilagus*, and large fibromas in their natural hosts.

CHAPTER 8

MYXOMATOSIS IN *ORYCTOLAGUS CUNICULUS*: PATHOGENESIS AND HISTOPATHOLOGY

We have just seen that in *Sylvilagus* spp. myxoma viruses, whether of Californian or South American type, produce localized benign fibromas, which in the case of particular associations of virus and host are readily transmitted by arthropods. When an infective arthropod bites a European rabbit, however, an entirely different disease is porduced, with widely generalized lesions and a very high mortality rate. The experience in Australia since 1950 and in Europe since 1952 has shown that this generalized disease can persist on a continental scale as an enzootic-epizootic infection of *Oryctolagus cuniculus*.

Proper understanding of these dramatic outbreaks of myxomatosis in Australia and Europe, and appreciation of the factors which have influenced the accompanying evolutionary changes of the virus, the host, and the disease which is the product of their interaction, requires an understanding of the disease process of myxoma virus infection in individual *Oryctolagus*.

This chapter will give an account of the pathogenesis and the histopathology of myxomatosis in *O. cuniculus*. In the following chapters consideration will be given to the immunological response in infected *Oryctolagus* and the effects of various physiological and environmental conditions on the disease process. Almost all laboratory experiments on myxomatosis in *Oryctolagus* have been carried out with the South American type of virus or derivatives of it.

THE PATHOGENESIS OF MYXOMATOSIS IN *ORYCTOLAGUS CUNICULUS*

Myxomatosis can be spread from one rabbit to another by a variety of methods, but under natural conditions by far the most common and important mode of transfer of the infection is by arthropod vectors. Transmission is mechanical, depending upon the carriage of virus on the biting mouthparts of the vector (see chapter 11). In consequence, the infecting dose is usually small and it is lodged in the skin, either in the epidermis or more usually in the dermis.

In considering the pathogenesis of myxomatosis in *Oryctolagus* as it affects the natural history of the disease in that host, we must therefore devote our attention to the sequence of events which follows lodgement of a small dose of virus in the skin of a rabbit. For the most part, our account is derived from the work of Fenner & Woodroofe (1953), who studied the distribution of virus in various organs of rabbits which had been infected by intradermal inoculation with small doses of virulent myxoma virus.

Their results are summarized in Fig. 4. There was a well-defined sequence of appearance of virus in different organs, and its subsequent multiplication in those organs. Virus was recovered from the inoculation site, and nowhere else, on the first day after infection. A day later the draining lymph node also yielded virus. By the third day, high titres were reached in the inoculation site and the draining lymph node, and virus was also found in the cellular elements of the blood, and in the spleen and lung. Next day generalization was complete and virus was recovered from all tissues examined, including the conjunctival washings and the skin distant from the inoculation site. It is noteworthy that at this time the only macroscopic sign of infection was a small lump in the skin at the inoculation site. Conjunctival swelling was apparent on the fifth day, and by the sixth day there were widespread early signs of generalization, consisting of swellings on skin distant from the inoculation site, blepharo-conjunctivitis, and anogenital oedema. Symptomatology, and the virus titres in all sites, were at their maximum on the eighth and ninth days, and death usually occurred on the tenth day.

As Mims (1964) has pointed out, in viraemia in poxvirus infections the virus is associated with formed elements in the blood, and in myxomatosis the most important cells appear to be the white blood cells, probably the lymphocytes. The serum never contains virus, but in the late stages of severe acute myxomatosis, as in severe variola in man (Downie, McCarthy, McDonald, MacCallum & Macrae, 1953), soluble antigens can be found in high concentration in the serum. Using the gel-diffusion precipitin test Mansi & Thomas (1958) demonstrated soluble antigens in the primary lesion from the third day, in the regional lymph node from the fifth day, and in the serum and most organs from the seventh day until death.

In myxomatosis of *Oryctolagus* due to attenuated strains the sequence of events and their time relationships were the same, but with greatly attenuated strains like neuromyxoma the maximum

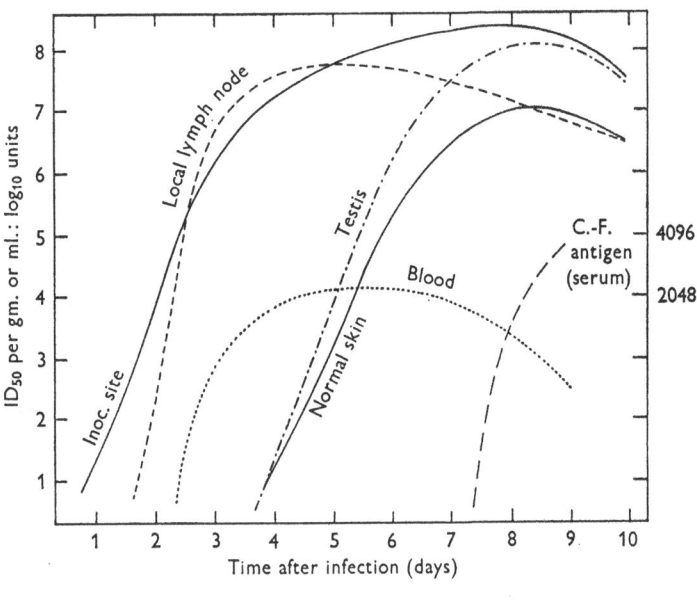

Fig. 4. The sequence of appearance and multiplication of virus in various organs after the intradermal inoculation of small doses of the standard laboratory strain of myxoma virus. The times of appearance of abnormal physical signs, and of soluble antigens, are indicated below the graph (from Fenner & Woodroofe, 1953; Mansi & Thomas, 1958). ± to + + + + +, arbitrary grading of severity of physical signs, 'normal skin', skin on a portion of the body distant from the inoculation site.

titres attained were much lower, viraemia was transient, free antigen did not appear in the blood, and generalization was much less extensive. Some significant comparative figures are shown in Table 15.

The cause of death in myxomatosis is obscure. Although the virus multiplies to reach a high titre in the skin in many parts of the body, studies by virus titration and by fluorescent antibody staining (Mims,

TABLE 15. *Maximum viral concentrations (expressed as log ID50 per gram) in different tissues and organs of rabbits inoculated intradermally with 20 ID50 of myxoma virus*

Assays were made every second day until death, or until the fourteenth day of the disease (data from F. Fenner, unpublished observations).

		Maximum titre and day of occurrence				
Virus	Virulence Grade	Inoculated skin	Regional lymph node	Testis	Ovary	Brain*
Standard laboratory strain	I	8·6 (8)†	7·7 (6)	7·3 (10)	0	0
Aust/Uriarra/2–53/1	IV	8·2 (12)	7·0 (8)	7·3 (12)	1·0 (8)	...‡
Neuromyxoma	V	7·5 (8)	6·0 (8)	6·4 (8)	0	0
England/Nottingham/4–55/1 (attenuated)	V	8·3 (8)	7·6 (8)	4·0 (12)	0	...
California/San Francisco/ 1950/1 (MSW)	I	7·6 (8)	7·8 (8)	6·7 (8)	4·0 (8)	5·0 (8)

* Brain tested on eighth day only.
† $10^{8·6}$ ID50 per gram, on the eighth day after inoculation.
‡ ..., Test not done.

1964) showed only a limited involvement of the adrenals, kidneys, spleen, liver, lung and brain. Thus death cannot be attributed to the growth of virus in a vital organ, although multiplication of Californian type virus in the brain (Table 15) constitutes an exception to this statement. Rabbits infected with Californian myxoma virus often show signs of involvement of the central nervous system.

THE HISTOPATHOLOGY OF MYXOMATOSIS
ORYCTOLAGUS CUNICULUS

Before the spread of myxomatosis through the wild rabbits of Australia and Europe, almost all laboratory investigations with these viruses had been concerned with the possibility that because myxoma virus (and more particularly the related virus of rabbit fibroma) produced pronounced proliferative changes in the skin of rabbits, infections with them might throw light on neoplastic processes. Since many of these early workers were histopathologists there are available numerous descriptions of the histopathological changes in infections with myxoma virus (Hurst, 1937a; Ahlström, 1940a) and fibroma virus (Hurst, 1938a; Ahlström, 1938), both in normal adult rabbits, in newborn rabbits (Duran-Reynals, 1940, 1945; Harel & Constantin, 1954), and animals treated with tar or carcinogens (Ahlström & Andrewes, 1938; Ahlström, 1940b) X-rays (Clemmesen, 1939) or hormones (Smith, 1952; Harel & Constantin, 1954), all of which increase the severity of the lesions.

No attempt will be made to traverse this field in detail, and in this section attention will be concentrated upon an account of the major histological lesions in myxoma and neuromyxoma infections (derived from Hurst, 1937a, b; Ahlström, 1940a), the histology of myxoma in fibroma-vaccinated rabbits (Hurst, 1938b), and a consideration of the nature of the inclusion bodies found in some cells infected with myxoma or fibroma virus.

The histology of myxomatosis due to virulent myxoma virus

Myxomatosis in *Oryctolagus cuniculus* is a generalized infection in which high titres of virus occur in many tissues throughout the body before there are any clinical signs except a lump at the inoculation site. Typical pathological lesions are found in most organs of the body, and are particularly associated with cells of the reticuloendothelial system. The term myxoma is somewhat misleading, for even the large lumps found in the skin are hardly true tumours; their size being due more to an accumulation of mucinous material than to cellular proliferation. There is, nevertheless, some cellular proliferation in the dermis, and in various other sites throughout the body, the characteristic cell being a stellate cell with a large nucleus and abundant cytoplasm, the so-called 'myxoma cell' (see Pl. VII).

Hurst (1937a) has drawn attention to what he regards as a unique lesion in myxomatosis, namely a proliferative endothelial change in the capillaries and small venules. The endothelial cells, like the reticular cells in the spleen, lymph nodes and thymus, proliferate and develop into large stellate cells which narrow the lumen or appear to bud off into the surrounding tissues, where they take the appearance of typical myxoma cells.

In spite of the profound clinical differences between myxoma and fibroma (in the adult domestic rabbit) there is a close relationship between them histologically (Ahlström, 1940*a*). In both the virus attacks the undifferentiated mesenchyma and in the early stages cellular proliferation takes place mainly from certain centres, composed of capillaries and venules with hyperplastic endothelium. Whereas in fibroma the skin tumours are due essentially to cellular proliferation, in myxomatosis the size of the skin lesions is determined by the degree of seromucinous exudation.

The histological changes in neuromyxoma infections

Neuromyxoma is the name given by Hurst (1937*b*) to the attenuated strain of myxoma virus he derived by serial intracerebral passage of virulent myxoma virus in rabbits. The clinical features of the disease caused by neuromyxoma are much milder than those due to virulent myxoma virus, but (especially with large doses) they are undoubtedly those of myxomatosis. However, Hurst considered that the microscopic picture was not, for the typical stellate myxoma cells were never seen. The newly formed cells in the dermis resembled more closely young fibroblasts, but differed from those seen in fibroma. Hurst's description of the gradual change in the histological appearances on passage was confirmed by Rhodes (1938) although the latter worker found no change in virulence on intracerebral passage.

The histological changes in myxoma lesions in fibroma-vaccinated rabbits

The skin reaction to the injection of myxoma virus is accelerated in fibroma-immune rabbits. The lesions are clearly demarcated from the surrounding normal skin and central necrosis appears earlier. By the seventh day a frequent appearance is a plaque with a deep purple centre surrounded by a pale ring, then by a thin haemorrhagic ring, and again by a pinkish oedematous zone, recalling the 'cocarde' of a strongly positive tuberculin response.

Microscopically, Hurst (1938*b*) found typical inclusions in the epithelial cells and vascular changes and myxoma cells in the dermis. The seromucinous exudate was less than in non-immune animals, and there was a massive cellular exudate not found in non-immune rabbits. By the fifth day there was extensive central necrosis involving the epithelium, dermis, and muscle, and later vascular granulation tissue walled off the dead area. Slight but typical myxomatous changes were found in the local lymph node, but not in other organs.

Inclusion bodies in myxoma- and fibroma-infected cells

In the early days of animal virology great importance was attached to the presence and type of inclusion body found in infected cells. More recently the correlation of light microscopic with electron microscopic studies of infected cells has clarified the significance of the different types of inclusion bodies.

There appear to be two types of cytoplasmic inclusion body in poxvirus infections. With suitable staining methods it is always possible to detect localized areas of cytoplasm in which virus multiplication is occurring. These virus 'factories' may be detected by fluorescent-antibody staining or autoradiography (Cairns, 1960), by electron microscopy (the viroplasm, Gaylord & Melnick, 1953), or as Feulgen positive areas, which stain reddish-purple with Giemsa. They have been termed 'B-type inclusions' by Kato, Takahashi, Kameyama & Kamahora (1959).

More prominent than these in sections stained by conventional methods are the eosinophilic inclusion bodies found in fowlpox, ectromelia and cowpox. These are not sites of active viral multiplication, and they may contain many mature virions (Barnard & Elford, 1931), or none at all (Boswell, 1947; Matsumoto, 1958). These have been designated 'A type inclusions' by Kato et al., and they are probably a rather late cellular response to viral infection.

Rivers (1930) described typical cytoplasmic inclusion bodies in the epithelial cells overlying skin tumours in myxomatosis of Oryctolagus. Hurst (1937a) confirmed this and reported that suitably stained myxoma cells also usually showed small inclusion bodies. In fibroma Shope (1932) described the presence of similar inclusion bodies in the epidermal cells in lesions in cottontail rabbits.

In addition very prominent inclusion bodies occur in fibroma cells in the dermis of cottontail fibromas (Fisher, 1953), in tumours of suckling domestic rabbits or adult domestic rabbits treated with X-rays or cortisone (Dalmat & Stanton, 1959), and in cultured cells

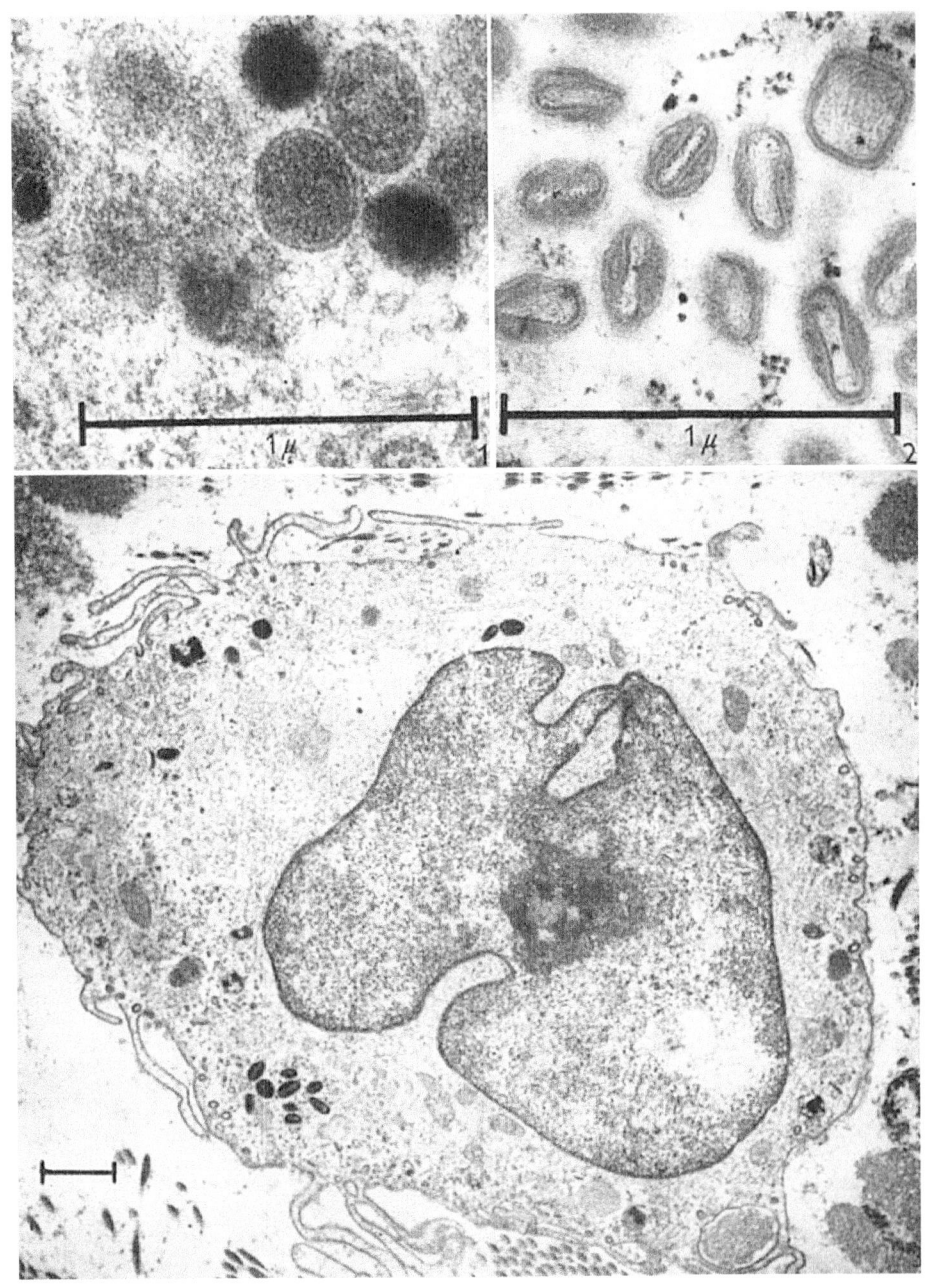

Plate VII

infected with fibroma virus (Renzoni & Castrucci, 1960). These inclusions are stained intensely by the periodic acid-Schiff method, and are probably glycoprotein (Fisher, 1953). Their exact association with the virions of fibroma is uncertain, but their presence is correlated with infectivity of the fibroma for mosquitoes (Dalmat & Stanton, 1959), a topic which is discussed more fully in chapter 11. If sought by Feulgen or Giemsa staining methods, or by fluorescent antibody staining, cells infected with myxoma or fibroma viruses always contain B type inclusions also (Kato & Cutting, 1959).

ELECTRON MICROSCOPY OF MYXOMATOUS LESIONS

Epstein *et al.* (1952) described the occurrence of particles varying in diameter from 100 to 650 mμ in thin sections of the stellate myxoma cells in the dermis of infected rabbits. By modern standards their photographs are so poor that interpretation of them will not be attempted here. As Pl. VII shows, thin sections of epidermal cells of primary lesions contain many typical virions. These are also seen in the myxoma cells in the dermis, and in secondary lesions the latter cells seem to contain virions more commonly than cells of the epidermis.

Pötz (1957) has described the electron microscopic appearances of the skin lesions in myxomatosis. Besides describing typical mature virions and developmental forms in the cytoplasm of cells of the epidermis and the dermis, she commented at length upon the disturbed condition of the collagen fibres. The specificity of the latter is uncertain.

PLATE VII. Electron micrographs of myxoma virus in rabbit skin. Lesion from rabbit infected intradermally with standard laboratory strain seven days earlier, fixed with osmium tetroxide, embedded in Araldite and sections stained on the grid with 1 % uranyl acetate. Preparations made by Dr E. H. Mercer and photographed with a Siemens Elmiskop I.

Fig. 1. Immature forms of myxoma virus in an epidermal cell.

Fig. 2. Mature myxoma virions in an epidermal cell. Note multiple membranes and staining of DNA.

Fig. 3. A 'myxoma cell' in the dermis. These are large rounded cells with very large lobulated nucleus, myxoma virions in cytoplasm, and numerous sinuous processes on cell surface.

ALTERATIONS OF HOST RESPONSE BY PHYSIOLOGICAL FACTORS

The response of rabbits to infection with myxoma and fibroma viruses is affected by a number of factors which alter the physiological state of the host animal. Some of these, which are of considerable epidemiological importance, are dealt with separately (active and passive immunity, chapter 9; ambient temperature, chapter 10). Here we will consider briefly a number of other factors which alter the host response to infection with these two viruses.

Age

Fenner & Marshall (1954) found that animals aged less than three weeks at the time of inoculation died on the fifth or sixth day after infection with a highly virulent strain of myxoma virus, with minimal signs of generalized myxomatosis. Even after very small infecting doses, there was often only a slight lump at the inoculation site and a slight thickening of the margin of the eyelids. If the life of these young animals was prolonged by passive immunity, they developed the same severe lesions as adult rabbits suffering from acute generalized myxomatosis. The resistance of young rabbits to the highly virulent strain of myxoma virus increased with increasing age (see Table 16).

TABLE 16. *The effect of age upon the response of non-immune rabbits to infection with strains of myxoma virus of differing virulence (data from Fenner & Marshall, 1954, 1957; F. Fenner, unpublished observations)*

	Strain of virus and its virulence grade		
Age at time of infection (days)	Standard laboratory strain, Grade I	Uriarra, Grade IV	Neuromyxoma, Grade V
9–14	$5\cdot4\pm0\cdot2$*	$11\cdot7\pm2\cdot1$ no S†	14, 17, 18, S, S, S, S
21–27	$6\cdot0\pm0\cdot3$	$17\cdot0\pm3\cdot9$, no S	All S
47–52	$7\cdot3\pm0\cdot3$	12, 14, 18, S, S, S, S, S, S	All S
120	$10\cdot8\pm0\cdot3$	$26\cdot2\pm12\cdot1$, 19 S out of 45	All S

* Survival time in days.
† S, survivor.

Attenuated strains also caused a more lethal disease in juvenile than in mature rabbits, but the increase in resistance with age which led to a prolongation of the survival time in recently weaned rabbits infected with highly virulent myxoma virus allowed many rabbits infected with an attenuated strain to recover. Neuromyxoma virus caused lethal infections only in rabbits infected when less than two weeks old. In older rabbits it caused the same mild non-lethal infection as was found in adult rabbits.

The influence of the age of domestic rabbits on their response to fibroma virus infection has attracted a good deal of attention. Duran-Reynals (1940) showed that large doses of fibroma virus produced a rapidly fatal inflammatory type of disease in newborn rabbits. Somewhat smaller doses caused generalized fibromatosis, which was usually fatal. His observations were later extended (Duran-Reynals, 1945) and confirmed (Harel & Constantin, 1954). In newborn rabbits fibroma virus could be recovered from the blood up to the thirteenth day after infection, and in fatal cases, from the viscera up to twenty-one days after infection. In normal adult rabbits, on the other hand, virus was never recovered from the blood and could be recovered from the viscera only between the fifth and eighth days (Hurst, 1937c; Duran-Reynals, 1945). In adult rabbits the antibody response was remarkably prompt and effective. Neutralizing antibodies were present to high titre on the fifth day after infection, and immunity to re-infection was present within twenty-four hours of the primary infection and was complete by the fifth day. In the young rabbits, on the other hand, immunity to re-infection was still incomplete by the twelfth day, and neutralizing antibodies were absent on the ninth day but had appeared a week later.

Treatment of adult rabbits with carcinogens or X-rays

Ahlström & Andrewes (1938) showed that after a single treatment of adult rabbits with tar, or benzpyrene, regression of the lesions produced by the intradermal or subcutaneous injection of fibroma virus was greatly delayed; and after intravenous injection generalized fibromatosis, sometimes fatal, was common. Clemmesen (1939) showed that X-irradiation had a similar effect. The modes of action of tar and X-rays do not seem to be the same, however, for while X-rays considerably delayed the onset of active immunity (eleven to fifteen days instead of five to six days), tar had no such effect on the response of non-immune rabbits to a slightly attenuated strain of myxoma virus. In fibroma-immune rabbits challenged with this strain the local lesions which developed at the inoculation sites were larger and persisted much longer than in untreated rabbits (Ahlström, 1940b). The mode of action of tar is not clear; it had no effect in infections of rabbits with highly virulent myxoma virus, IA fibroma virus, or vaccinia virus (Ahlström, 1940b). Recently it has been shown that 20-methylcholanthrene inhibits interferon production (De Maeyer & De Maeyer-Guignard, 1964). This effect seems to be common to several carcinogenic aromatic hydrocarbons, and may explain why tar and benzpyrene enhance fibroma virus infections.

Hormonal effects

Sprunt (1932) found that pregnant rabbits reacted differently from non-pregnant females to infection with myxoma virus. The size of the primary skin lesions, and the frequency and size of secondary skin lesions, were less in the pregnant animals. On the other hand, involvement of the liver, spleen, and lungs was more common and more extensive. The survival time was unaffected by pregnancy, but about half the pregnant animals aborted.

In rabbits infected with virulent myxoma virus, treatment with cortisone led to a striking absence of malaise and anorexia, and a diminution in the mucopurulent discharge from the mucous membranes, but the mean survival time was unchanged (Smith, 1952). In fibroma-infected rabbits cortisone had an effect like X-rays, that is it led to larger and more persistent skin tumours. Harel & Constantin (1954) found that massive doses of cortisone caused adult rabbits to react to fibroma virus in much the same way as tarred rabbits, that is they suffered severe generalized fibromatosis. They concluded, like Duran-Reynals, that immaturity or cortisone led to a modified response because of alterations in immunological reactivity. It is possible, however, that cortisone acts by some mechanism other than the inhibition of the immunological response, perhaps by suppressing interferon production (Kilbourne, Smart & Pokorny, 1961), for Jacotot, Vallée & Virat (1962) found that concomitant cortisone treatment enhanced the immunity to myxomatosis produced by the inoculation of adult rabbits with fibroma virus.

TABLE 17. *Some relationships between host, environmental, and viral factors as they affect the disease picture*

		Type of disease			
Virus	Local tumour only	Acute generalized inflammatory and proliferative, lethal	Acute generalized lethal. Little proliferation	Acute generalized lethal. Little proliferation	Milder generalized proliferative sometimes lethal
South American myxoma from *Sylvilagus brasiliensis*	In *Sylvilagus brasiliensis*	In normal adult *Oryctolagus*		In normal juvenile *Oryctolagus* In adult starved or pregnant *Oryctolagus*	In genetically resistant *Oryctolagus* In passively immune *Oryctolagus* In actively immunized *Oryctolagus* (after fibroma virus infection) In normal *Oryctolagus* held at high environmental temperatures *In normal *Oryctolagus* inoculated with large doses of stored virus, containing few infective particles
South American myxoma attenuated by natural transmission in wild *Oryctolagus*	—	In normal juvenile *Oryctolagus* In normal *Oryctolagus* held at low environmental temperatures	—	—	In normal adult *Oryctolagus*
Californian myxoma from *Sylvilagus bachmani*	In *Sylvilagus bachmani*	—	In normal adult *Oryctolagus*		—
Fibroma from *Sylvilagus floridanus*	In *Sylvilagus floridanus* In normal adult *Oryctolagus*	In newborn *Oryctolagus*	—		In adult *Oryctolagus* treated with tar or cortisone

Except for case marked with asterisk (*), small doses of freshly prepared, active virus inoculated intradermally. Several combinations other than those shown are possible with attenuated South American or virulent Californian myxoma virus.

8

F & R M

The complex interactions between virus and host are summarized in Table 17, which illustrates some of the ways in which variation in genetic and physiological factors can modify the host response and thus the clinical picture.

Concurrent infections with myxoma virus and other viruses

Syverton & Berry (1947) found that cells of a Shope papilloma could be superinfected with several viruses, including myxoma virus. Indeed cells of the papilloma appeared to have a selective affinity for myxoma virus introduced at a distant site.

Semliki Forest Virus has an inhibitory effect on the development of lesions due to both fibroma and myxoma viruses (Ginder & Friedewald, 1951, 1952). The intramuscular injection of Semliki Forest Virus together with myxoma virus greatly modified the response of rabbits to the latter. Generalized myxomatosis always developed but several rabbits recovered. The highest recovery rates (75%) were recorded in rabbits inoculated with myxoma-cell passaged Semliki Forest Virus and myxoma virus. Although Ginder & Friedewald discussed viral interference as the mechanism of this inhibition, the problem needs re-investigation, with especial emphasis on the possible role of interferon.

THE EFFECT OF MYXOMATOSIS ON FERTILITY

An invariable sign of myxomatosis in *Oryctolagus*, except with highly attenuated strains, is inflammation and swelling of the ano-genital region. In the males this involves the prepuce, and the scrotum and testes are always inflamed and swollen. As seen in Fig. 4, the virus titre in the testis itself reaches very high levels in infections with virulent myxoma virus, and with attenuated strains also the testis is a highly vulnerable organ (Table 15), doubtless because of its exposed position and consequent somewhat lower temperature (see chapter 10).

In some male animals which are apparently recovering from acute myxomatosis rupture of the scrotum occurs in the fourth week of the disease, and often leads to death. If such animals recover, the male external genitalia are sometimes deformed due to scarring, and atrophy and fibrosis of the testes are very common. This severe involvement of the male genitalia might be expected to decrease the fertility of buck rabbits which recovered from myxomatosis, and thus diminish the contribution they could make to the genetic resistance of subsequent generations. In the course of breeding experiments with rabbits which had recovered from infection with the moderately virulent KM 13 strain Sobey & Turnbull (1956) did in fact find abundant evidence of partial or complete male sterility among the recovered animals. Six months after their recovery ten out of twenty bucks examined were sterile and some of the others were only intermittently fertile, in the sense that motile sperm were sparse in their ejaculates. One-third of the matings of the latter group failed to result in conception, and the mean litter size of their successful matings (3·16) was also low. As might be predicted from the degree of involvement of the testes, the severity of the disease in bucks which recovered had a marked influence on their fertility. Semen tests made six months after recovery showed motile sperm in the ejaculates of each of six animals which had sustained a relatively mild infection, but in only four out of thirteen animals which had recovered from severe disease.

Sobey & Turnbull suggested that most of the rabbits which had suffered severe infections would not have survived in nature. However, the only assessment available of the fertility of recovered wild bucks did show small differences between normal and myxomarecovered rabbits (Poole, 1960). At Merricumbene, for example, 72 % of non-infected bucks were fertile, as judged by the sperm content of the epididymal tubules, compared with 52 % of recovered animals.

The titre of virus found in the ovaries during acute myxomatosis was very low, except in some cases of infection with Californian type virus (Table 15), so that recovered females would not be expected to show diminished fertility due to direct damage to the ovaries. Sobey & Turnbull encountered two infertile females amongst thirteen animals tested, but found no loss of fecundity amongst the fertile does.

During field investigations in northern New South Wales, Marshall, Dyce, Poole & Fenner (1955) recorded data on the reproductive state of female rabbits which were subsequently correlated with their serological status. It was impossible in this way to obtain information on completely infertile does, but the data shown in Table 18 suggest that recovery from myxomatosis did not reduce either the fertility or fecundity of female rabbits. The slightly larger numbers of embryos found in the recovered animals can probably be ascribed to their greater age, for litter size increases with repeated pregnancies (Brambell, 1944). Studies of the reproductive capacity

TABLE 18. *The reproductive capacity of recovered and uninfected does (data from Marshall, Dyce, Poole & Fenner, 1955, data for August and October 1952)*

	Lactating	Pregnant	Mean no. of embryos	Percentage pregnant or lactating
Recovered	57	104	5·9	76
Uninfected	54	113	5·4	83

of female rabbits in two other areas in New South Wales, by corpora lutea counts, showed no obvious difference between uninfected wild rabbits and those which had recovered from myxomatosis (Poole, 1960).

SUMMARY

In their natural hosts the viruses of the myxoma-fibroma group produce localized fibromas. In *Oryctolagus cuniculus* the myxoma viruses produce a generalized infection, the spread of the virus throughout the body following the same pattern as in other generalized poxvirus infections. There is relatively little multiplication of the virus in the internal organs compared with the skin, and the testis. The 'rash' of myxomatosis consists of swellings which develop especially at the muco-cutaneous junctions (in the eyelids, nose, and anogenital region), and to a lesser extent on the pinnae, legs, and on the body generally.

Histologically these swellings are due to a large extent to an accumulation of a mucinous material together with some degree of cellular proliferation in the dermis. The characteristic lesion consists of proliferation of the endothelial cells of the small capillaries and venules, from which the 'myxoma cells', which are stellate cells with a large nucleus and abundant cytoplasm, bud off into the surrounding tissues. The sequence of spread of virus to distant organs and tissues is similar in myxomatosis due to attenuated strains, but histologically the lesions are more proliferative and contain less sero-mucinous exudate.

In juvenile rabbits highly virulent strains of myxoma virus cause death within about six days, with minimal external signs of infection. Attenuated strains are also more rapidly lethal in very young rabbits than in mature animals, but neuromyxoma virus causes a mild infection even in rabbits only a month old.

Fibroma virus, which produces only a localized skin tumour in normal adult rabbits, causes a lethal generalized infection in newborn rabbits. Generalized fibromatosis may occur in adult rabbits after treatment with tar, cortisone, or X-rays.

Infertility is common in male rabbits which have recovered from severe infections with myxomatosis, due to the extensive multiplication of the virus in the testis and the external genitalia.

CHAPTER 9

MYXOMATOSIS IN *ORYCTOLAGUS CUNICULUS*: THE IMMUNOLOGICAL RESPONSE

Three host factors may play roles of varying importance in determining the outcome of viral infections in susceptible, non-immune animals, namely, interferon, the immune response, and innate resistance (which may depend to an unknown extent upon the first two factors and upon other undefined physiological mechanisms). Specific acquired immunity is of major importance in protecting recovered animals from re-infection. The role of interferon and the importance of innate resistance will be considered elsewhere (chapters 13 and 14). In this chapter we are concerned with the immune response.

Both active and passively acquired immunity affect the epidemiological pattern of viral infections. In man, there appears to be only one antigenic type of the viruses which cause measles, rubella, and smallpox; and recovery from any of these infections is followed by lifelong immunity to the clinical manifestations of re-infection. With a long-lived animal like man the endemic-epidemic pattern of these diseases is governed to a large extent by active immunity in the herd. The importance of active immunity in the epidemiology of myxomatosis is limited by the short life-span of wild rabbits (see chapter 4), especially in areas where there are annual brief epizootics rather than intermittent exposure throughout the year. However, in certain marginal areas in Australia the longevity of rabbits appears to be much greater, and under these circumstances immune rabbits constitute up to 30 % of the population (K. Myers and D. C. Regnery, personal communication, 1963).

Passive immunity, which under natural conditions means maternal transfer of antibodies, may play an important part in protecting newborn infants from viral infections, as exemplified by herpes simplex

(Anderson & Hamilton, 1949). Passive immunity may be of some epidemiological importance in myxomatosis, depending to a large extent upon the seasonal incidence of infection and upon the virulence of the strains of myxoma virus circulating in nature.

SEROLOGICAL TESTS IN MYXOMATOSIS

Several different serological tests have been used to study the antigenic relationships between myxoma virus and other viruses (see chapters 5 and 7), and to follow the immune response in infected animals. Antigens obtained from lesions of infected rabbits, or from infected chorioallantoic membranes, consist of the virions and smaller viral products or components (the soluble antigens). Different serological tests may be used to measure the response to the surface or internal antigens of the virion, to the soluble antigens, or to both.

Serological reactions involving the virions

Two *in vitro* and one *in vivo* reaction can be carried out with suspensions of washed virions, namely, agglutination and complement-fixation (CF), and neutralization tests. Rivers & Ward (1937) showed that myxoma- and fibroma-immune sera agglutinated washed myxoma virions. Ledingham (1937), using a micro-agglutination test, showed reciprocal relationships between virions of myxoma and fibroma and the appropriate antisera. Rabbits which had recovered from infection with fibroma or myxoma viruses produced agglutinins to suspensions of both. He noted the early development of agglutinins (within three days of the inoculation of fibroma virus) and the presence of agglutinins in the serum of moribund myxoma-infected rabbits. Teixeira & Smadel (1941) showed that myxoma-immune serum, or antiserum produced against the soluble antigens of myxoma, would fix complement in the presence of washed virions. Heating the latter at 56° for thirty minutes had no effect on their reactivity.

Immune sera are also able to neutralize the infectivity of myxoma and fibroma viruses. Neutralization can be demonstrated by the inoculation of serum-virus mixtures into the skin of rabbits, scoring the result by death or survival (Lush, 1939) or by the development of a local lesion (Parker & Bronson, 1941).

The most convenient methods for assay of neutralizing potency are the reduction of pock count (Lush, 1937; Fenner, Marshall & Woodroofe, 1953) or plaque count (Woodroofe & Fenner, 1962, 1965). Typical results are shown in Table 19. It is not possible to compare the results of such assays directly with quantal assays in the rabbit, but with the same serum and virus the 50 % infectivity endpoint with a dose of 30 ID50 was obtained with a serum dilution of 1/500, and with a dose of 300 ID50 with a serum dilution of $\frac{1}{2}$.

TABLE 19. *Neutralization of the infectivity of myxoma virus by a rabbit antiserum, tested by the reduction of the pock count on the chorioallantoic membrane and by the reduction of plaque counts on rabbit kidney cell monolayers (G. M. Woodroofe, unpublished observations)*

			Chorioallantoic membrane		Rabbit kidney cell monolayers	
Serum from	Serum dilution	Virus dilution	Mean pock count	Percentage reduction	Mean plaque count	Percentage reduction
Normal rabbit	1/1	10^{-4}	50	—	146	—
Recovered rabbit	1/1	10^{-2}	28	99·4	66	99·6
	1/10	10^{-2}	65	98·7	65	99·6
	1/100	10^{-2}	Hundreds	—	Hundreds	—
	1/1000	10^{-3}	55	89	110	92·4
	1/100,000	10^{-4}	45	0	150	0

The pock neutralization test is probably the simplest to perform, but since plaque counts are more reproducible than pock counts, the plaque inhibition test is more accurate.

Serological reactions involving the soluble antigens

Antigen-antibody reactions involving the soluble antigens can be demonstrated either by precipitation or complement-fixation tests. The Rockefeller Institute workers used tube precipitation tests and demonstrated specific precipitins in sera from fibroma- and myxoma-immune animals, and in rabbits immunized with the soluble substances (Rivers & Ward, 1937; Rivers *et al.* 1939). Recently precipitin reactions have been studied by the Ouchterlony method (see chapter 5). This has led to the recognition of a number of different soluble antigens, and to the demonstration of clearcut antigenic differences between the South American and Californian types of myxoma virus, as well as between both of these and fibroma virus. The gel-diffusion precipitin test is very convenient for carrying out serological surveys.

The most comprehensive study of complement-fixation reactions with myxoma and fibroma viruses is that of Shaffer (1941). He demonstrated the presence of specific antibodies in the sera of rabbits which had recovered from infection with myxoma and fibroma, and also in animals which had been inoculated with suspensions of heat-inactivated virions. Cross-fixation occurred in all cases, titres with the homologous antigen being the higher. Preparations freed of virions by filtration were shown to have the same antigenic potency as the original crude tissue suspensions. Antigen for complement-fixation tests can be produced from infected chorioallantoic membranes (Lush, 1939), but Mansi (1957a) showed that suspensions of skin lesions of rabbits infected with virulent myxoma virus were more potent and less liable to be anti-complementary.

ACTIVE IMMUNITY IN MYXOMATOSIS

Classical myxomatosis is almost invariably lethal in laboratory rabbits, so that until the recent appearance of attenuated strains (see chapter 13) investigation of active immunity was impossible. There are now a few studies on the development of active immunity in infections with attenuated strains of myxoma virus, and somewhat more on active immunity to fibroma and myxoma in rabbits inoculated with fibroma virus. A few experiments with inactivated myxoma vaccines have been reported.

The development of antibodies in rabbits infected with myxoma virus

Susceptible laboratory rabbits infected with highly virulent myxoma virus fail to develop antibody whether this is tested for by complement fixation, gel-diffusion precipitation, or egg neutralization (Fenner & Woodroofe, 1953; Mansi, 1957b; Mansi & Thomas, 1958). However, Fenner & Woodroofe found that the serum of such rabbits became anticomplementary on the fifth day, and this activity increased until death. There was a large amount of circulating soluble antigen from the eighth day onwards. The anticomplementary activity was interpreted as an index of the presence in the serum of an antigen-antibody complex.

Several studies have been made of the time sequence of antibody production in rabbits infected with attenuated strains of myxoma virus. Fenner & Woodroofe (1953) found that rabbits infected with small doses of neuromyxoma virus showed anticomplementary activity on the seventh day, neutralizing antibody on the tenth day and complement-fixing antibody on the eleventh day. Peak titres of neutralizing antibody were reached by the fourteenth day, whereas the titre of CF antibody continued to rise until the twenty-eighth day.

TABLE 20. *The appearance of soluble antigens, and antibodies to them, in the serum of rabbits at various times after the intradermal inoculation of a small dose of myxoma virus (data from F. Fenner, unpublished observations)*

Strain of virus	Virulence Grade		Days after infection with 10 ID50					
			4	6	8	10	12	14
Standard laboratory strain	I	Antigen	−	.	+	+	.	.
		Antibody	−	.	−	−	.	.
California/San Francisco/ 1950/1 (MSW)	I	Antigen	−	+	+ +	.	.	.
		Antibody	−	−	−	.	.	.
Aust/Uriarra/2–53/1	IV	Antigen	−	.	−	.	+	−
		Antibody	−	.	−	.	+	+ +
England/Nottingham/4–55/1, attenuated	IV	Antigen	−	.	−	.	−	+
		Antibody	−	.	−	.	+	+ +
Neuromyxoma	V	Antigen	−	.	−	.	−	−
		Antibody	−	.	−	.	+	+ +

Table 20 records the results of an experiment in which gel-diffusion tests were used to demonstrate the presence of soluble antigens, and antibodies to them, in the serum of rabbits at various stages after infection with small doses of virulent and attenuated strains of myxoma virus. With the highly virulent strains soluble antigen was found at the end of the first week, and antibody did not appear before the animals died. With the moderately attenuated strains low concentrations of soluble antigen were found at the end of the second week, sometimes together with antibody, and in neuromyxoma only antibody was detected.

Extensive tests with English field strains of myxoma virus, mostly of Grade II or Grade III virulence, showed that on the eleventh day after inoculation it was fairly common to find antigen in the serum, and rare to find antibody (Fenner & Chapple, 1965). Sometimes both occurred together. There was no absolute correlation between the presence of either and the fate of the infected rabbit, although animals which died before the end of the third week had usually shown antigen and rarely showed antibody. Animals which died later, or survived, had often lacked demonstrable antigen and antibody, or had antibody (either alone or together with antigen) in their serum. An occasional survivor had serum antigen only on the eleventh day.

TABLE 21. *Antigen and antibody in the serum and local lesion of rabbits infected with an attenuated strain of myxoma virus (England/Nottingham/4–55/1), determined by gel-diffusion precipitation tests (data from Mansi & Thomas, 1958)*

Days after infection ...		3–5	6–30	31–40	41–50	Recovered
Antigen	Primary lesion	+	+	±	–	...*
	Serum	–	–	–	–	–
Antibody	Primary lesion	–	–	+	+	...
	Serum	–	+	+	+	+

* No lesion remains.

In a long term study, Mansi & Thomas (1958) found that in rabbits infected with an attenuated strain of myxoma virus (England/Nottingham/4–55/1 (attenuated)), antibodies to different soluble antigens appeared at different times. In contrast to our results, they did not find soluble antigen in the serum, but antibody to one of the antigens appeared at the end of the first week, followed later by antibodies to the others (Table 21). The antibodies which were detected last persisted for the shortest period (one to nine months), whereas the antibody which appeared first was still present in the serum of some rabbits for as long as 34 months after the infection.

Active immunity to myxomatosis produced by inactivated vaccines

McKee (1939) and Hyde (1939a) studied the active immunization of rabbits with heat-inactivated myxoma virus (heated at 60° C. for thirty minutes). Five intradermal inoculations at three-to four-day intervals of either heated myxoma virus alone, or heated myxoma virus plus a living culture of pneumococcus type III, resulted in the production of high titre complement-fixing antibody and the development of some resistance to a highly virulent strain of myxoma virus, indicated by increased survival time and, in a few cases, recovery. Pneumococcus type III produces a sharp inflammatory response in the skin of the rabbit, and the inclusion of this agent in the vaccine slightly enhanced its efficiency.

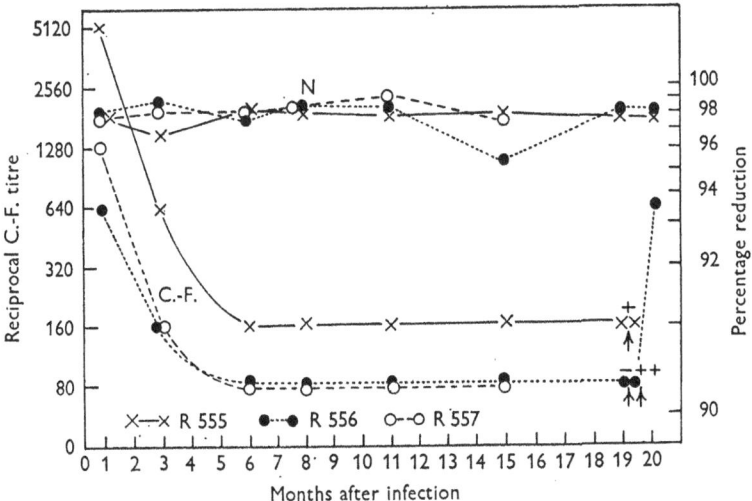

Fig. 5. The persistence of complement-fixing (C-F) and neutralizing (N) antibodies to myxoma virus in wild rabbits which had recovered from natural infection contracted about one month before the first titrations. Nineteen months after infection two survivors (R 555 and R 556) were challenged with 10 ID50 of myxoma virus, and ten days later R 556 with 10^6 ID50. The serological response is shown (from Fenner, Marshall & Woodroofe, 1953). $-$, $+$, $++$ indicate local reactions to challenge inoculations.

More recent workers have had entirely negative results, using formolized vaccine (Jacotot & Vallée, 1953 b) or virus inactivated by heat, phenol, nitrogen mustard or crystal violet (McKercher, 1952).

Resistance to re-infection in rabbits which have recovered from myxomatosis

Fenner *et al.* (1953) found that antibodies persisted for a prolonged period in wild rabbits which had recovered from severe natural infections (Fig. 5). Such animals did not develop generalized myxomatosis after challenge infection with virulent myxoma virus, but when the interval between the initial infection and the challenge inoculation was prolonged a localized lump developed at the inoculation site (Fig. 6). These lumps developed more rapidly than in normal rabbits inoculated with similar doses of virus, they were always clearly demarcated from the adjacent skin, and sometimes they showed the 'cocarde' characteristic of a severe hypersensitivity reaction. After reinoculation of immune rabbits the titre of the complement-fixing antibodies rose when a lump developed at the

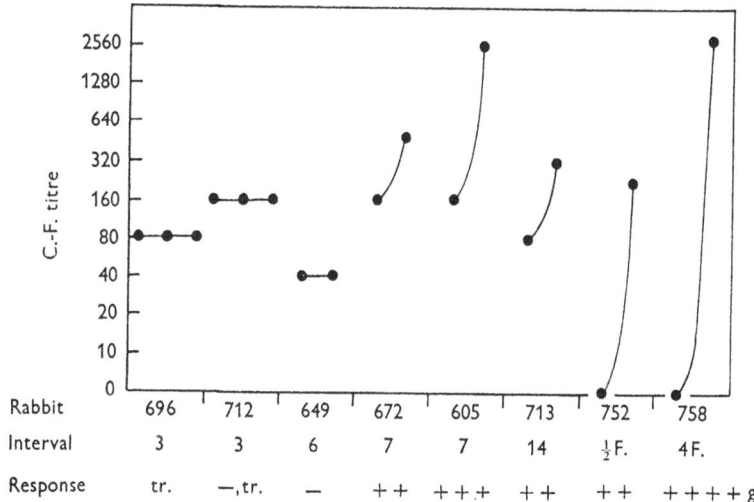

Fig. 6. The response of wild rabbits which had recovered from myxomatosis acquired naturally or from fibroma infection (752 and 758) at the indicated periods (interval in months) before challenge inoculation with myxoma virus (from Fenner *et al.* 1953). − to + + + + are arbitrary measures of the extent of the local reaction at the site of intradermal inoculation of the virus. tr., trace; g, generalized.

inoculation site, but not otherwise, and the extent of the rise was correlated with the severity of the lesions produced by the challenge inoculation. This was particularly evident in challenge infections of fibroma-vaccinated rabbits (Fenner & Woodroofe, 1954).

Recovery from myxomatosis, with all strains of myxoma virus that have been examined, confers almost complete protection against re-infection with any other strain (Table 22). Neutralization tests on the chorioallantoic membrane also indicate complete cross-immunity between strains of myoma virus obtained from California, Brazil, Australia and Europe (Fenner & Marshall, 1957).

125

In the course of experiments upon passive immunization of young rabbits with myxoma-immune serum, Fenner & Marshall (1954) recorded three survivors among twenty rabbits which had been infected with virulent myxoma virus after having been inoculated intraperitoneally with a dose of immune serum equivalent to one

TABLE 22. *Cross-immunity tests between strains of myxoma virus from different parts of the world (from Fenner & Marshall, 1957; F. Fenner, unpublished observations)*

Strain of virus causing initial infection	Strain with which recovered rabbits were challenged	No. of rabbits	Response: nodule at inoculation site		
			Nil	Small	Large
Aust/Corowa/12–52/2 (KM 13)	Standard laboratory strain	6	0	6	0
	Brazil/Campinas/1949/1 (Lausanne)	6	1	4	1
	Brazil/Campo Grande/12–53/1	6	3	3	0
	U.S.A./San Francisco/1950/1 (MSW)	5	5	0	0
Neuromyxoma	Standard laboratory strain	20	7	10	3
	Brazil/Campinas/1949/1 (Lausanne)	2	0	2	0
	U.S.A./San Francisco/1950/1 (MSW)	2	0	2	0
England/Nottingham/4–55/1, attenuated	Standard laboratory strain	5	0	5	0
	U.S.A./San Francisco/1950/1 (MSW)	5	1	4	0
U.S.A./San Diego/1949/1, attenuated	Standard laboratory strain	6	3	3	0

fortieth of their body weight. Two of these animals were challenged again seven months later. Each survived but suffered generalized myxomatosis which was much more severe in the rabbit which had experienced the milder initial infection. This result accords with many others which suggest that the degree of active immunity depends, to a large extent, on the intensity of the initial antigenic stimulus.

Active immunity in fibroma

Shope's fibroma virus constitutes the most widely used vaccine for the control of myxomatosis in domestic rabbits. Although complement-fixing antibodies cannot be detected until about two weeks after infecting laboratory rabbits with the virus (Fenner & Woodroofe, 1954) resistance to re-infection with fibroma virus develops very rapidly (Clemmesen, 1939; Duran-Reynals, 1945). After the intradermal inoculation of 10^3 ID50 of Boerlage fibroma virus the complement-fixing titre to homologous antigen was always slightly higher than to myxoma antigen, and demonstrable antibody persisted, at a low titre, for one year.

Dalmat (1958 *b*) investigated the serological response of laboratory rabbits and *S. floridanus* to infection with Boerlage fibroma virus. Fibroma-neutralizing antibodies appeared on the seventh day, reached a peak at seven weeks, and disappeared entirely by the fifth month. Re-infection at this stage led to a somewhat different response in the two species of rabbits. In laboratory rabbits there was an accelerated response, the tumour appearing on the second day, reaching maximum size by the fifth day, and then regressing. In the cottontails small lumps appeared by the tenth to twelfth day and then regressed. There was a concurrent increase in neutralizing antibodies in all animals. The failure of antibodies to persist in cottontail rabbits made it impossible to use serum surveys to study the natural history of fibromas in eastern U.S.A. Marshall *et al.* (1963) came to similar conclusions concerning the use of serum surveys in myxomatosis of the Californian brush rabbit.

Although some early investigators failed to demonstrate neutralizing antibody for myxoma virus in fibroma-immune sera, there is no doubt that these do occur. Perhaps the most comprehensive tests were those of Fenner & Woodroofe (1954) who found that each of six rabbits inoculated with 10^3 ID50 of Boerlage fibroma virus developed significant titres of neutralizing antibody by the third week. Titres remained high for 36 weeks and fell slowly thereafter.

Active immunity to myxomatosis following infection with fibroma virus

The existence of myxomatosis in places where domestic rabbits were raised for laboratory or commercial use, first in Brazil and California, and later in Australia and Europe, led to the demand for an effective vaccine. Until recently, the outcome of this was very similar to the situation which developed with human smallpox vaccination. In neither case did inactivated vaccines prove effective, nor have attenuated strains of variola or myxoma virus been used for vaccination. In both cases a naturally occurring related virus has been utilized.

Shope (1932) showed that fibroma virus, recovered from cottontail rabbits in eastern U.S.A., would protect domestic rabbits against myxomatosis, and the reciprocal protection was later demonstrated in cottontails (Shope, 1932; Shope, 1936a; Berry & Lichty, 1936). In early experiments on the protection of domestic rabbits against myxomatosis by vaccination with fibroma virus the challenge inoculations were made a few weeks after vaccination (Hyde, 1936; Hurst, 1938b; Shope, 1938), and under these conditions vaccination was invariably successful. For practical purposes, however, a single vaccination should protect for a prolonged period, and no evidence was available on the duration of protection.

Fenner & Woodroofe (1954) showed that the duration of the protection varied greatly according to the strain of fibroma virus used for vaccination. Protection conferred against small doses of virulent myxoma virus by the OA strain had declined greatly by the third month after vaccination. The Boerlage strain of fibroma virus, which produced much larger tumours in domestic rabbits, produced substantial protection as long as a year after vaccination. Complete protection was rare; usually a small lump developed at the site of inoculation of myxoma virus and underwent an accelerated evolution and decline, but generalization did not occur. Fenner & Woodroofe concluded that the intradermal inoculation of 10^3 ID50 of Boerlage fibroma virus constituted a safe and effective method of vaccination of domestic rabbits against myxomatosis. In view of Duran-Reynal's (1945) report of generalized fibromatosis in suckling rabbits they recommended that only animals older than two weeks should be vaccinated.

Immunity develops very rapidly. Hurst (1937c), for example, found that with an interval as short as two days between vaccination and challenge the course of myxomatosis was modified, and some animals suffered a modified infection even when the two viruses were inoculated simultaneously. Hyde (1939b) and Jacotot, Vallée & Virat (1955c) confirmed these results. Vaccination with fibroma virus has been successfully used for the protection of laboratory rabbits in Australia. The outbreak of myxomatosis in Europe led to a demand for vaccination on a greater scale. Rowe, Mansi & Hudson (1956) describe experiments concerned with the practical use of fibroma vaccine. They concluded that Boerlage fibroma virus, inoculated intradermally, gave serviceable protection; but recommended revaccination at intervals of six months if myxomatosis was still a threat.

In France, Jacotot, Vallée & Virat (1955c, 1958) noted the variable effects of fibroma vaccination. Protection did not appear to be closely correlated with the size of the fibroma produced by vaccination. Some rabbits showed substantial immunity many months after vaccination; in others protection against myxomatosis was slight. As might be expected, protection was more effective and more prolonged when attenuated myxoma virus was used for challenge inoculations. Recently, Jacotot *et al.* (1962) showed that small doses of cortisone, administered to adult rabbits after they had been inoculated with small doses of fibroma virus, enhanced the size of the fibroma tumours and also greatly increased the resistance of the rabbits to infection with myxoma virus. Jacotot *et al.* suggested that this greater protection was due to an enhanced antibody response associated with the larger amount of fibroma virus produced in these rabbits.

Due to the fact that mosquito transmission is rare in Britain, natural infections of domestic rabbits with myxomatosis have been very rare there, and fibroma vaccine is not now used. Outbreaks in domestic rabbits have been common in France, and fibroma vaccine has been used there on a large scale. Owing to the great reduction in the numbers of wild rabbits myxomatosis amongst domestic rabbits is now much rarer than it was at the time of the great epizootics (1953–55), and the use of fibroma vaccine has declined from about ten million to perhaps a million doses annually. Routinely, breeding stock is inoculated twice each year. The satisfactory results now reported are in part due to the reduced incidence of myxomatosis amongst wild rabbits, and the reduced severity of the challenge to which vaccinated rabbits are exposed (H. Jacotot, personal communication, 1962).

Vaccination with attenuated strains of myxoma virus

Attenuated strains of myxoma virus have not yet been used in practice for the vaccination of domestic rabbits. Small doses of neuromyxoma virus produce little sickness in rabbits older than fourteen days, but large doses of this virus may produce undesirably severe symptoms. McKercher & Saito (1964) have produced a highly attenuated variant of Californian type myxoma virus which produces only trivial lesions even with large doses (see page 77). We have confirmed the effectiveness of the resistance conferred to challenge infection with virulent myxoma virus, but long-term experiments have yet to be carried out.

PASSIVE IMMUNITY IN MYXOMATOSIS

Early investigators (Martin, 1936; Hyde & Gardner, 1939) failed to find any evidence that immunity to myxomatosis, or to infection with fibroma virus, could be passively transferred from mother to young, although Hyde (1936) had demonstrated a slight degree of passive immunity to myxomatosis by inoculating adult rabbits on several occasions with blood from immune animals. Jacotot, Vallée & Virat (1954b) re-investigated this problem, using highly virulent myxoma virus for the challenge infections, and likewise failed to demonstrate the transfer of immunity from mother to young. This result, if valid, would be unique, for there is abundant evidence that neutralizing antibodies are present in the serum of myxoma-immune rabbits, and much evidence of the *in utero* transfer of antibodies in the rabbit (Brambell, Hemmings & Henderson, 1951).

Using more sensitive criteria Fenner & Marshall (1954) demonstrated unequivocally the existence of passive immunity in myxomatosis, both by inoculating immune serum into young rabbits and by testing the progeny of myxoma-immune does. The successful results of these workers depended upon their use of small doses of virus for challenge inoculation, and the assessment of the results not only in terms of survival or death, but by consideration of the survival time of fatal cases. Recently Harel (1956) has shown that female rabbits hyperimmunized with fibroma virus transferred immunity to their young. He found CF and neutralizing antibodies in the young offspring of the hyperimmunized mothers, and twenty young rabbits inoculated at ages of a few hours to twenty-one days were found to be resistant to infection, whereas most control animals of this age developed fatal generalized fibromatosis. The resistance disappeared in two to three months.

The following account of passive immunity is derived from Fenner & Marshall (1954). The titres of CF and neutralizing antibodies in the sera of newborn rabbits borne by immune mothers are the same as those of the mother. They fall off fairly rapidly so that little antibody can be detected at seven weeks.

The offspring of does which were immune as the result of infection with fibroma by myxoma virus, or neuromyxoma virus alone, were challenged by interrupted feedings of two *Aedes aegypti* mosquitoes per rabbit, the mosquitoes having previously probed through the skin

TABLE 23. *The resistance of progeny of immune does to infection with the standard laboratory strain of myxoma virus by interrupted mosquito biting (excluding repeated attempts at infection) (data from Fenner & Marshall, 1954)*

No. of effective mosquito bites	9 to 11-day-old progeny of		27-day-old progeny of	
	Immune does	Normal does	Immune does	Normal does
0/2	6	0	4	0
1/2	9	1	8	0
2/2	4	6	3	6
Total effective bites				
Total bites	17/38	13/14	14/30	12/12

lesions of a rabbit infected with the highly virulent standard laboratory strain of myxoma virus. Very young non-immune rabbits die within five or six days of infection with minimum doses of virulent myxoma virus, with very slight signs of generalized myxomatosis (see chapter 8). The offspring of immune mothers were infected by mosquito bite much less effectively than were normal rabbits of the same age (Table 23). When they were infected, all of them died, but survival times were invariably longer than those of normal juvenile rabbits (Table 24). Fenner & Marshall give histories of two siblings which failed to become infected on three successive weekly attempts by mosquito bites and a further intradermal inoculation of 15 ID50 of virulent virus. One of them was fatally infected when inoculated with 300 ID50 when 49 days old.

Since 1953, highly virulent strains of myxoma virus have not been common in the field in Australia (chapter 13). It was therefore of some importance to determine the effect of maternal antibody on infection of immature rabbits with the somewhat attenuated viruses which occur commonly in the field. The offspring of normal and immune does were challenged with small doses of a strain which produces about 60 % mortality in adult rabbits. All of the eleven 28-day-old control rabbits died, the mean survival time being 17 days (range 11–25 days). All the 37 test kittens had demonstrable neutralizing antibody in their sera at the time of challenge infection.

TABLE 24. *Increase in survival time due to passive immunity after mosquito-bite infection with the standard laboratory strain of myxoma virus (data from Fenner & Marshall, 1954)*

Age at infection (days)	Survival time (days) ± standard error		
	Normal kittens	Immune kittens	Increase
9–11	5.4 ± 0.2	8.8 ± 0.3	3.4 ± 0.4
21–27	6.0 ± 0.3	10.1 ± 0.4	4.1 ± 0.4
47	7.3 ± 0.3	10.7 ± 0.3	3.4 ± 0.4

Twenty-one died, with a mean survival time of 20·5 days (range 11–41), seven recovered after a mild or a severe attack of myxomatosis, and nine were not affected. The latter were reinfected when 45 days old, when all of them still had demonstrable neutralizing antibody. Five suffered severely and two of these died; the other four recovered after a mild infection.

There is no doubt, therefore, that passive immunity is sufficient to protect young rabbits from infection with the strains of virus which are now common in the field in Australia. Protection may either be complete, when it leaves the individual susceptible to infection as an adult; or partial, when it protects a young rabbit from death and allows the animal to develop a high degree of active immunity. Whether maternal immunity is of practical importance depends primarily on the time of the year when epizootics occur and the virulence of the viruses which then circulate.

SUMMARY

Rabbits infected with either myxoma or fibroma virus produce antibodies against the virions and soluble antigens of both myxoma and fibroma. These can be detected by neutralization or agglutination tests, which involve the virions, and by complement-fixation and precipitin tests, which involve the soluble antigens.

In rabbits infected with myxoma virus antibodies to different soluble antigens develop at different rates, which allows the simultaneous demonstration of antigen and antibody in the serum and tissues of infected rabbits by gel-diffusion precipitin tests. After recovery from myxomatosis rabbits show a high degree of resistance to challenge infection. This is not always absolute, for after prolonged intervals challenge infection frequently produces a localized tumour, and rabbits which survive a mild initial infection may subsequently suffer from generalized myxomatosis. Inactivated myxoma virus will induce antibody production and a slight degree of protection against challenge infection.

Infection of domestic rabbits with fibroma virus rapidly induces active immunity to myxomatosis, but complete resistance is of short duration. Challenge infections at intervals of several months produce responses which vary from a local lesion to fatal, but usually modified, myxomatosis.

Immune does transfer antibody to their foetuses *in utero*, and this may completely protect the young animals from mosquito-bite infection. If passively-immune young rabbits are infected with a virulent strain of myxoma virus they die after a slightly prolonged infection, and passive immunity may protect young rabbits from otherwise fatal infections with attenuated strains of myxoma virus.

CHAPTER 10

MYXOMATOSIS IN *ORYCTOLAGUS CUNICULUS*: ENVIRONMENTAL EFFECTS

The complex interactions between host and parasite which are manifested by the development of disease are affected by many factors. In the previous chapter we considered the effects of active and passive immunity on myxomatosis in *Oryctolagus* and in subsequent chapters special consideration will be given to the effects on the disease of genetic changes in the virulence of the virus (chapter 13), and changes in the genetic resistance of *O. cuniculus* (chapter 14). Here we will consider what may be called environmental effects, that is effects which alter the disease picture with a given combination of virus virulence and host resistance. A major environmental factor in myxomatosis which probably has considerable practical importance is temperature. Other environmental factors which merit consideration include malnutrition and the effects of other infections.

TEMPERATURE

Arising out of his investigations of the effect of temperature on the growth of poliovirus in cultured cells, Lwoff has emphasized the vital role which body temperature may have played in the evolution of viral infections (Lwoff, 1959; Lwoff & Lwoff, 1960). Important evidence supporting Lwoff's thesis comes from observations on the behaviour of myxoma and fibroma viruses, both in cultured cells and in the intact animal.

Experiments in rabbits

Earlier experiments of Thompson (1938) were extended by Parker & Thompson (1942). Thompson found that fibroma infections were markedly inhibited or completely prevented by maintenance of rabbits in incubators operating at 35–38° C., which raised skin temperatures of the rabbits from the usual 33–36° to 38–40° C., and rectal temperatures from 38·7–39·4° to 39·4–40·5° C. Under similar conditions six out of thirteen rabbits survived the inoculation of highly virulent myxoma virus, which killed all rabbits maintained at room temperature. Three of the survivors had shown no sign of infection, and when challenged they died of myxomatosis.

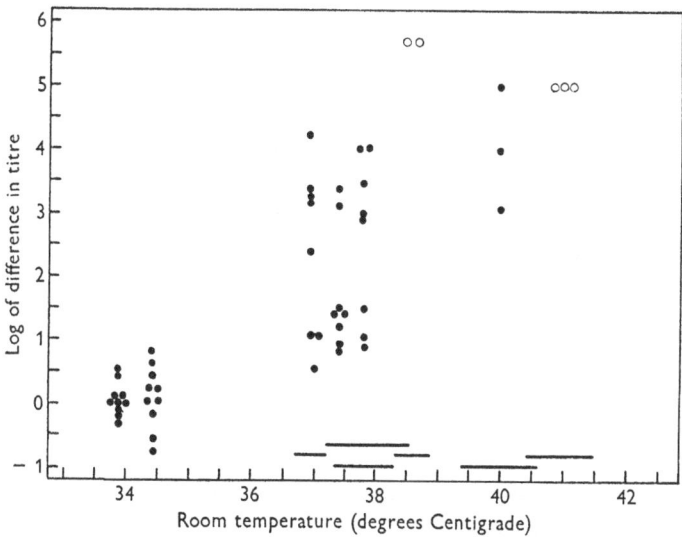

Fig. 7. Effect of environmental temperature on infection with myxomatosis. Each circle represents the result obtained with a single animal. The titre of virus in the control was taken as base (with the scatter shown on the left of the diagram). The diminished titre found in rabbits maintained at temperatures higher than 35° C. is plotted as a positive logarithm. (Log control titre − experimental titre = log difference). Open circles indicate that no lesions appeared in the experimental animals (from Parker & Thompson, 1942).

Parker (1940) had shown that myxoma virus gave a one-particle response when tested by the intradermal inoculation of graded doses into rabbits, and there were only slight inter-rabbit variations in susceptibility. With rabbits maintained at ambient temperatures greater than 37° C. the results were variable (Fig. 7); although replicate titrations in any one animal gave comparable results.

Some rabbits maintained at the higher temperatures failed to become infected. Two such animals which developed no lesions in seventeen days at a mean room temperature of 41° C. were found to have large skin tumours at some of the inoculation sites five days after the temperature was reduced to 36·5° C. The room temperature was again raised and the animals recovered. At temperatures slightly above normal lesions appeared which were circumscribed and more clearly demarcated from the surrounding skin than in the controls. They reached maximum size in six to eight days and then began to regress, healing after the separation of a secondary slough of necrotic tissue. If secondary lesions appeared they were few in number and small in size.

Most of the animals (23/28) held at temperatures of 36–42° C. survived infection. The responses of the survivors to re-inoculation differed greatly. Some animals were completely resistant but showed a strong allergic reaction to the inoculation of large doses of virus. This was manifested by large areas of oedema (3 × 5 cm. in diameter), which developed rapidly and then subsided, usually within 72 hours of the inoculation. One unexpected result was that, of four animals convalescent from generalized infection, two succumbed on re-inoculation, while of seven with no evident secondary lesions five recovered.

Marshall (1959) re-investigated the effect of ambient temperature on the course of myxomatosis paying particular attention to its possible epidemiological importance. For this reason he studied the effect of temperatures which were altered each day so that for eight hours the animals were at the usual room temperature (20–22° C.), and the other sixteen they were at 4° C. (cold room) or 37–39° C. (hot room). These conditions approximate to the diurnal changes found in winter and summer in many rabbit-infested areas of Australia, although as pointed out in chapter 4 the diurnal temperature variation is much less in the rabbit burrow than it is above ground. The rabbits were infected with a standard dose (20 ID50) of a somewhat attenuated strain of virus, which produced about 60% mortality in adult laboratory rabbits maintained under the usual laboratory conditions (at a temperature of 20–22° C.). Marshall's results, summarized in Table 25, indicate the profound effects of both high and low environmental temperature on the response of rabbits to myxomatosis. Plate VIII shows animals from the cold room and the hot room, photographed 20 days after infection with the attenuated strain of virus. Observations were made on the level of viraemia, and the appearance of soluble antigens and circulating antibody, in some rabbits infected under the conditions described. Figure 8 shows that

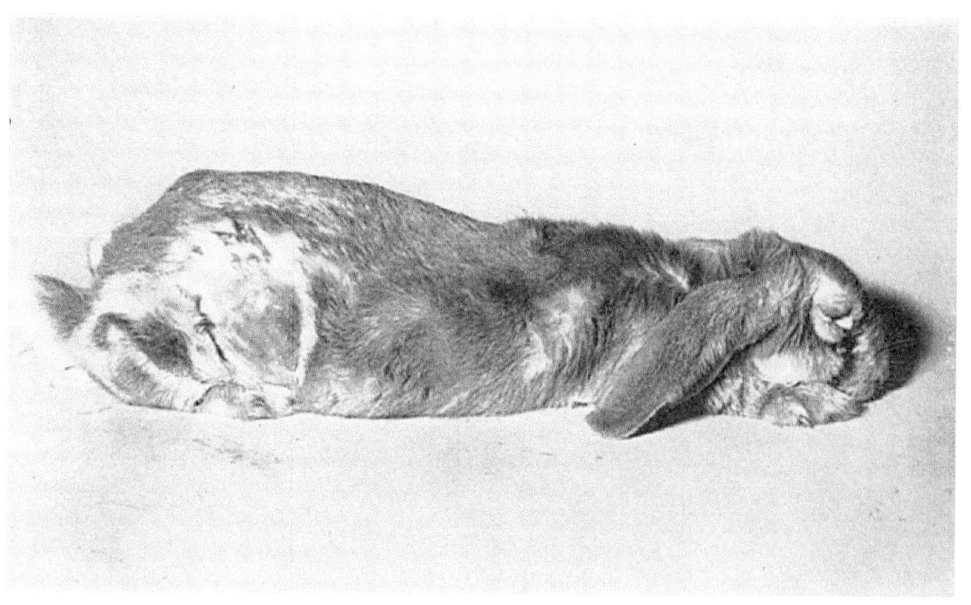

Fig. 1

Fig. 2

PLATE VIII. The effects of high and low temperatures on the symptomatology of myxomatosis (from Marshall, 1959).
Fig. 1. Rabbit infected with KM 13 myxoma virus and showing symptoms typical of those held in the cold room. Photographed twenty days after inoculation of 150 ID 50 of virus.
Fig. 2. Rabbit infected with KM 13 myxoma virus and showing symptoms typical of those held in the hot room. Photographed twenty days after inoculation of 150 ID 50 of virus.

viraemia appeared earliest, and reached the highest levels, in the rabbits exposed to low temperatures, whereas a slight viraemia was seen, for two days only, in the rabbits subjected to elevated environmental temperatures. Circulating soluble antigen appeared at the time of maximum viraemia but was not found at all in animals kept in the hot room. Circulating antibody, on the other hand, was detected in the hot room rabbits from the eleventh day onwards, and

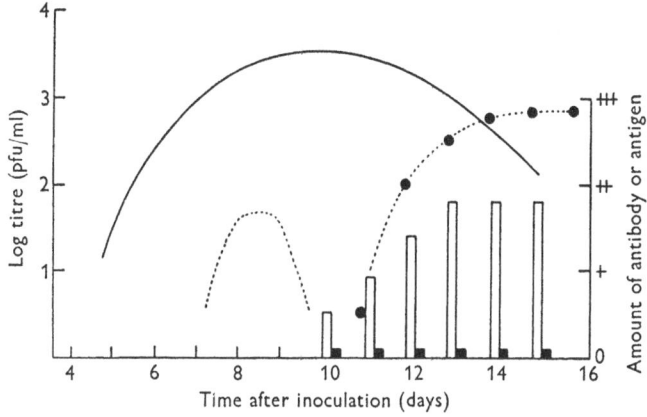

Fig. 8. Viraemia, and serum antigen and antibody, in rabbits held for sixteen hours out o each twenty-four at 4° C. (cold room) or 37°–39° C. (hot room) (modified from Marshall, 1959). Curves on left indicate viraemia: ——— in cold room rabbits, ········· in hot room rabbits. ●······● Antibody in serum of hot room rabbits. White columns = antigen in serum of cold room rabbits; black columns = no antigen in serum of hot room rabbits.

rapidly increased to high levels. Only one out of the five rabbits in the cold room (the only animal which ultimately survived) developed circulating antibody. The other four continued to circulate soluble antigen only up to the fifteenth day, and died a few days later.

Thus rabbits exposed to temperatures at about freezing point for sixteen hours out of each twenty-four developed severe myxomatosis and usually died. They showed high and prolonged viraemia and failed to develop circulating antibody. Rabbits infected with the same dose of the same virus, but exposed to environmental temperatures higher than normal, developed trivial viraemia, failed to circulate soluble antigen, and had circulating antibody from the eleventh day onwards. In these rabbits the lesions at the inoculation sites were localized and healed rapidly, and generalization was slight. The control rabbits, at standard room temperature, developed slightly less severe myxomatosis than those in the cold room, with less viraemia, and they developed circulating antibodies a day later than those in the hot room. The disease picture was more variable in this group than in either of the others.

TABLE 25. *The mortality and symptomatology of myxomatosis (20 ID50 inoculated intradermally) in rabbits held for 16 hours daily at temperatures of 0–$4°$ C. (cold room), 20–$22°$ C. (mild room) and 37–$39°$ C. (hot room)*

For the other eight hours each day the temperature was allowed to rise or fall to ambient temperature ($15°$–$22°$ C.). Data from Marshall (1959).

Virus strain	Virulence Grade		Cold room	Mild room	Hot room
Aust/Corowa/ 12–52/2A	III	Mortality	36/39	23/35	2/23*
		Symptoma-tology	Severe, progressive generalized disease	Variable	Mild with early demarc-ation of lesions
		Viraemia	High and prolonged	Moderately high and prolonged	Low and transient
		CF. antigen in serum	Present in 4/5 from day 10 to day 15 (absent from one survivor)	Present in 2/4 on day 12	Absent
		CF. antibody in serum	Absent except in one survivor (from day 12)	Present in 3/4 from day 13	Present in 3/5 from day 11
Brazil/ Campinas/ 1949/1	I	Mortality	—	6/6	6/6
		Survival times	—	10, 11, 11, 11, 11, 13	9, 9, 10, 12, 12, 12
Nil (control rabbits)		Mortality	0/36	—	0/31

* Seven deaths not due to myxomatosis omitted.

Evidently high and low ambient temperatures affect both the immune response and the degree of multiplication of myxoma virus, both in the skin (there are large changes in skin temperatures, particularly in the animals exposed to cold) and in the internal organs.

In Marshall's experiments rabbits inoculated with the highly virulent Lausanne strain of myxoma virus and held in the hot room responded no differently from the control rabbits, all dying between the ninth and thirteenth days (Table 25). He did not determine whether this point of difference from the results of Parker & Thompson (1942) was due to the strain of virus used, or the fact that the latter workers maintained their rabbits at sustained instead of fluctuating high ambient temperatures.

Minor differences in environmental temperature are probably responsible for seasonal differences in the mortality rate noted by both Mykytowycz (1956) and Sobey (*in litt.*). Mortality rates have been higher than expected in winter and lower than expected in summer.

Figure 9 shows the seasonal differences observed by Mykytowycz in a series of contact infections carried out in an unheated animal house in Canberra with the attenuated strain Aust/Uriarra/2–53/1. After contact infection this strain caused negligible mortalities except

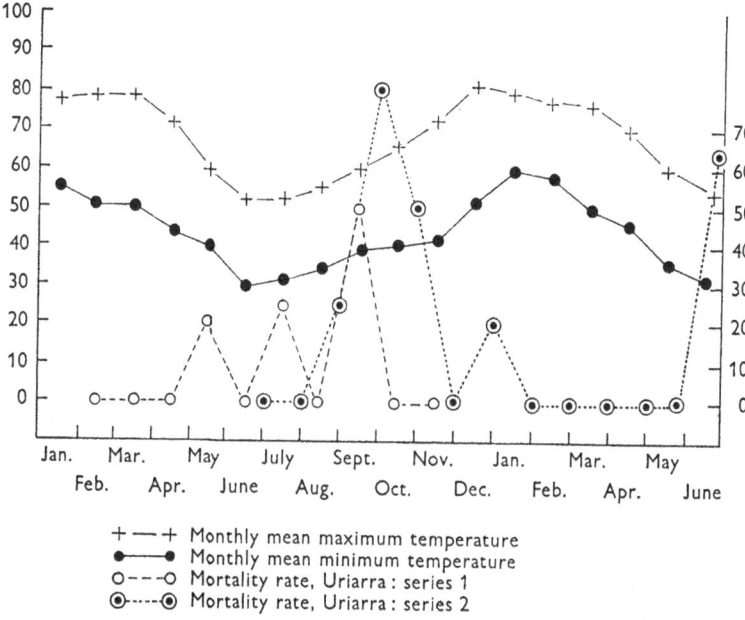

+ — + Monthly mean maximum temperature
●——● Monthly mean minimum temperature
○- - -○ Mortality rate, Uriarra: series 1
◉····◉ Mortality rate, Uriarra: series 2

Fig. 9. Seasonal fluctuations in the mortality rate of rabbits maintained in an unheated animal house and infected by contact with the attenuated Uriarra strain of myxoma virus (modified from Mykytowycz, 1956).

during the winter months, when the mean minimum temperature varied between 0 and 5° C. In early experiments on genetic selection carried out in an unheated animal house in Sydney, where the winter temperatures are much higher than in Canberra, W. R. Sobey (personal communication) also found that myxomatosis was more severe, and mortality rates higher, than during the warmer times of the year. He noticed that a short spell of cold weather would be followed by unexpected deaths from severe myxomatosis.

During his more recent experiments, Sobey (*in litt.*) has used an animal house which is heated in winter, so that the temperature never falls below 22° C. There was still a seasonal effect, which must be attributed to the sparing effect of higher summer temperatures.

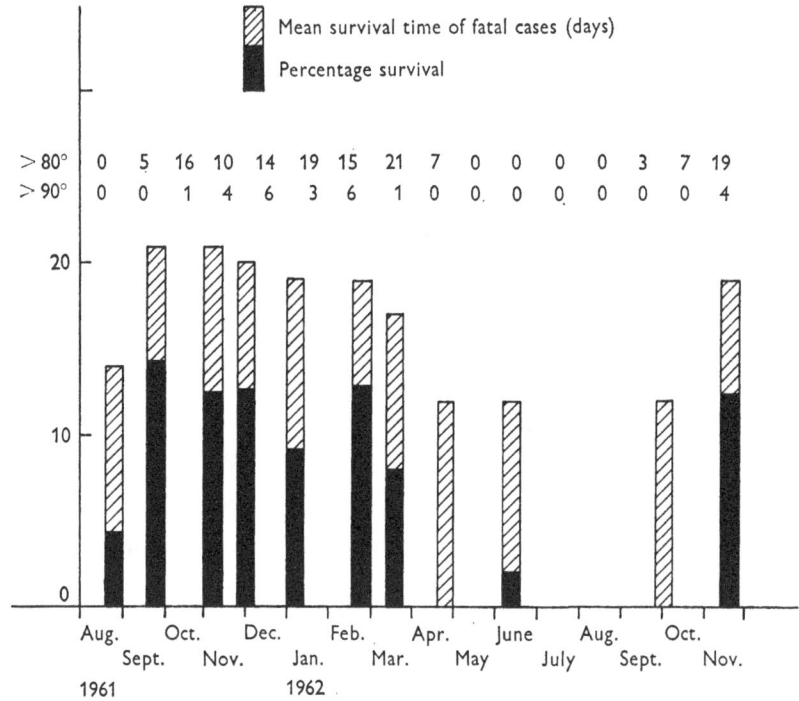

Fig. 10. Seasonal fluctuations in the mortality rate and mean survival times (of fatal cases) of rabbits showing some genetic resistance, which had been infected with the virulent standard laboratory strain of myxoma virus. The animal house was heated in winter so that its temperature never fell below 72° F. (22° C.), but it was not cooled in summer. Figures indicate the numbers of days each month when the maximum daily temperature in the animal house was between 80° and 90° F. and over 90° F. (data from W. R. Sobey, unpublished observations).

Typical results are shown in Fig. 10, which illustrates the mortality rate and mean survival times in rabbits with some genetic resistance, after challenge with the virulent standard laboratory strain of myxoma virus.

Precise experiments are still to be performed, but it is probable that high and low temperatures exert their critical effects during the first few days of infection.

140

Experiments in developing eggs and cultured cells

In the intact animal it is not possible to dissociate the effects of elevated or lowered environmental temperatures on host response and viral multiplication. The immune response (but not some other host responses, like interferon production) can be eliminated by using the developing egg or tissue cultures; and a number of experiments with myxoma and fibroma viruses have been carried out with these systems.

Using Maitland-type cultures of chick embryo tissue Thompson & Coates (1942) found that the growth of myxoma virus was much the same at 32, 35 and 38° C., but was greatly reduced at 40° C. Developing eggs incubated at 39° C. after infection produced virtually no virus, compared with the large yields obtained from eggs incubated at 35–37° C. (page 78).

Kilham (1959) found that in cultures of rabbit kidney cells virulent myxoma virus multiplied at 40° C., though not quite as well as at 38° C., whereas the growth of fibroma was somewhat inhibited at 38° C. and prevented at 40° C.

Changes in virulence associated with growth at high
or low environmental temperature

Marshall's (1959) experiments showed that fluctuating high or low environmental temperatures, of a type which might be encountered in nature in Australia, could have important effects upon the response of individual rabbits to infection with the slightly attenuated virus strains found in the field. One therefore might expect that, with rabbits of the same level of genetic resistance, and viruses of the same degree of attenuation, summer epidemics would be less lethal than those occurring in midwinter. Casual field observations and records of experimental inoculations in unheated animal houses indicate that in the winter infections with attenuated strains usually cause high mortalities. There is also some evidence (Marshall & Douglas, 1961; Fig. 10) that under hot summer conditions the mortality amongst genetically more resistant rabbits infected with moderately or highly virulent strains of virus is less than amongst similar rabbits maintained at a constant temperature of 70° F. in the laboratory.

More important than this immediate physiological effect, however, is the possible effect of temperature in selecting mutants of different virulence. Lwoff & Lwoff (1960) propounded the hypothesis that in the course of the cold season, 'cold' attenuated strains of myxoma virus might be selected rather than 'hot' virulent strains. Elsewhere (chapter 11) evidence will be presented that once attenuated strains have emerged they will be selected for by mosquito transmission. It would be important, therefore, to determine whether lowered temperatures would favour the emergence of attenuated strains of myxoma virus, but no satisfactory experiments on this have been carried out.

MALNUTRITION

Rabbits dying of acute myxomatosis usually eat well until shortly before death. Houlihan & Derrick (1945) investigated the effect of malnutrition, which resulted in a weight loss of about 25 %, on the response of rabbits to infection with virulent myxoma virus. Skin lesions at sites of intradermal inoculation were delayed in appearance in the malnourished rabbits, and were smaller than in control animals, and there were few signs of generalization. Nevertheless, all these rabbits died in the usual time (seven to thirteen days after infection). Starving rabbits would be poor reservoirs of infection, but otherwise these observations do not appear to have any epidemiological importance.

SIMULTANEOUS INFECTIONS WITH MYXOMA VIRUS AND OTHER AGENTS

A small amount of evidence is available on the synergistic and antagonistic effects of certain infections on the course of myxomatosis. In chapter 8 we described experiments by Ginder & Friedewald (1951, 1952) which showed that with both fibroma and myxoma viruses concurrent infections of cultured cells, or of intact animals, with Semliki Forest virus greatly modified the responses. The effects they observed included the reduction of the mortality rate of myxomatosis from the usual 100 % to levels varying in different experiments from 25 to 62 %. E. L. French and F. Fenner (unpublished experiments) showed that in large doses Murray Valley encephalitis virus, which sometimes occurs in areas of Australia where myxomatosis also occurs, had similar effects. However, small doses transmitted by mosquito failed to affect the course of myxomatosis which had also been produced by mosquito bites, so that the phenomenon is unlikely to be of epidemiological importance.

Many rabbits which do not die from acute myxomatosis get snuffles, due to Pasteurella, and this undoubtedly contributes to the mortality in chronic cases.

Mykytowycz (1956, 1959 b) has reported some interesting interactions between myxomatosis and endoparasite infestation. He found (Mykytowycz, 1956) that some heavily parasitized wild rabbits died prematurely after infection with an attenuated strain of myxoma virus, with symptoms of myxomatosis less severe than would have been expected. Later Mykytowycz (1959 b) carried out faecal counts of oocysts and nematode eggs in some of the wild rabbits used in the experiments on genetic resistance reported by Marshall & Douglas (1961). There was a significant fall in the oocyst-ova count at the time of virus generalization (seventh to twenty-first day). The ova count then rose in some animals, and all of these rabbits died. Autopsy showed that these fatal cases were associated with very high levels of infestation with *Graphidium strigosum*. Snuffles and endoparasite infestation are probably only two of several sorts of concurrent infection which may play a part in determining death in myxoma-infected rabbits, especially in animals which survive the acute disease.

SUMMARY

The response of *Oryctolagus* to infection with myxoma virus is affected by the genetic resistance of the host, the virulence of the virus, and by immunological factors. If all these variables are standardized important environmental effects are revealed. The most dramatic and the one which probably has considerable epidemiological importance, is high or low environmental temperature. The former greatly diminishes the severity of symptoms, whereas low temperatures accentuate the disease.

Concurrent infections have variable effects, usually exacerbating the disease, sometimes diminishing its severity. Malnourished rabbits show negligible symptoms but die as rapidly as the controls.

CHAPTER 11

MECHANISMS OF TRANSMISSION
OF MYXOMATOSIS

Since viruses do not multiply outside living cells, transfer of a virus from one host to another involves transfer from the cells of that host to those of the other, without the opportunity for multiplication in the inanimate environment. Broadly speaking, there are three ways in which this may be accomplished. In 'contact infections' the virus infects cells which, by their breakdown, release virus on the surface of the body, or into the respiratory or alimentary tracts, and thus into the environment. In such cases the virus can usually infect a susceptible host through intact mucous membranes, sometimes only through traumatized skin or mucous membranes.

The second method involves a vector, usually another animal, such as an insect, rarely an inanimate object such as a hypodermic syringe. In these cases virus need not be shed into the environment, but is acquired from the blood, or from the infected cells through which the biting parts of the vector, or the needle, may probe. Sometimes, as with the arboviruses, the virus multiplies in the insect vector and is injected with the saliva at a subsequent feeding; sometimes virus merely contaminates the mouth parts, from which it may be dislodged during feeding operations.

The third method of transfer of virus from one host to another, as far as we are aware of no importance in myxomatosis, is via the egg, as in avian lymphomatosis (Rubin, Cornelius & Fanshier, 1961), or *in utero*, as in lymphocytic choriomeningitis of mice (Traub, 1939). In such cases contact infections may also occur between infected and susceptible adult animals. Often more than one mode of transfer exists, for example in fowlpox the diphtheritic form may spread as a respiratory tract infection, whereas the epitheliomatous form is spread by mosquitoes.

Infections with myxoma virus cause very different symptoms in *Sylvilagus* and *Oryctolagus*. In *Sylvilagus* both myxoma and fibroma

144

viruses produce localized tumours in the skin, whereas in *Oryctolagus* myxoma virus produces widespread lesions of the skin and superficial mucous membranes, with the discharge of large quantities of virus in the conjunctival and perhaps in other secretions. It is therefore convenient to consider separately the modes of transfer of myxomatosis in *Sylvilagus* and *Oryctolagus*, although in doing this we do not wish to imply that there are any fundamental differences in the transmission mechanisms.

TRANSMISSION OF MYXOMA AND FIBROMA IN *SYLVILAGUS*

Many examples are known of natural fibroma infections in *Sylvilagus* captured in various parts of eastern U.S.A. and southern Canada, and in all cases the lesions consisted of one or a small number of tumours of the skin, commonly on the foot. Laboratory experience shows that each lump is the result of independent local introduction of the virus.

Only single cases are known of natural infections of *Sylvilagus* with South American or Californian myxoma virus (Aragão, 1943; Marshall & Regnery, 1960), but extensive experiments by Regnery & Marshall (see chapter 7) with Californian myxoma virus indicate that in adult *Sylvilagus* of several species it produces localized skin tumours only. Although there are differences in the response of different species of *Sylvilagus*, the symptomatology of fibromas and myxoma in their natural *Sylvilagus* hosts is almost identical. Except possibly in very young animals there is no generalization and at most a slight and transient viraemia. The skin lesions do not break down and release virus in the scabs (as happens in human smallpox), and since congenital infection can be excluded, the most likely mechanism for virus transfer involves a vector which can pick up virus from the infected cells of the skin tumour.

South American myxomatosis

Aragão (1943) showed that mosquitoes which fed on the myxomatous skin lesions of *Sylvilagus brasiliensis*, but not those which fed on adjacent normal skin, could transfer virus to other rabbits. He further showed that *Aedes aegypti* mosquitoes could transmit myxomatosis up to seventeen days after feeding through a skin tumour on

S. brasiliensis. Assay of virus in several parts of the mosquito showed that between the sixth and seventeenth days virus was found only on the proboscis, and Aragão concluded that transmission was mechanical.

Fibroma in cottontail rabbits

The results of experiments by Kilham & Woke (1953), Kilham & Dalmat (1955) and Dalmat (1959), on the transmission of fibroma in cottontail rabbits will be discussed in some detail because myxoma and fibroma present very similar epidemiological problems in these animals, and a good deal more information is available on fibroma than on myxoma in *Sylvilagus.*

Impressed by the frequent occurrence of natural cottontail fibromas on the feet, Shope (1949) suggested that soil nematodes might be concerned in transmission. Experiments by Rendtorff & Wilcox (1957) with larvae of *Nippostrongylus muris* were negative. Subsequently, Dalmat (1959) reported that in an enzootic area in Maryland, tumours were found as often on the nose and around the eyes as on the legs and feet; and in a field experiment R. E. Shope (personal communication, 1961) found that domestic rabbits exposed in an enzootic area in New Jersey, on a platform above ground level, developed many fibromas on their shaved flanks. In view of the evidence now available on arthropod transmission the nematode vector hypothesis does not warrant further study.

Since there is no viraemia and virus is found only in the tumours (Kilham & Fisher, 1954) the arthropods tested were all fed on or probed through the skin overlying these lesions. Results are summarized in Table 26. There is general agreement that there is no vector specificity of the sort found with the arboviruses, for example: any arthropod which will bite two rabbits in succession is a potential vector of cottontail fibroma. With the insects listed in Table 26 transmission was effected both by interrupted feeding and by refeeding at various intervals up to five weeks after the infective feed. Experiments with chiggers (trombiculid mites) gave negative results, but the life cycle of these arthropods is such that transmission could have occurred only if the virus were transovarially transmitted in the chigger.

Most experiments have been carried out with mosquitoes, and dissections show that virus is invariably associated with the head and mouth parts, mainly the stylets and clypeus (Dalmat, 1959). Mechanical transfer of virus from contaminated mouthparts certainly occurs; whether there is also multiplication in the mosquito is a question which will be discussed elsewhere in this chapter.

TABLE 26. *Arthropods which have been tested for their ability to transmit fibroma or myxoma virus in* Sylvilagus

Arthropod	Virus	Comment	Reference
Mosquitoes			
Aedes aegypti	Fibroma	Positive results recorded from interrupted feeding and after delays of up to 5 weeks	Kilham & Woke (1953); Kilham & Dalmat (1955); Dalmat (1959)
	Californian myxoma	Positive results with *S. bachmani* only	Grodhaus, Regnery & Marshall (1963)
Ae. aegypti, Ae. scapularis	South American myxoma	Positive results with *S. brasiliensis* from interrupted feeding and after delays of up to 17 days	Aragão (1943)
Ae. triseriatus, Culex pipiens, Anopheles quadrimaculatus, Culex quinquefasciatus	Fibroma	Positive results from cottontail tumours	Kilham & Dalmat (1955); Dalmat (1959)
Anopheles freeborni, Aedes sierriensis, Culiseta incidens, Culex tarsalis	Californian myxoma	Positive results from *S. bachmani* tumours	Grodhaus *et al.* (1963)
Fleas			
Cediopsylla simplex, Odontopsyllus multispinosus, Hoplopsyllus affinis	Fibroma	Transmission achieved with difficulty from cottontail tumours	Kilham & Woke (1953)
Bugs			
Rhodnius prolixus, Triatoma infectans, T. phyllosoma pallidipennis	Fibroma	Effective vectors, both by interrupted feeding, and after delays of at least 25 days	Dalmat (1959)
Cimex lectularius	Fibroma	Effective vectors, both by interrupted feeding, and after delays of at least 25 days	Dalmat (1959)
Mites			
Trombicula splendens	Fibroma	Negative. Single individuals do not feed twice on mammals	Dalmat (1959)

In considering the natural history of cottontail fibroma, and by implication that of myxomatosis in *Sylvilagus*, it is important to know when cottontail fibromas become infective and how long they remain

10-2

so. Rabbits captured as adults in enzootic areas may be partially immune from prior infection, and this probably accounts for the refractory state and evanescent tumours described in such animals by Dalmat (1959). In mature susceptible animals lesions appear seven to eight days after infection, reach their maximum size two weeks later, and may persist for three to five months before crusting and regressing. In young animals the tumours develop earlier and persist longer. Kilham & Dalmat (1955) reported the case of an animal inoculated when one month old, in which the tumours appeared after five days and persisted for a year. Earlier reports suggested that tumours were rarely infectious before the fourth week. Infectivity is certainly highest from this time until regression, but Dalmat (1959) has reported that in twenty-two infected rabbits examined by refeeding experiments the tumours of six were first infectious during the second week after inoculation, three during the third week, three in the fourth week, and the remainder after the end of the fifth week. Tumours remained infectious until they were so crusted that the mosquitoes could no longer feed on them.

In the rabbit infected as a juvenile animal, mentioned earlier, positive results were obtained in transmission experiments carried out from the third through the tenth month. The possibility that such persistent tumours might effectively serve as an over-wintering reservoir of virus was suggested by the capture of a naturally-infected cottontail in Michigan in December. This animal had three fibromas, and mosquitoes fed on these during the months of December, January, March and April transmitted the disease to other cottontails. However, there remain some puzzling features about the over-wintering of fibroma virus (see chapter 15).

Californian myxomatosis

Our knowledge about the transmission of Californian myxomatosis is derived entirely from the work of Regnery & Marshall. In the outbreak at Palo Alto, California, in which virus was obtained from the lesion on the leg of a wild-caught *Sylvilagus bachmani* (Marshall & Regnery, 1960), a similar virus was also recovered from each of two batches of wild-caught *Anopheles freeborni* mosquitoes. Grodhaus *et al.* (1963) carried out serial passage of Californian myxoma virus in

148

Sylvilagus bachmani by the bites of *Anopheles freeborni*, and also showed that each of the four other species of mosquitoes adequately tested would serve as a vector. The duration of infectivity of infected mosquitoes was not investigated. The earliest successful transfer was made nine days after the inoculation of *S. bachmani* with Californian myxoma virus, and the latest was made on the eightieth day. There was no opportunity to examine the persistence of infective tumours in juvenile *S. bachmani*.

TABLE 27. *Failure to transfer viruses other than Californian myxoma virus from* Sylvilagus bachmani *by means of mosquito bites (modified from Marshall & Regnery, 1963)*

	Virus strain		
	Attenuated Australian (KM 13)	Virulent South American (Lausanne)	Virulent Californian
Ratio $\dfrac{\text{titre in } O.\ cuniculus}{\text{titre in } S.\ bachmani}$	1000	1	1
Mosquito bite transfers from lesions in S. bachmani	. . .*	0/20	16/20
Virus titre†; skin slices from S. bachmani	. . .	10^4	10^8

* . . ., not done.
† Infectious doses per gram of tissue.

Considerable interest attaches to experiments by D. C. Regnery & I. D. Marshall (personal communication, 1962) on the attempted serial passage of Californian myxoma virus in several species of *Sylvilagus*. The initial donor was *S. bachmani*, and transfers were attempted by interrupted feeding of *Aedes aegypti*. Tumours were readily produced in four other species of *Sylvilagus* by mosquitoes infected from *S. bachmani*. However, attempts to passage the disease by means of the mosquitoes feeding through these tumours were uniformly unsuccessful (see Table 12). Thus the Californian virus can be maintained by serial mosquito bite infection only in *S. bachmani*, and not in other species of *Sylvilagus*.

Having established the high degree of specificity of the host animal, Marshall & Regnery (1963) examined the specificity of the virus strain. Parallel titrations of three strains of myxoma virus in *Oryctolagus cuniculus* and *S. bachmani* showed that *S. bachmani* was much less susceptible than *O. cuniculus* to the production of local tumours by the attenuated Australian virus, but equally sensitive to the others (Table 27). However, attempts at mosquito transfer of South American myxoma virus were uniformly negative, whereas the tumours produced by the Californian virus in *S. bachmani* gave the expected positive results.

TRANSMISSION OF FIBROMA AND MYXOMA IN *ORYCTOLAGUS*

Fibroma in Oryctolagus cuniculus

Unlike myxoma, fibroma virus produces in *Oryctolagus* the same sort of benign localized tumour that is found in its natural host, *S. floridanus*. Such tumours differ from those in cottontail rabbits, however, in two respects, (*a*) they regress rapidly, usually within three weeks of inoculation, and (*b*) mosquito transmission from fibromas in normal adult *Oryctolagus* has been achieved very rarely. Day, Fenner, Woodroofe & McIntyre (1956) for example, recorded five positive results out of 1627 bites by 619 mosquitoes which had probed through or fed on fibromas in *Oryctolagus*, and Dalmat (1959) recorded ten positives (in animals which had more persistent tumours than usual) out of 'many hundreds' of attempts. Dalmat & Stanton (1959) obtained no positives in tests on fibromas in 200 adult domestic rabbits.

However, fibromas which are infectious for mosquitoes can be produced in *Oryctolagus* by treatments which prolong the survival of the tumours. Infection of suckling rabbits is one such method, and treatment of adult rabbits with X-rays (Clemmesen, 1939) or carcinogens (Ahlström & Andrewes, 1938) another. Dalmat (1958*a*) found that persistent tumours produced by these treatments were invariably infectious for mosquitoes when they were tested after the twenty-sixth day (a time at which fibromas in normal *Oryctolagus* had always disappeared). The evolution of lesions in suckling domestic rabbits was more rapid than in either cottontail or treated adult domestic rabbits, and infectivity for mosquitoes developed more rapidly in such animals.

Figure 11, from Dalmat & Stanton (1959), summarizes the rela-

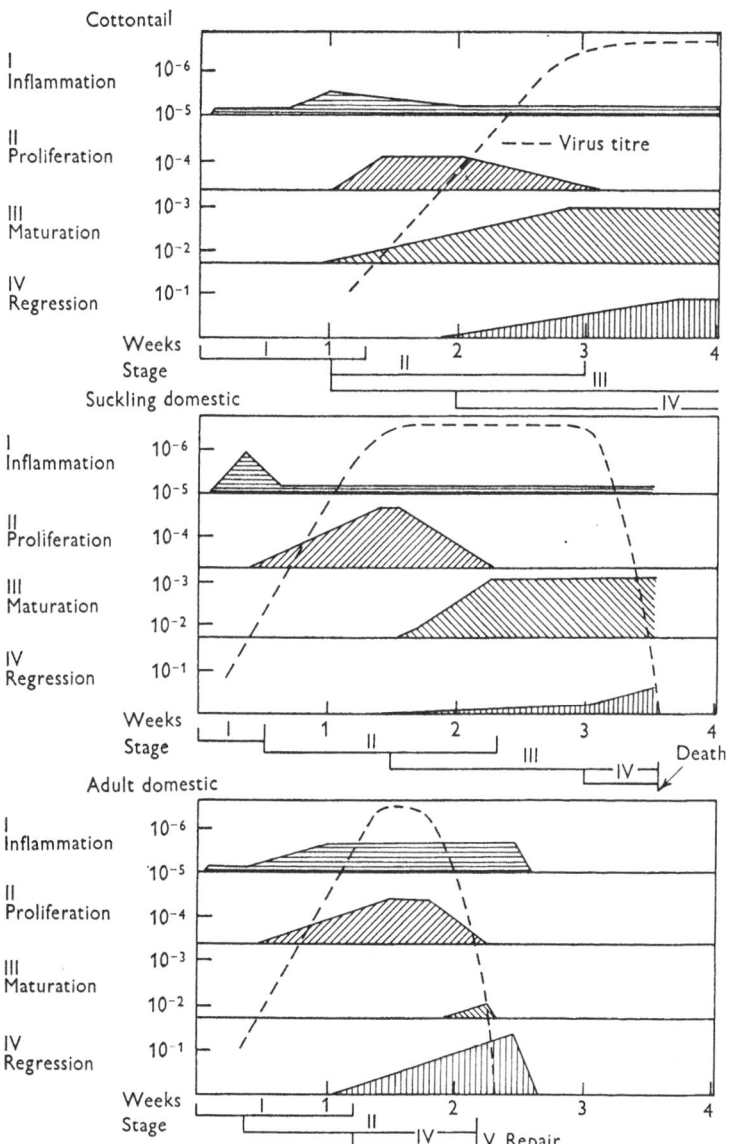

Fig. 11. Diagrammatic pattern of connective tissue response and virus titre after the intradermal injection of fibroma virus into cottontail rabbits, suckling domestic rabbits, and adult domestic rabbits (from Dalmat & Stanton, 1959). Stage I, acute inflammatory stage; Stage II, proliferative stage; Stage III, maturation stage (only at this stage are the lesions regularly infective for probing mosquitoes); Stage IV, regression; Stage V, repair.

tionship between infectivity for mosquitoes and the rate of development and the persistence of fibromas in cottontail, normal adult domestic, and suckling domestic rabbits. The reason for these differences in transmissibility lay not in the titre of virus in the lesions, for this was the same in all cases, but in the 'accessibility' of virus to the probing mosquitoes.

Myxomatosis in Oryctolagus cuniculus

Although myxoma virus evolved in association with species of *Sylvilagus*, it is best known for the disease it produces in *Oryctolagus cuniculus*, and it is with this disease that most work on transmission has been carried out. As described in chapter 8, myxomatosis in *O. cuniculus*, produced by virus as it comes from *Sylvilagus brasiliensis*, is a generalized disease of great severity, in which both viraemia and excretion of virus occur. Hence modes of transfer other than by vectors have to be considered, although it will be shown that in myxomatosis of *Oryctolagus* also transfer by insect vectors is the most important mode of transmission.

Contact infection

Severe myxomatosis in *Oryctolagus* is accompanied by a profuse ocular discharge, rich in virus (van Rooyen, 1937; Fenner & Woodroofe, 1953), and the skin lesions often ooze virus-containing fluid. The question of contact infection therefore arises in a way which does not occur with myxomatosis of *Sylvilagus*. In the laboratory, indeed, myxomatosis is usually regarded as a highly infectious disease, even in the absence of winged vectors. Infection can undoubtedly occur by the direct contact of excreted virus with scratches on the skin or superficial mucous membranes, as may be produced during social or sexual activity. The importance of this mode of contact transmission under field conditions will be considered in the next chapter.

Mechanisms of contact infection other than the entry of virus through abrasions are difficult to evaluate. Mykytowycz (1958b) held infected and susceptible rabbits in flank-to-flank contact, with their heads in opposite directions, for periods of $3\frac{1}{2}$–24 hours in boxes designed to minimize the possibility of transmission by the respiratory or conjunctival routes. Eight of the twenty-four wild rabbits so treated contracted myxomatosis. Since the very common rabbit ecto-parasites *Haemodipsus ventricosus* (a blood-sucking louse) and *Cheyletiella parasitivorax* (a mite) will transmit myxomatosis, although with low efficiency (Mykytowycz, 1958b), they may have been responsible for the observed 'contact' transmissions. However, three out of the seventeen rabbits which had been treated with insecticide and chloroform vapour contracted myxomatosis after contacts with similarly treated infected rabbits of up to twenty-four hours in the boxes, but the effectiveness of this treatment in freeing the rabbits of ectoparasites is open to question.

Mykytowycz also showed that the disease could be transferred by the sexual activity of the infected bucks, and suggested that virus contaminating the feet could be deposited in the sensitive conjunctival sac during the process of face washing.

All investigators have found that oral infection is of negligible importance. Infection by the respiratory route may occur but appears to be unusual. Martin (1936) rarely observed infection in susceptible rabbits kept in cages six inches away from diseased rabbits, and Mykytowycz (1958b) found that susceptible animals held nose to nose, but not in contact, with sick rabbits for twenty-four hours failed to become infected.

Infection by the respiratory tract or conjunctival sac may occur more readily if rabbits are exposed to virus aerosols for a prolonged period, as in the experiments of Calaby (quoted by Mykytowycz, 1958b), when 30% of rabbits were infected in warrens into which a virus aerosol had been blown. Bull & Mules (1944) found that infection sometimes occurred if sick and healthy rabbits were held in a tunnel so arranged that a steady stream of air moved from the infected to the healthy rabbits.

P. J. Chapple and E. A. Boulter (personal communication, 1963) have carried out some interesting experiments on infection via the respiratory tract, using the Henderson apparatus (Henderson, 1952). When this was adjusted to produce dried particles about $1\,\mu$ in diameter, many of which penetrate into the alveoli, they found that the 'Glenfield' strain (Aust/Dubbo/2–51 of Fenner & Marshall, 1957) was highly infectious by the respiratory route, calculated doses of $1\cdot5$ pfu (equivalent to 15 rabbit ID50) infecting all exposed rabbits. In parallel experiments the KM 13 strain (Aust/Corowa/12–52/2) failed to produce infection in any rabbits which had inhaled as much as 15 pfu. The Glenfield strain had been chosen for test because of W. R. Sobey's observation (personal communication, 1962) that it spread within an animal house much more readily than other strains he had used, including the standard laboratory strain. Chapple & Boulter confirmed the ready spread of the Glenfield virus within an animal house.

Experiments in which virus suspensions were instilled into the respiratory tract will not be reviewed since this is a highly artificial procedure which has little relevance to natural transmission.

In summary, then, infection does not occur by feeding, occurs rarely by the respiratory route, and more commonly by the physical contact of infected animals. Infection by contact is much less efficient than transmission by vectors, and its importance in nature lies in its possible role in maintenance of the disease during the periods when vectors are rare or absent.

Infection by arthropod vectors

The importance of insect vectors in myxomatosis was first appreciated by Aragão (1943), who had shown as early as 1920 that the disease could be transmitted between *Oryctolagus* by the cat flea. Torres (1936) first showed that mosquitoes could act as vectors. Since then a large number of arthropods have been tested in the laboratory, and in several cases transmission has been effected by wild-caught insects, or virus has been recovered from them.

Table 28 summarizes the results: every species of mosquito which has been adequately tested is capable of transmitting myxomatosis of *Oryctolagus*, as have other biting flies, fleas, ticks, mites and lice. The only requirement for transmission appears to be the capacity of the same arthropod to bite two rabbits in succession. The lack of specificity, the fact that the interrupted feeding is usually highly efficient, and the absence of an extrinsic incubation period, indicate that transmission is mechanical, at least in the early stages after the infective feed. Whether 'biological' transmission (which implies multiplication of the virus in the vector) may also occur is a problem to be discussed in detail later. Its possible importance lies not in the fact that mechanical transmission would not suffice to explain the epizootic spread of myxomatosis among wild *Oryctolagus*, but that multiplication of virus in an invertebrate vector could be an important mechanism in the selection of virus variants.

Under natural conditions in Australia mosquitoes appear to be by far the most important vectors of myxomatosis (chapter 12), whereas

TABLE 28. *Arthropods which have been shown to be implicated in the transmission of myxomatosis between infected and normal* Oryctolagus

Diptera	Comment	Reference
Culex fatigans	Laboratory*	Torres (1936) and others
C. pipiens australicus	Field (inoculation)	Ratcliffe (unpublished observations)
C. annulirostris	⎰ Field	Mykytowycz (*in* Myers, 1954)
	⎱ Laboratory	Fenner, Day & Woodroofe (1956)
Aedes aegypti	Laboratory	Aragão (1943) and others
Ae. scapularis	Laboratory	Aragão (1943) and others
Ae. alboannulatus	Laboratory	Bull & Mules (1944) and others
Ae. camptorhynchus	Laboratory	Bull & Mules (1944)
Ae. imperfectus	Field (inoculation)	Dyce & Marshall (unpublished observations)
Ae. notoscriptus	Field (inoculation)	Dyce & Marshall (unpublished observations)
Ae. cinereus	Laboratory	⎫
Ae. cantans	Field	⎬ Muirhead-Thompson (1956a)
Ae. annulipes	Field	⎭
Ae. theobaldi	Field (inoculation)	⎫ Myers, Marshall & Fenner (1954)
Anopheles annulipes	⎰ Field (inoculation)	⎭
	⎱ Laboratory	Fenner *et al.* (1956)
An. stephensi	Laboratory	Jacotot, Toumanoff, Vallée & Virat (1954)
An. atroparvus	Field	Muirhead-Thompson (1956b)
Simulium melatum	Field	Mykytowycz (1957)
Austrosimulium furiosum	Field	Myers (unpublished observations)
Fleas		
Ctenocephalides felis	Laboratory	Aragão (1920)
Echidnophaga myrmecobii	Laboratory	Bull & Mules (1944)
Spilopsyllus cuniculi	Field	Lockley (1954)
Lice		
Haemodipsus ventricosus	Laboratory	Mykytowycz (1958b)
Mites and ticks		
Cheyletiella parasitivorax	Laboratory	Mykytowycz (1958b)
Ixodes ricinus	⎧ Laboratory	Shanks, Sharman, Allan, Donald, Young & Marr (1955)
	⎨ Field	Chapple (personal communication, 1962)
I. reduvius	Laboratory	Allan & Shanks (1955)

* 'Laboratory' indicates that transmission has been demonstrated in laboratory experiments. 'Field' indicates that wild-caught arthropods have transmitted infection. 'Field (inoculation)' indicates that suspensions of wild caught arthropods have been shown to contain myxoma virus by inoculation.

in Britain the rabbit flea appears to be the major vector (chapter 17). In the next sections of this chapter the substantial amount of experimental work that has been carried out on the mechanism of mosquito transmission will be reviewed. Relevant information will be taken from experiments with fibroma as well as myxoma virus.

MOSQUITO TRANSMISSION OF MYXOMATOSIS

The source of the virus

In myxoma and fibroma of adult *Sylvilagus* there is no viraemia, and even in juveniles viraemia is rare and transient. The only substantial source of virus for a mosquito are the cells of the tumour and the overlying skin. In myxomatosis of *Oryctolagus*, on the other hand, there is a viraemia, especially in infections with virulent strains (see chapter 8). However, even in myxomatosis associated with high viraemia the significant virus, from the point of view of mosquito transmission, is that in the infected cells of the skin. Fenner, Day & Woodroofe (1952) allowed *Aedes aegypti* to feed either on the ears or the local skin lesions of rabbits infected six to eight days earlier, at a time when large amounts of virus were circulating in the blood stream. Both groups of mosquitoes imbibed demonstrable virus with the blood meals. Virus could be demonstrated in the head and mouth parts, as well as the gut, of the mosquitoes which had bitten through the local lesions, but only in the gut of those which had fed through the normal skin of the ear. Furthermore, feeding of both groups of mosquitoes on normal rabbits at frequent intervals for a period of four weeks resulted in transmission only by the mosquitoes which had fed through the skin lesions. Other experiments (Fenner & Woodroofe, 1953; Fenner, Day & Woodroofe, 1956) showed that mosquitoes feeding through normal skin never became infectious, whereas those feeding either on the primary lesion or on a well-developed secondary lesion of the face, or elsewhere, almost always acquired infectivity.

The fact that fibromas in normal domestic rabbits failed to infect probing mosquitoes, whereas those in suckling domestic rabbits, or adult domestic rabbits treated with carcinogens were infective, led to attempts to determine with somewhat greater precision the cellular location of virus and its relation to mosquito transmission. As is shown in Fig. 11, the peak titre of virus in the fibromas of these different groups of rabbits was the same. A histological study of the lesions led Dalmat & Stanton (1959) to correlate infectivity for mosquitoes with what they regard as a 'mature' fibroma. This was characterized by a stroma which was packed with fibroma cells containing large cytoplasmic inclusions and cytoplasmic inclusion bodies in the epithelium overlying the fibroma. Since the large inclusions in the fibroma cells are apparently not associated with virus (Fisher, 1953), Roberts (personal communication, 1962) utilized fluorescent antibody staining to analyse the situation more precisely. He found that whereas almost every cell in the hypertrophied epidermis of a domestic rabbit infected when five days old contained antigen, infected epidermal cells were very rarely seen in the fibromas of adult *Oryctolagus*, in which antigen occurred in scattered dermal cells only. In myxomatosis of adult *Oryctolagus*, on the other hand, many cells containing antigen occurred throughout both the dermis and the epidermis.

The location of the virus in infective mosquitoes

There is widespread agreement that mosquito infectivity is correlated with the presence of virus on the mouthparts and head. Mosquitoes known to be infective have never been found to contain virus in their salivary glands, and neither myxoma virus (Day *et al.* 1956) nor fibroma virus (Dalmat, 1959) multiplies after being inoculated into the haemocoele. Separate titration of the mid-gut and other parts of the abdomen and thorax in mosquitoes which had fed on viraemic blood through normal skin showed that under these conditions virus was associated with the mid-gut only, and had disappeared entirely by the end of two days. Excreta tested one day after the blood feed contained virus, and it is presumed that all imbibed virus was either excreted or inactivated.

The small size of the component parts of a mosquito's proboscis, and the small and variable virus load of different mosquitoes, made it difficult to determine precisely the distribution of virus associated with the head and mouthparts. By separately treating the heads and the probosces of groups of infective mosquitoes, Day *et al.* (1956) found that myxoma virus occurred in higher concentration in the head than in the proboscis immediately after the infective feed, but three days later the maximum concentration was on the proboscis. Dalmat (1959), on the other hand, found the maximum concentration of fibroma virus on the stylets and clypeus, only a small amount being associated with the head capsule and none with the palpi. Virions may well be spread secondarily to various parts of the mosquito during its 'beak-cleaning' activities, and techniques of dissection are probably not precise or clean enough for the accurate localization of virus on components of the head and mouth parts.

Filshie (1964) has recently succeeded in demonstrating myxoma virions on the maxillae, in the food canal, on the outside of the hypopharynx, but not in the salivary canal, of mosquitoes which have probed through skin lesions (see Pl. IX).

The duration of infectivity

Table 29 summarizes the results of experiments on the duration of infectivity of mosquitoes which have probed through the skin lesions of *Sylvilagus* or *Oryctolagus* infected with either myxoma or fibroma virus.

The very prolonged infectivity of some mosquitoes (e.g. hibernating *An. atroparvus*—lines G and H), and the finding that positive results are often obtained during the third, fourth and fifth weeks after the infective feed in mosquitoes kept at 80° F., have naturally led to suggestions that in addition to mechanical transmission (the existence of which is not questioned), there may also be some multiplication of virus in the mosquito, and subsequently a stage of 'biological' transmission. This proposition is susceptible, with difficulty, to proof : impossible to disprove.

In those cases of arbovirus infection in which there is early mechanical transmission and subsequently prolonged biological transmission (review, Chamberlain & Sudia, 1961), there is a period between the time of early mechanical transmission and later biological transmission when the mosquitoes are non-infective—the extrinsic

---

PLATE IX. Head and mouth parts of *Aedes aegypti*.

Fig. 1. Longitudinal section of head. Magnification × 200. Photograph by Dr M. F. Day. *br*, brain; *fg*, foregut; *fr.g*, frontal ganglion; *m*, muscles; *mp*, mouth parts, *o*, omnatidia; *sog*, suboesophageal ganglion; *sp*, salivary pump.

Fig. 2. Transverse section of mouth parts. Fixed in 0·6 % $KMnO_4$, embedded in Araldite, and photographed by phase contrast. Magnification × 1000 (from Filshie, 1964).

Fig. 3. Myxoma virion on a maxilla. Fixed in 0·6 % $KMnO_4$, embedded in Araldite, stained on grid with uranyl acetate, and photographed with Siemens Elmiskop I (from Filshie, 1964).

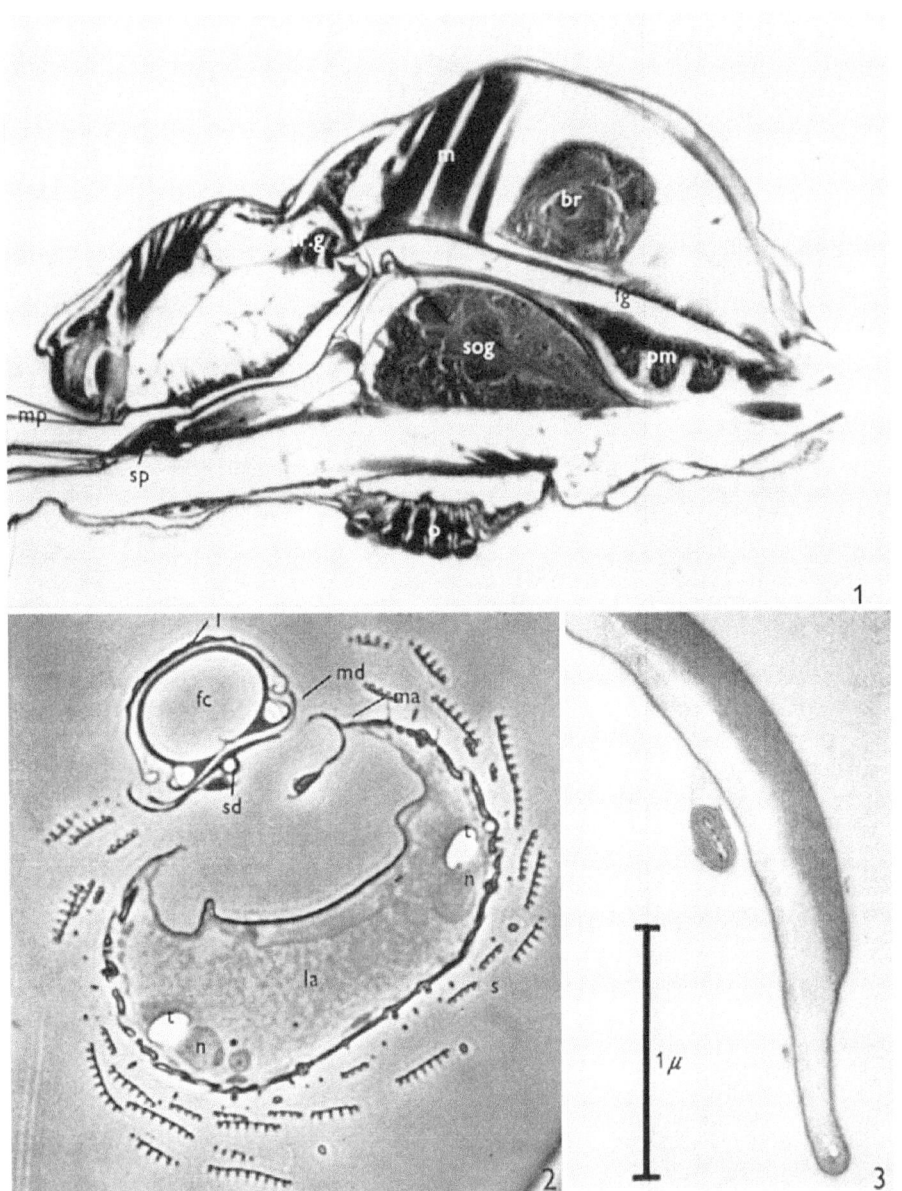

1

2

3

1μ

PLATE IX

incubation period. Infectivity is subsequently associated with high titres of virus in the salivary glands, and infection is due to the injection of infective saliva. If biological transmission does occur with myxoma and fibroma viruses it is quite different from this, for at all times infectivity is associated with virus on the head and mouth parts, the salivary glands do not contain virus, and there is no conclusive evidence for an extrinsic incubation period.

Failure of either pins or 'dead heads' to remain infective for more than a few days (Fenner *et al.* 1952; Andrewes, Muirhead-Thompson & Stevenson, 1956) is no evidence that virus must multiply in the mosquito, for conditions on the mouthparts of living insects are vastly different from those on the surface of a pin or a dead insect. Suspended in the way in which they are probably maintained on the mouthparts of a mosquito both myxoma and fibroma viruses are very stable (see chapter 5). If the initial virus load is large enough, such stability could account for the observed persistence of infectivity both in hibernating mosquitoes (low temperatures and few or no 'wipe off' feeds), and insects kept in an insectary (higher temperatures and frequent 'wipe off') feeds. The shape of the 'decay curve' for the ability to transmit revealed in Table 29, especially in the detailed experiments summarized in lines A, B and C, constitutes strong evidence for the exclusively mechanical nature of transmission.

The other objections to biological transmission are the complete lack of vector specificity, even for prolonged transmission, and the difficulty of visualizing what cells in the head or mouthparts could serve as a site for multiplication and release of virus in such a way that continued infectivity was achieved (see Pl. IX).

On balance it seems to us that mechanical transmission, which all workers agree occurs during the early stages after the infective feed, is probably the only mechanism of transmission, and that multiplication does not occur. Apart from its intrinsic interest, this conclusion is important in that it excludes multiplication in an arthropod vector as a mechanism which might select virus variants.

Objections have been raised to our use of the term 'flying pin' to characterize the role of the mosquito in the transmission of myxomatosis (Fenner *et al.* 1952). We agree with these objections, if the expression is taken literally. Obviously the complicated mouthparts of a living mosquito are quite different from a pin, and feeding is very

159

different from the insertion of a pin into the surface skin of a rabbit. However, we believe that the expression was useful to drawing attention to three important features of mosquito transmission of myxomatosis (a) non-specificity of the vector, (b) absence of an extrinsic incubation period, and (c) the importance of virus in the superficial cells of the tumour and infected skin, and the irrelevance of viraemia, for the infection of arthropods.

Quantitative aspects of the mechanical transmission of myxomatosis by mosquitoes

Assuming that transmission is mechanical, virus acquired during probing of the mosquito through skin lesions being 'wiped off' in the skin of a susceptible rabbit, certain aspects of the efficiency of the process may be examined. Of major importance are the biting preferences of potential vectors, their numbers, their range of movements, and their survival in nature. These matters will be discussed in detail in the next chapter.

Experimentally it is possible to compare the suitability of skin lesions as sources of virus infections for mosquitoes, either as a function of their age (see Fig. 11) or as a function of the degree of multiplication of different strains of myxoma virus in the skin. It is also possible to compare the virus loads acquired from the same lesions by individual mosquitoes, and by mosquitoes of different species, and to estimate the average loss of virus due to subsequent probing and feeding, and to the passage of time.

(i) Infectivity for mosquitoes as a function of virus strain

In fibroma important differences occur in the accessibility of virus to probing mosquitoes due to host factors such as species and age of the host, and age of the tumour. Infectivity is not related only to the overall virus concentration in the tumour. In myxomatosis of *Oryctolagus* these factors are of minor importance, but the strain of virus is of great importance in determining the amount of virus in the superficial layers of the skin, and the length of time that rabbits remain alive and in a state infectious for mosquitoes. Except for neuromyxoma and the attenuated Californian viruses, there do not seem to be substantial differences in the maximum titres of myxoma virus in the skin with strains of differing virulence (Figs. 12, 13),

Reference line	Virus	Donor host	Species	Maintenance Temperature (°F.)	Assay method	0	1	2	3	4	5	6	
A	Myxoma	*Oryctolagus*	*Aedes aegypti*	80	Mosquito probes on *Oryctolagus*, expressed as positives out of total tests	3/5	0/2	0/4	0/3	1/4	0/3	0/2	
B	Myxoma	*Oryctolagus*	*Ae. aegypti*	80		9/11	14/17	15/41	9/21	4/21	5/27	3/31	4
C	Myxoma (Uriarra strain)	*Oryctolagus*	*Ae. aegypti*	80		67/100	55/136	9/60	16/75	25/77	4/46	4/42	1
D	Myxoma	*Oryctolagus*	*Ae. aegypti*	80	Titration of mosquito pools in rabbits, expressed as ID50/mosquito	130	—	80	—	—	—	—	16
E	Myxoma	*Oryctolagus*	*Ae. aegypti*	70	Titration of mosquitoes in rabbits, expressed as positives out of total tests	—	—	—	9/22	—	—	—	
F	Myxoma	*Oryctolagus*	*Anopheles annulipes*	70		—	—	—	4/8	—	—	—	
G	Myxoma	*Oryctolagus*	*An. atroparvus*	32–36	Mass feeding on single rabbits, expressed as positive or negative out of number of mosquitoes tested	—	—	—	—	—	—	—	
H	Myxoma	*Oryctolagus*	*An. atroparvus*	32–70		—	—	—	—	—	—	—	
I	Myxoma	*Oryctolagus*	*An. atroparvus*	60–70	Titration of pools of mosquitoes in rabbits	—	—	—	10^{-2}	—	—	—	1C
J	Myxoma	*Oryctolagus*	*An. atroparvus*	60–70		—	—	—	10^{-2}	10^{-3}	—	10^{-2}	1C
K	Fibroma	*Sylvilagus*	*Aedes aegypti*	80	Mass feeding on single cotton tails, expressed as fibromas to mosquitoes	18/15	15/17	—	—	18/32	—	—	1
L	Fibroma	*Sylvilagus*	*Ae. aegypti*	80		—	—	—	—	—	—	—	
M	Fibroma	*Sylvilagus*	*Ae. aegypti*	80		—	—	—	12/120	—	—	—	5
N	Fibroma	*Sylvilagus*	*Ae. aegypti*	80	Titration of pools of mosquitoes in rabbits	—	—	—	10^{-4}	—	10^{-3}	—	1C
O	Fibroma	*Sylvilagus*	*Ae. aegypti*	80		—	10^{-4}	—	—	10^{-2}	—	—	
P	Fibroma	Suckling *Oryctolagus*	*Ae. aegypti*	80		10^{-4}	10^{-2}	10^{-2}	10^{-3}	10^{-3}	—	—	
Q	Myxoma	*Oryctolagus*	—	60–80	Survival in skin of rabbit tested by rabbit inoculation, expressed as positives out of total tests	5/5	—	—	—	—	—	—	

TABLE 29. *Experiments on the persistence of infectivity of mosquitoes which had probed or fed through the skin lesions of domestic rabbits or cottontails infected with myxoma or fibroma virus*

	Days after infective feed													
8–11	12–14	15–18	19–21	22–25	26–28	50–60	79	105	149	163	213	220	Reference
7/18	3/14	0/7	0/6	1/11	0/3	—	—	—	—	—	—	—	Fenner, Day & Woodroofe (1
3/55	3/28	1/9	—	—	—	—	—	—	—	—	—	—	Fenner et al. (1952)
1/20	0/1	—	—	—	—	—	—	—	—	—	—	—	Fenner, Day & Woodroofe (1
—	0·5	—	—	0·5	3	—	—	—	—	—	—	—	Day, Fenner, Woodroofe & McIntyre (1956)
9/10	—	1/10	—	1/10	2/13	—	—	—	—	—	—	—	Day et al. (1956)
6/8	—	3/10	—	0/10	2/12	—	—	—	—	—	—	—	Day et al. (1956)
—	+/95	—	+/95	—	—	+/35	+/12	+/5	−/1	—	—	—	Andrewes, Muirhead-Thomps Stevenson (1956)
—	—	—	—	—	—	+/50	—	+/5	—	+/25	−/3	+/2	Andrewes et al. (1956)
10^{-1}	—	$10^0/10^{-1}$	—	$10^0/10^{-1}$	$10^0/10^{-1}$	—	—	—	—	—	—	—	Andrewes et al. (1956)
$10^{-1}/10^{-3}$	10^{-2}	10^{-2}	10^{-2}	19^{-2}	10^{-1}	—	—	—	—	—	—	—	Andrewes et al. (1956)
—	13/16	—	—	—	—	—	—	—	—	—	—	—	Kilham & Woke (1953)
—	0/11	—	9/5	—	—	—	—	—	—	—	—	—	Kilham & Dalmat (1955)
6/45	6/25	—	—	—	—	—	—	—	—	—	—	—	Dalmat (1959)
10^{-3}	10^{-3}	10^{-4}	—	—	—	—	—	—	—	—	—	—	Dalmat (1959)
10^{-1}	10^{-2}	—	—	—	10^{-2}	—	—	—	—	—	—	—	Kilham & Dalmat (1955)
—	—	—	—	—	—	—	—	—	—	—	—	—	Dalmat (1959)
—	—	—	2/2	—	3/4	2/2	3/3	2/8	1/2	—	0/3	2/2	Jacotot, Vallée & Virat (1955

although there are considerable variations between different sites on a single rabbit and between different rabbits infected with the same strain of virus (Fenner *et al.* 1956). There are, however, great differences in the rate of disappearance of the virus as a source of infective feeds, either due to death of the host or healing of the skin lesions.

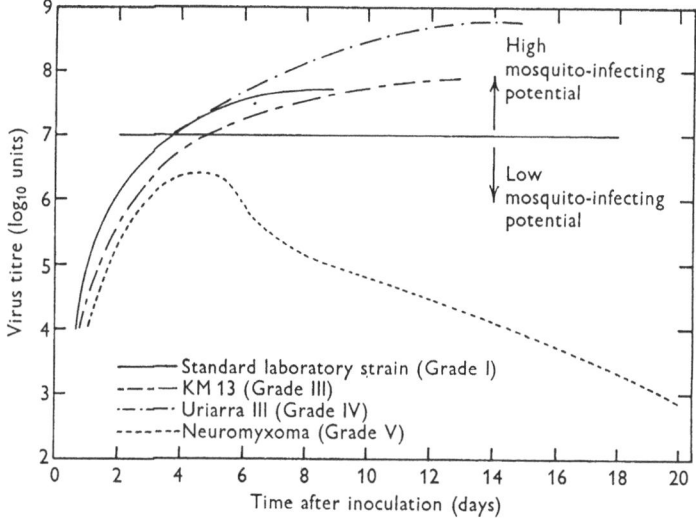

Fig. 12. Changes with time in the titre of virus in slices of skin taken from the surface of lesions produced by the intradermal inoculation of rabbits with large doses of the standard laboratory strain of myxoma virus and three strains of diminished virulence (virulence grade indicated, see chapter 13) derived from it (modified from Fenner, Day & Woodroofe, 1956).

Tests of mosquito infectivity carried out with several strains of virus are shown in Table 30. Skin lesions of most strains are infectious for mosquitoes until the rabbit dies or until they become crusted, regress, and finally heal. With neuromyxoma and the attenuated Californian strain the virus titre fails to reach a high level, and regression sets in so early that the mosquito transmission is rare; whereas the highly virulent strains kill rabbits by the tenth or eleventh day, or even earlier with virulent Californian strains, and thus remove them as a source of virus for mosquitoes, although it should be recorded that in the laboratory mosquitoes will probe the skin of infected rabbits for several hours after their death. Moderately virulent strains, like Uriarra III or KM 13, allow rabbits to survive for a prolonged time, yet the skin lesions do not heal rapidly and they remain infective for mosquitoes for an extended period.

(ii) *Loss of virus from contaminated mouthparts*

If myxoma virus does not multiply in the mouthparts of mosquitoes infectivity will fall due to two factors, (*a*) loss of virus which is wiped

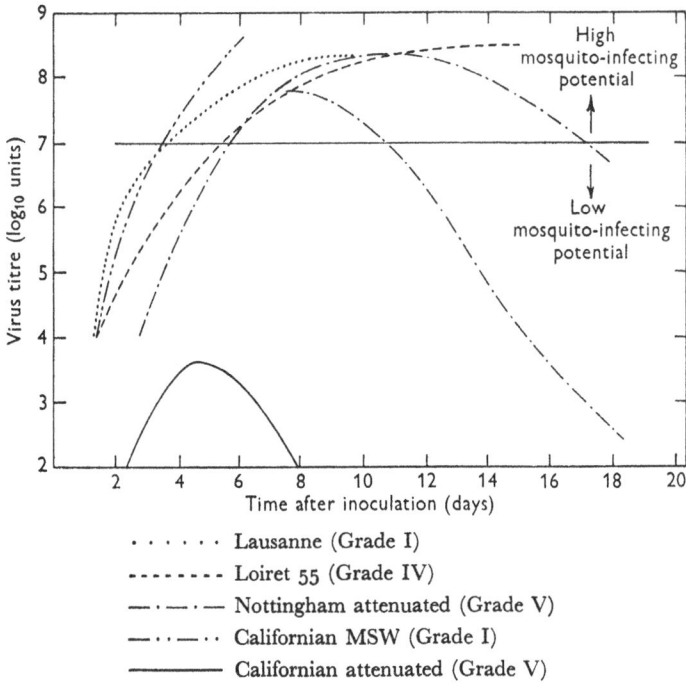

Fig. 13. Changes with time in the titre of virus in slices of skin taken from the surface of lesions produced by the intradermal inoculation of rabbits with large doses of the prototype virulent European strain (Lausanne) and two attenuated derivatives of this; and the prototype virulent Californian virus and an attenuated Californian strain (from Day, Fenner, Woodroofe & McIntyre, 1956; F. Fenner, unpublished observations).

off when the mosquito probes or feeds, or (*b*) loss due to other causes like thermal inactivation. Day *et al.* (1956) devised a system of multiple probing to give data from which calculations could be made of the median minimal initial virus load, the average loss of virus per probe, and daily loss of virus from causes other than probing. Large numbers of *Aedes aegypti* were infected by probing through skin lesions and each mosquito of a group of about twenty-five was induced to probe on nineteen marked sites on the shaved backs of rabbits. The procedure was repeated with a replicate batch of mosquitoes (maintained in an insectary at 27° C.) on the second, third, and fourth days after the infective feed. Typical results are shown in

TABLE 30. *The relative efficiency of mosquito transmission of neuromyxoma and the standard laboratory and Uriarra III strains of myxoma virus*

Mosquitoes probed through the primary skin lesions of rabbits infected with the appropriate virus at various intervals after inoculation, and then fed on marked skin sites on normal rabbits. The first feeds were usually made immediately after the infective probe, and subsequent feeds were taken at intervals of 1, 2 or 3 days (from Fenner *et al.* 1956).

Day of disease in donor rabbit	Neuromyxoma		Uriarra III		Standard strain*	
	No.	%	No.	%	No.	%
Up to 4	0/33	0	—	—	0/4	0
5	0/11	0	—	—	4/16	25
6	0/8	0	3/7	43	9/15	60
7	0/14	0	2/13	15	12/20	60
8	1/12	8·3	11/18	61	7/11	64
9	1/16	6·3	3/16	19	10/14	71
10	2/16	12·5	11/17	65	11/12	92
11	4/20	20	9/16	56	—	—
12	1/13	7·7	13/18	72	—	—
13	1/10	10	9/19	47	—	—
14	2/16	12·5	11/17	65	—	—
15	—	—	13/17	76	—	—
16	—	—	10/14	71	—	—
17	—	—	16/18	89	—	—
18	—	—	15/18	83	—	—
Totals	12/169	7·1	126/208	60·6	53/92	57·6

* Data from Fenner *et al.* (1952).

Table 31, from which it can be seen that individual mosquitoes may induce several infections if they probe several times, a fact clear from earlier experiments with both myxoma virus (Fenner *et al.* 1952) and with fibroma virus (Dalmat, 1959). Calculations by McIntyre on similar data, obtained in experiments carried out with several strains of myxoma virus, allowed estimates to be made of some of the parameters of mosquito infectivity, with the results shown in Table 32.

(iii) *The effects on the virus load of feeding on an immune rabbit*

It was thought that taking a blood feed on an immune rabbit might neutralize virus present on the mouthparts, and thus affect the

TABLE 31. *The results of successive probes on the backs of susceptible rabbits by Aedes aegypti which had previously probed through skin lesions produced by intradermal inoculation of a rabbit with 5×10^4 rabbit-infectious doses of the Lausanne strain of myxoma virus 7 days earlier*

The virus titre of the skin lesions was $10^{7.8}$ rabbit-infectious doses per gram. Separate groups of mosquitoes used at 0, 2, 3 and 4 days. Experiment conducted at 27° C. and 60–80 % relative humidity. Mosquitoes used 3 and 4 days after infectious probing released into cage and given sugar and water on the second day (data from Day et al. 1956)

Days after infective probe	1	2	3	4	5	6	7	8	9	10	11	12	13	14	15	16	17	18	19	Totals
														Probing no.						
0	20/24*	15/24	13/23	6/23	9/23	10/23	10/22	6/22	8/21	7/21	6/20	4/20	6/20	3/20	2/19	4/19	2/19	4/19	7/18	142/400
2	16/23	13/23	10/23	12/23	9/22	7/22	9/22	8/21	8/21	4/20	0/20	4/19	1/19	2/19	1/19	2/18	5/18	2/15	2/15	115/382
3	14/26	10/24	9/23	7/23	4/22	4/22	9/21	5/20	6/20	5/20	3/19	5/19	0/19	3/18	2/18	3/18	1/18	1/17	0/16	91/383
4	10/22	12/22	10/22	7/22	7/22	6/20	5/20	3/19	1/19	3/19	2/19	2/19	2/18	2/17	4/17	2/15	2/15	1/14	1/13	83/355

* Numerator = number of probes resulting in production of a local lesion; denominator = total number of probes.

infectivity of mosquitoes. An experiment carried out to test this showed that a blood meal on an immune rabbit had no greater effect on the infectivity of a mosquito, when tested two days later, than had a blood meal on a normal animal, that is any loss of virus was due to wipe-off and not neutralization (Fenner *et al.* 1956).

TABLE 32. *Characteristics of the transmission of four strains of myxoma virus by* Aedes aegypti *(data from Day et al. 1956)*

Virus strain	Median minimal initial virus load	Amount of virus lost on successive probes (percentage of load)	Amount of virus lost per day (percentage of load)
Standard:			
Expt. I	21·6	17	15
Expt. II	4·8	22	—
Expt. III (a)	1·3	19	45
(b)	3·1	—	—
Neuromyxoma	0	—	—
KM 13	1·2	35	—
Lausanne	7·8	12	20

(iv) *The virus load acquired by different mosquitoes*

Consideration of the scattered distribution of virus-laden cells in the skin, especially in the dermis, and the differences in probing and feeding behaviour of different mosquitoes, suggested that there would be differences in the load of virus acquired by different individual mosquitoes feeding on the same skin lesion, and perhaps even larger differences between different species of mosquito. This expectation was borne out by the very different performance of replicate mosquitoes in experiments involving many successive probes.

The size and variability of the virus load on the head and mouth parts of infected mosquitoes was assayed directly (F. Fenner, unpublished results). Fifty *Aedes aegypti*, each in a separate tube, were allowed to probe several times through the skin lesions of a rabbit which had been inoculated intradermally with the standard laboratory strain of virus six days earlier, and killed just before exposure to the mosquitoes. Next day the head and mouth parts were dissected off each of the 40 surviving mosquitoes, placed in diluent, treated for one minute with the Mullard ultrasonic drill, and assayed by the intradermal inoculation of rabbits. All specimens yielded virus, the amount varying between 10 and 2000 infectious units per mosquito. The modal virus load was about 100 infectious units, the amount of virus being between 40–200 infectious units in 30 of the 40 mosquitoes. Three yielded only 10–20 infectious units each, six yielded about 500, and one about 2000.

Although *Aedes aegypti* is an excellent laboratory animal it is not a natural vector of myxomatosis, at least in Australia, and it is a small mosquito compared with some of the important natural vectors like *Anopheles annulipes* and *Culex annulirostris*. The latter mosquitoes were much better vectors than *Aedes aegypti*, the results of a serial probing experiment showing that only 8% of probes of *Aedes aegypti* were positive compared with 35 % of these by *Anopheles annulipes* and 37 % of these by *Culex annulirostris*. The considerable differences in the structure of the maxillary stylets of several species of mosquito are illustrated by Day *et al.* (1956).

TRANSMISSION BY FLEAS

As long ago as 1920 Aragão showed that the cat flea, *Ctenocephalides felis*, could transmit myxomatosis between *Oryctolagus*, but it was not a very efficient vector. Like most mammals, the wild European rabbit carries many ectoparasites. Prominent among these is the rabbit flea, *Spilopsyllus cuniculi* (Dale) which Rothschild (1953) regarded as an unusually common flea in Britain, prior to myxomatosis. (The effect of myxomatosis on flea numbers and habits is discussed in chapter 17.) The rabbit flea did not survive the transport of *O. cuniculus* to Australia and New Zealand, and has never occurred in those countries.

Although *O. cuniculus* in Australia is not infested with *S. cuniculi*, in certain areas it carries large numbers of the stickfast flea, *Echidnophaga myrmecobii*, which usually infests certain marsupials. In early field trials of myxomatosis in an arid part of South Australia, Bull & Mules (1944) found a correlation between the degree of infestation of rabbits with *E. myrmecobii* and the spread of myxomatosis. Since *E. myrmecobii* is a sessile species it was assumed that it transferred the virus to another rabbit when fleas left a rabbit which had died from myxomatosis. In laboratory experiments, Bull & Mules showed that fleas left the carcass of an infected host in batches from time to time up to 36 hours after death. A proportion of them found and attached to a new host which contracted myxomatosis. Having fed on a first infected host some fleas would transfer to and infect a second and a third host.

Flea larvae fed on flea faeces which contained virus were shown to harbour the virus, but when they became adults they did not transmit the disease. They concluded that *E. myrmecobii* was a mechanical vector, and that virus did not multiply in the flea.

During the first few years of its occurrence in France myxomatosis was a disease of violent summer epizootics, and with the overriding importance of mosquitoes as the vectors of myxomatosis in Australia in mind it was not surprising that at first attention in Europe was concentrated upon these vectors. But various factors, notably the apparent restriction of the spread of myxomatosis in Britain by fences, and the occurrence of obvious cases throughout the year, directed attention to the rabbit flea. Shanks had shown that the rabbit flea was an effective vector, and this was confirmed by Lockley (1954). Rothschild (1953) had emphasized the peculiar suitability of the rabbit flea for maintaining infection through the winter, and Lockley (1954) had shown that fleas would remain viable on the body of a rabbit dead from myxomatosis for at least six days, in mid-winter, and if then transferred to a new host the latter would contract myxomatosis. Brown, Allan & Shanks (1956) further showed that rabbits released in deserted burrows 50 days after the occupants had died of myxomatosis were infected, after 70 days they became infested with fleas but did not get myxomatosis, and after 90 days they did not become infested or infected. In an attempt to assess the role of fleas in maintaining infection in deserted burrows Chapple & Lewis (1964) carried out experiments with fleas which were enclosed with a small bundle of rabbit fur in polythene-stoppered glass tubes. These were buried underground, out of doors, and removed at intervals. They found that with fleas which had previously fed through lesions of a rabbit with myxomatosis two out of three which subsequently fed on a normal rabbit 84 days later were infective, as was one out of two which fed after 112 days. This experimental evidence confirms the opinion of Brown *et al.* (1956), based on field observations, that fleas may act as a reservoir of infection for several months after rabbits have deserted a burrow.

It is only very recently that it has been possible to breed *S. cuniculi* in captivity, due to the unexpected dependence of the female flea upon the blood of a pregnant rabbit, if successful egg production is to occur (Mead-Briggs & Rudge, 1960). Investigations have now been commenced into the bionomics of the flea, and several results of importance in understanding its vector potential have already emerged. Wherever they are placed on the animal, most fleas rapidly migrate to the head and many become firmly attached to the ears. Experiments with marked fleas (Mead-Briggs, 1964) showed that movement between rabbits, and from the open pasture on to rabbits, was much greater and more rapid than had hitherto been suspected, a fact of considerable epidemiological importance.

Few experiments have been carried out on the mechanism of flea transmission of myxomatosis. Muirhead-Thompson (1956c) found that serial transfer of infected fleas from one normal rabbit to another, at intervals of two to three days, resulted in infection of only the first rabbit parasitized. The fleas lost their infectivity rapidly and showed no sign of regaining it, that is there was no biological cycle. Fleas subjected to short, controlled feeds on several rabbits in succession infected the first, second, and third recipient animal, as long as a week after the infective feed. Muirhead-Thompson found that fleas which fed on the top of the ear of a rabbit with advanced myxomatosis (where there were no skin lesions) failed to become infected, whereas those which fed on the head acquired the virus; a result in accordance with similar experiments on mosquito transmission (Fenner & Woodroofe, 1953).

Scanty though experimental work has been, there seems to be no reason to expect that the mechanism of transmission of myxomatosis by fleas will be different in principle from that established for mosquitoes, that is mechanical transfer from infected skin lesions, with loss of virus from infected fleas by wipe-off and inactivation. However, differences in the behaviour of mosquitoes and fleas may be of profound epidemiological and evolutionary importance. Natural selection of virus survival depends upon transmissibility, and the viral properties selected in transmission by a highly mobile mosquito acting as the vector may be quite different from those favoured by the relatively sessile flea. This will be discussed further in chapter 18.

TRANSMISSION BY INANIMATE VECTORS

Since steel pins will serve to transmit myxomatosis, it is conceivable that inanimate vectors might, under exceptional circumstances, play a role in nature. Only one case has been proven, namely that described by Dyce (1961). An epizootic of myxomatosis broke out in a colony of wild rabbits maintained in an enclosure for studies of their social habits (Mykytowycz, 1961). No insect vector could be found, and suspicion fell upon the dense litter of thistle, *Cirsium vulgare*, which blanketed the enclosure and formed almost the only food available to the rabbits. When two groups each of five laboratory rabbits were driven through the thistles they were seen to suffer numerous small abrasions from the spines of the thistles. Six of ten animals thus exposed contracted myxomatosis.

In the early days of the French epizootics statements were made in the popular and quasi-scientific press that the ingestion of hay 'contaminated' by wild rabbits with myxomatosis was an important mechanism whereby domestic rabbits became infected with myxomatosis. This is not the case; the activity of mobile arthropod vectors is unquestionably much the most important mode of introduction of infection into colonies of domestic rabbits. However, Borg and Bakos (1963) recovered myxoma virus from the beaks and claws of two buzzards (*Buteo buteo*) and a crow (*Corvus corone cornix*). They suggested that occasionally predatory birds might afford a means of carrying the virus over long distances.

SUMMARY

A pre-occupation of virologists with laboratory manipulations has led to the neglect by many of them of the great importance of mechanisms of transmission in determining the pattern of evolution of virus diseases, and indeed the very nature and structure of the virions. Our studies with myxomatosis, and those of other workers with some of the plant viruses, have led us to recognize the importance of mechanical transmission by arthropods as a major mechanism of transmission of several animal and many plant viruses. Among animals, it is reasonable to consider mechanical transmission by arthropods as a probable means of transfer of all those viruses which produce localized skin tumours, such as the various fibromas and papillomas of wild and domestic animals.

Because of the major importance of malaria and yellow fever, much more attention has been devoted to 'biological' transmission of viruses by mosquitoes and other arthropods than to mechanical transmission. Confining our attention to mosquito vectors, it is worth emphasizing the major points of difference between these two modes of transmission (Table 33).

TABLE 33. *Major differences between mechanical and biological modes of transmission of animal viruses by mosquitoes (data from Fenner & Day, 1952)*

	Mechanical	Biological
Source of virus	Skin lesions	Blood stream
Specificity of mosquito vectors	None	High
Interrupted feeding	Positive	Negative
Extrinsic incubation period	None	Present
Arthropod transmission the only natural mechanism	No	Yes
Multiplication of virus in the mosquito	No	Yes

In this chapter we have shown that both in its natural *Sylvilagus* hosts in the Americas, and in the novel situation which arose when myxoma virus was introduced into populations of *Oryctolagus* in Australia and Europe, natural transmission of myxoma viruses is due principally to the activity of insect vectors. All the criteria relating to mechanical transmission set out in Table 33 are satisfied. The epidemiological and evolutionary consequences of this will be dealt with in chapters 12 and 18.

CHAPTER 12

TRANSMISSION IN THE FIELD: AUSTRALIAN OBSERVATIONS

THE ORGANIZATION AND SCOPE OF THE FIELD STUDIES

When, in December 1950, myxomatosis revealed itself as well established in the wild rabbit population and spreading rapidly, it was already realized that the likelihood of the virus proving effective as an agent of control depended primarily on whether the indigenous fauna of blood-feeding arthropods would provide vectors in sufficient abundance in the various regions where rabbits were important. Even then there were clear indications that insect transmission of the infection was mechanical, so that any species which habitually fed on rabbits might qualify as a vector. It seemed probable, therefore, that many species might be found to play a useful role in the natural spread of the disease. However, as most blood-sucking insects have aquatic larvae and much of the rabbit's habitat in Australia is but poorly provided with surface waters, it also seemed probable that the spread of myxomatosis—at a useful level of intensity, anyway— might be subject to quite severe spatial limitations.

Within a matter of days after the first natural outbreak of myxomatosis was reported it was obvious that the virus had encountered some highly effective vectors along the river frontages of the central portion of the Murray–Darling system. Once doubt concerning the ability of the infection to become firmly established in the field had been dispelled, and the latent potentialities of the situation had been appreciated, it was clear that the vector aspects of the epidemiology of myxomatosis in Australia presented an ecological problem of considerable interest and probable complexity.

An investigation into the identity of the vectors involved in the initial spread of the infection along the Murray River near Albury was undertaken without delay. Later, as the distribution of myxomatosis became more and more extensive and as staff was recruited for

171

the work, a number of additional regional field studies were organized. Investigations with a useful degree of continuity over a period, and which could be termed intensive or reasonably so, were initiated in the following areas between 1951 and 1955 (see map following p. 371):

New South Wales—the Riverina[1] (at least three areas, representative of river-frontage, foothills and plains: Lake Urana, in the plains, became the most important field station for the virological research programme); the Moree district (from which observations were extended to a river-frontage area near Texas on the Queensland border); New England (Armidale district: three areas at different altitudes); Colo Vale and the Araluen Valley (both in the hills adjacent to the south coast).

Victoria—around Ouyen and across the South Australian border to Peebinga (in the Mallee); Goroke (northern Wimmera); Yarram (south Gippsland).

South Australia—South-Eastern district.

Western Australia—Donnybrook and Darkan (in the south-west).

The various study areas were selected as being representative of distinct climatological and topographical regions in which the impact of the rabbit on primary production was important. The concentration of effort in these relatively few localities meant that vast areas were omitted from the scope of regular observations; and although quite a number of extensive surveys were made from time to time to collect information from further afield, substantial tracts of country still remained from which virtually no relevant data were obtained.

Not long after the entomological investigations started in 1951, it became apparent that mosquitoes were likely to prove by far the most important vectors in Australia; and the observations made in subsequent years confirmed this early indication (see below, under Simuliidae and Fleas). In those areas where the vector situation proved difficult to clarify, continuing attention had to be given to potential vectors other than mosquitoes; but most of the field investigators were able to confine their attentions almost exclusively to the Culicidae. While this simplified the programme, a great deal was still left to be done for remarkably little was known about the ecology and

---

[1] The Riverina is the region enclosed by the Murray and Murrumbidgee Rivers: the foothills of the Dividing Range form its natural boundary on the east.

behaviour of the mosquitoes occurring in the rabbit-infested regions. The identification of the species which were playing, or were capable of playing, a significant vector role was only one of the objectives of the field investigations. An overall assessment of the value of myxomatosis as an agent of rabbit control was required and this involved a study of the factors underlying local and seasonal variations in disease activity.

Naturally enough the various regional investigations developed different emphases and special approaches in response to local conditions, but they all followed the same essential pattern. The incidence of myxomatosis and any significant changes in the rabbit populations were recorded, while the details of the background entomological picture were filled in. The work was of course by no means confined to epizootic periods, although the main dividends came from observations made during outbreaks. The luck of the seasons favoured some investigators more than others: in more than one area no epizootic worthy of the name occurred during two to three years of intensive field work.

NATURE OF EVIDENCE SOUGHT: TECHNIQUES

The first prerequisite of a vector is that it should feed on rabbits. There need be no overriding preference for the rabbit as a blood source, but it must be attacked with a reliable frequency. Although it provided some useful data, direct observation of insects attracted to rabbits is of limited value because of the light factor and also because of the disturbing effect of a human observer in the immediate vicinity. Most of the data on mosquito feeding habits were obtained from precipitin tests and the use of rabbit-baited traps.

Mr D. J. Lee of the School of Public Health and Tropical Medicine, Sydney, not only played an active part in some of the field studies but arranged for his staff to carry out precipitin tests on blood spots obtained in all the study areas. The reliability of the test data as an indication of the relative importance of different blood sources will depend, of course, on the degree to which the collection of the specimens is free from bias. Naturally engorged mosquitoes are not always easy to come by in the field, and some of the collections were known to have a marked bias. Due attention was paid to this when interpreting the results of the tests (see in particular Lee, Clinton & O'Gower, 1954). The commoner Australian mosquitoes are apparently not specialized feeders, but normally obtain blood meals from a fair range of hosts. Although some may be preferentially attacked, it seems likely that availability would be more important than preference in most ecological situations.

The rabbit-baited traps used in the field vector studies were developed early in the history of the investigations by Myers (1956). The bait animal was secured to a wooden base board; and with the aid of a rope and pulleys (enabling the operator of the traps to avoid his person competing, as bait, with the pilloried rabbit) a gauze-covered cone was lowered on to the board at intervals. The insects feeding upon, and many of those attracted to, the bait rabbit were caught in this way. Myers also developed two others types of trap designed to collect insects emerging from rabbit burrows in the evening and entering them in the morning. The three types were commonly referred to as 'drop', 'warren-in' and 'warren-out' traps respectively.

The traps just described provided invaluable data in some of the study areas, but they had certain limitations. The warren-in and warren-out traps could only be employed profitably where mosquitoes harboured in rabbit burrows; and reliability of the drop traps for determining the relative extent to which different mosquito species fed on rabbits depended on there being no marked difference in their readiness to approach and settle on an animal held below a suspended frame. When the drop traps were tried out in areas where the mosquito density was very much lower than in the Riverina, where the trap was developed and first used with success, there were indications that some species were collected very much less readily than were others—were, in fact, 'trap shy'. Where these species were involved almost complete dependence had to be placed on the results of precipitin tests. Naturally engorged adults, providing material for these tests, were obtained by searching daytime resting places and by catches made on human and other types of bait (which often contained a significant percentage of insects containing blood from a previous meal that was fresh enough for test purposes). The relative productivity of the two collecting techniques varied from species to species.

Information on the feeding habits (particularly host preference data) can be expected to provide a valuable pointer to the identity of field vectors. More critical evidence must then be sought; and it is furnished by (a) the recovery of virus from insects collected in an outbreak area, and (b) the correlation between epizootics and the activity of particular species of mosquitoes.

TABLE 34. *Recoveries of myxoma virus from mosquitoes collected in outbreak areas*

Area and date	Collector	Proportion of batches positive	Total no. of insects collected (or average size of batch)
	Culex annulirostris		
Corowa, 1951	Mykytowycz*	1/1	168
Lake Urana, 1952	Myers†	1/7	(50)
Araluen, 1955	Dyce‡	2/8	75
	Anopheles annulipes		
Lake Urana, 1951, 1952	Myers†	8/20	(50)
Corowa, 1952	Myers†	7/17	(50)
Araluen, 1955	Dyce‡	1/22	176
	Culex pipiens australicus		
Bacchus Marsh, 1952	Ratcliffe‡	1/1	4
	Aedes theobaldi		
Lake Urana, 1952	Myers†	1/3	(50)
	Ae. alboannulatus and *Ae. rubrithorax*		
Colo Vale, 1953	Lee, Dyce & O'Gower (1957)	2/20	943
	Ae. alboannulatus		
Araluen, 1955	Dyce‡	1/9	33
	Ae. notoscriptus		
Araluen, 1955	Dyce‡	1/6	40
	Ae. imperfectus		
Araluen, 1955	Dyce‡	1/1	34

* Myers (1954). † Myers *et al.* (1954). ‡ Unpublished observations.

For the virus-recovery tests batches of the candidate species were used to prepare material for inoculation into susceptible rabbits. With the more abundant mosquitoes it was often possible to employ pools of 50 or more insects. The results obtained from the tests are presented under the appropriate species headings below, and are summarized in Table 34.

The number of virus recoveries made by those involved in myxomatosis vector studies in Australia compares favourably with the results of comparable attempts with insect-borne diseases in other parts of the world. It must be admitted, however, that they disappointed some of our field workers who, over-optimistically perhaps, had hoped that information from this source would provide a clear indication of the relative importance of different species contributing to transmission in a complex vector situation, for example at Colo Vale and Araluen. From 1954 on effort was directed mainly to the collection of ecological and behavioural data on candidate species, for correlation with disease performance.

In some of our study areas it was by no means easy to establish a clear correlation between virus and vector activity. This was due in no small part to the difficulty in assessing mosquito abundance with adequate precision. Every species is likely to present its own problem in this respect. Mosquitoes habitually attracted to man are handled relatively easily, as a satisfactory index of their abundance can be obtained from catches made on human bait, if environmental factors affecting the insects' activity are taken into consideration. At the other extreme, and presenting the toughest problem, is a species such as *Culex pipiens australicus*, which is not a man-biter and though feeding readily on rabbits happens to be trap-shy. It will be readily appreciated that an estimation of the relative abundance of two or more species of mosquitoes presents something of a problem when different criteria have to be used for each species as indices of density.

In retrospect one can say that most of the conclusive findings of the field investigations came fairly quickly. The early work was carried out in areas where the ecological situation turned out to be relatively simple, with one or two outstandingly efficient vectors temporarily dominating the mosquito fauna—in fact liable to occur in much greater numbers than were necessary to ensure intense outbreaks of the disease. Their sharply defined seasonal incidence, and the levels of their abundance, greatly facilitated the collection of relevant quantitative data. Work carried out later rarely produced the clearcut results obtained in the Riverina and western and north-western Victoria, and often revealed a complex picture that defied clarification mainly because the recording techniques available were not sufficiently delicate and precise. Those who had to deal with less co-operative material in such situations were forced to recognize that important relationships often worked themselves out in both rabbit and mosquito populations at a sub-observational level. However, the information they obtained on a wide variety of species advanced knowledge of the ecology and behaviour of the Australian Culicidae to a notable degree.

THE VECTOR INSECTS[1]

Even the extra-tropical portion of Australia has a fairly rich mosquito fauna, but the situation in the rabbit-infested regions proved much less complex than the mere length of the list of locally occurring mosquitoes might suggest. In the first place, the majority of species recorded from any locality will be uncommon or rare and as such can be dismissed as potential vectors of any significance. In the second place, the working approach was simplified by broad differences in regional ecology, exemplified in New South Wales by conditions on the eastern (coastal) and western (inland) sides of the Divide.

In this chapter, and in the early chapter dealing with the biology of the rabbit (chapter 4), the term Divide has been used as a convenient abbreviation for the Great Dividing Range. The Divide proper is of course a line, delimiting the watersheds of the streams flowing eastward into the Pacific. (To the south, in Victoria, it swings westward, in hilly rather than mountainous country, separating the northward-flowing tributaries of the Murray from the headwaters of the shorter, southward-flowing rivers.) The Range on the other hand is sufficiently wide to constitute a region in its own right. It tends to fall away more gently inland than towards the coast. Its backbone is steep and forest covered for the most part; but the highland belt includes two or three substantial areas of tableland, almost completely cleared and utilized mainly for sheep grazing, where the

---

[1] Some of the information given in this section was common entomological knowledge before the vector studies began. Most of it, however—particularly that presented under the species headings—derives from the pooled observations of the C.S.I.R.O. field team (J. H. Calaby, A. L. Dyce, B. V. Fennessy, K. Myers, R. Mykytowycz and E. J. Waterhouse) and of G. W. Douglas and D. J. Lee. The former started as a member of the C.S.I.R.O. team, continuing his investigations as a member of the Victorian State Service. The latter, an officer of the Commonwealth Department of Health, was jointly responsible with Dyce for the work at Colo Vale. As a result of recent revisions (e.g. Klein & Marks, 1960), the names of some of the mosquitoes mentioned here differ from those used in the publications cited.

rabbit has been a serious problem (though on the whole less serious than at lower altitudes further inland).

As a biological divide the Range is of great importance, for it separates the two main zoogeographical provinces of the Australian continent—the south-eastern coastal Bassian province from the inland Eyrean. Allowing for the transition zone of slopes and foothills, these two provinces or regions differ in topography, climate, vegetation and fauna; although differences in the mosquito fauna on either side of the Divide are masked to some extent by the fact that several of the more adaptable and abundant species are common to both.

The Riverina, where so much of the field work on vectors was carried out, is typical of that part of the inland Eyrean region that enjoys reasonably good rainfall. The ecology of the locally occurring mosquitoes is determined by the topography of the land more than by any other factor (see Pl. XI, fig. 1). Most of the region is level plain; and even where the land is gently undulating, or where hills occur, wide flat valleys will be found. Even moderately heavy falls of rain will leave local accumulations of surface water persisting for a week or more. (In more hilly country, of course, the run-off is channelled into tributary streams, and is quickly carried away.) The flatness of the terrain has permitted the rivers to shift their courses over the ages; and thus they now have an associated system of lagoons and swamps in which the water, if not permanent, at any rate persists for extended periods. While the lagoons (or billabongs as they are commonly termed) depend to some extent on local precipitation for their maintenance, they are normally filled by floods coming down the rivers, caused by melting snow or heavy rainfall in the mountains.

These two distinct breeding habitats—the temporary water of the plains and the persistent water of the swamps and lagoons—are exploited by different groups of mosquitoes. The main breeders in 'casual' ground water are species of $Aedes$ belonging to the subgenus $Ochlerotatus$ of which the most abundant in the Riverina are $Ae.$ $(O.)$ $theobaldi$ and $Ae.$ $(O.)$ $sagax.$ They lay resistant eggs, and can complete their development in any water that persists for eight to ten days. They occur in great numbers when conditions are favourable, and are the most important day-biting pest mosquitoes of the inland. One of the most unexpected results of the field investigations was

Fig. 1

Fig. 2

PLATE X. Fig. 1. Netting fence in central New South Wales separating rabbit-free and rabbit-infested land. (By permission of *Country Life*, Sydney.)
Fig. 2. Lake Urana study area, showing portion of the dry lake bed with sandhill harbouring rabbits in foreground (from Myers, Marshall & Fenner, 1954).

Fig. 1

Fig. 2

PLATE XI. Fig. 1. Dry watercourse (foreground and distance) in western New South Wales, a region with low and erratic rainfall and sluggish drainage. The higher ground in the middle distance carries scattered cypress-pines (*Callitris*): foreground trees mainly *Eucalyptus bicolor*, a species restricted to areas subject to flooding (from Frith, 1964).

Fig. 2. The Merricumbene (Araluen) study area in the coastal hills of New South Wales, typical of a region drained by persistent streams and providing a variety of vector-breeding habitats (from English, Mackerras & Dyce, 1957).

the discovery that the *Ochlerotatus* group could be almost completely excluded from responsibility for myxomatosis transmission (see under species headings below), some combination of factors apparently reducing the likelihood of an infective feed being followed within the life span of the average individual by a feed on a susceptible rabbit.

The most abundant of the mosquitoes breeding in the persistent water habitat are *Anopheles annulipes*, *Culex annulirostris* and *Culex pipiens australicus*. All three are active only at dusk, or after. *An. annulipes* and *C. annulirostris* were shown to be vectors of outstanding importance: in a susceptible rabbit population a high-grade[1] epizootic would almost invariably follow the appearance on the scene of either species in adequate numbers.

The belt typified by the Riverina merges into the arid zone, further inland, where hardly any entomological observations were made during myxomatosis outbreaks. The few items of relevant information obtained from the semi-desert are included in the notes below under species headings.

On the coastal side of the Divide, and in comparable areas in the south and west of the continent, the country is in general more hilly and the tree cover more dense (see Pl. XI, fig. 2). Although the annual rainfall is higher, the general level of mosquito abundance is much lower than it is, in season, west of the ranges. (Thus when mosquitoes could be collected in hundreds in the Riverina study areas, ten to twenty might be collected in the same time, using the same techniques, at Colo Vale.) Of the considerable number of species likely to be found in any area—over thirty were recorded at Colo Vale and Araluen—it is rare for any to achieve overwhelming relative abundance. The wide range of habitats is exploited by a variety of species each with its definite, though sometimes only slight, preference for a particular type of breeding place. Mosquitoes seem to be less mobile in this region than they are inland; and it is not unusual for an outbreak of myxomatosis to develop very patchily,

---

[1] As there is no way of assessing mortalities in a wild rabbit population with precision, the convention was adopted of classifying epizootics into three grades which could be recognized by ordinarily careful observation. A 'grade 1' outbreak left only a very sparse population of survivors, resulting from a mortality of over 90%. A 'grade 2' outbreak caused a substantial, readily observed, mortality in the local population (resulting, in practice, from kills of anything between 50 and 90%) from which recovery could be rapid. A 'grade 3' outbreak was characterized by observed myxomatosis activity with no obvious effect on rabbit density.

working through local groups of rabbits and reducing them to low numbers before appearing in warren colonies quite close by (A. L. Dyce, personal communication).

One, and sometimes both, of the two main vectors of the inland (i.e. *An. annulipes* and *C. annulirostris*) occur in every district on the seaward side of the Divide. Of the aedine mosquitoes, species of the subgenus *Finlaya* are usually dominant ecologically, even though there might be almost as many species of *Ochlerotatus* on the local list. Occasionally one of the *Ochlerotatus* species which is normally rare will enjoy a season of exceptional abundance, and may provide indications of playing a vector role (see notes below under Miscellaneous Mosquito Species). In areas close to the coast, or providing some brackish-water habitat, the viciously biting *Ae. (O.) vigilax* or *Ae. (O.) camptorhynchus* may appear in numbers at certain times of the year.

It was in the coastal and near-coastal study areas that field workers found the greatest difficulty in identifying vectors and providing a satisfactory explanation for the course of outbreaks, even though they might be clearly defined and markedly seasonal. The fact that several of the locally occurring mosquitoes were feeding on rabbits could readily be demonstrated, but the onset of an epizootic rarely coincided with observed peaks in the activity of any one of them. However, at Yarram, on the Victorian coast, Douglas was able to show that *An. annulipes* was the main vector in the well-defined seasonal outbreaks in that area.

Inland of the Divide, the identity of the main vector or vectors in outbreak was usually obvious, and recoveries of the virus from mosquitoes collected in the field tended merely to confirm a self-evident fact. In the coastal belt, with its more complex mosquito situation, virus recoveries were expected to be of greater significance; and a special effort was made to obtain them, both at Colo Vale and Araluen. In the former area, in the late summer of 1953, material from twenty-eight pools of five species of mosquitoes were tested (Lee, Dyce & O'Gower, 1957). Twenty of the pools consisted of a mixture of *Aedes alboannulatus* and *Ae. rubrithorax*, which it was impracticable to separate, at the time, under field working conditions. Only two of these pools, each consisting of about 100 insects, yielded positives: the remainder of the twenty were negative, as were all the pools made up of other species. Subsequent collections at Colo Vale all gave negative results.

At Araluen the attempt was much more productive, and Dyce (*in litt.*) succeeded in recovering virus from five species during a late summer epizootic in 1955. They included *C. annulirostris* and *An. annulipes*, of which the former was not predominant numerically and the latter was well past its peak at the time of the outbreak. These recoveries, and the associated data on species abundance and feeding behaviour, left little doubt that transmission resulted from the combined activity of several species—with *C. annulirostris* probably the most important, but not predominantly so. A comparable state of affairs, with the relative importance of individual vector species varying from area to area, would seem to be the rule throughout the eastern coastal belt.

Culex annulirostris *Skuse*

Distribution

This is a warm-country species. It is absent from Tasmania and the south-eastern highlands. In Victoria it does not range, as a common insect, far south of the Divide; and it is virtually absent from south-eastern South Australia. Locally common in south-western Australia.

Vector status

Culex annulirostris is a very important and effective vector of myxomatosis, and was the first species shown to be transmitting the infection during an epizootic. R. Mykytowycz (unpublished observations) exposed three caged laboratory rabbits to the biting of this mosquito near Corowa, on the Murray, in January 1951: two of the animals became infected. At the same time and in the same area he recovered virus from a single large pool of the species. Myers recovered virus from one out of seven batches collected at Lake Urana during November and December, 1952; and in April 1955, as has already been mentioned, Dyce (*in litt.*) recovered virus from this species at Araluen. Two out of eight batches of *C. annulirostris*, totalling 75 individuals, were positive, the comparable figures for *Anopheles annulipes* being 1/22, total 276.

At the time of the initial spread of myxomatosis along the frontages of the Murray, *C. annulirostris* was much more abundant than *An. annulipes*—as was to be expected at the height of summer—and Myers' survey of disease activity revealed that intense and persistent outbreaks were confined to the close proximity of water, that is to the zone beyond which the *Culex* was not found to range. In the years that followed, when and where it was dominant in the local fauna, the correlation between the course and extent of epizootics and the activity of this mosquito became so obvious that no further effort was made to recover virus from it.

Feeding behaviour

From the results of precipitin tests Lee *et al.* (1954) concluded that *C. annulirostris* had 'a strong tendency to attack rabbits', and that 'cattle and rabbits are probably the predominant blood sources'. The considerable data collected since the publication of these results fully confirm the conclusion that rabbits are a preferred blood source. A dusk and night biter, this mosquito attacks man readily; and as it can be collected from human bait, and in rabbit-baited drop traps, indices of its population density are easily obtained.

Relevant ecology

Culex annulirostris breeds in permanent or persistent water, preferably where there is aquatic or emergent vegetation. Its main stronghold within the rabbit's range is provided by the river-frontages, swamps and lagoons of the Murray–Darling system—from the plains of Queensland in the north and the foothills of the eastern ranges, down to the Murray mouth in South Australia. In the southern part of its range, in Victoria, in addition to the breeding grounds associated with the tributary streams of the Murray, the stockwater dams of the north-west (Mallee) and the extensive system of channels which feed them produce large numbers by about February.

To the north, towards the Queensland border, *C. annulirostris* may be found at any time of the year where conditions are favourable. Elsewhere it is typically active in mid and late summer when, in southern Australia, the open country is normally dry. Under these conditions it rarely ranges far beyond the vegetation fringing its breeding grounds, being sensitive to low atmospheric humidity; and those myxomatosis outbreaks that are based on its activity reflect this very obviously. However, when the exceptional season turns up, with water lying and persisting over the countryside in summer, it is quick to take advantage of the situation. It then not only breeds in almost any persistent ground water, but becomes widely ranging in its flight behaviour. This was observed by Myers in 1954–55 in his foothill-valley study area near Albury; by Waterhouse in New England in the same season; and by Douglas near Goroke, in western Victoria, in 1955 and 1956. Even in the semi-desert areas of the inland the species manages somehow to maintain a foothold, and will breed up and disperse quite widely during the occasional favourable season, when it is liable to stimulate myxomatosis activity.

In the neighbourhood of its permanent breeding grounds, which are reticulated over a substantial part of south-eastern Australia, *C. annulirostris* is the most reliable and probably the most important vector. Elsewhere epizootics are mainly initiated and maintained by other species. Even in its most favoured habitats, *C. annulirostris* will almost invariably be associated with *Anopheles annulipes* and *C. pipiens australicus*. The three species attain peak abundance at somewhat different times, however, and it will usually be obvious which is the main vector in any particular outbreak. In a typical '*annulirostris* outbreak' the vector is usually present in superabundance, and the fact that other species (which may include simuliids—see below) may assist in transmission is probably irrelevant. However, it sometimes happens that an epizootic may depend on different vectors in its successive phases or in different parts of the outbreak area.

Anopheles annulipes *Walker*

Distribution

Australia-wide, occurring in every area where the rabbit is found.

Vector status

Anopheles annulipes is an effective vector of first ranking importance. Probably more rabbits are infected by it in the aggregate than by *C. annulirostris*. Its significance was not revealed in the first season of myxomatosis activity, 1950–51; but in the years that followed its role in transmission was clearly established. Virus was recovered from one out of two batches of adults collected at Lake Urana in December

1951, and 7/18 collected in November and December 1952; and from 7/17 batches collected at Corowa in December 1952 (Myers, Marshall & Fenner, 1954). A clear correlation between disease outbreaks and peaks of *An. annulipes* was established in the Riverina during the 1951–52 and 1952–53 seasons; and when surveys were carried out in the 1951–52 summer in Victoria and New South Wales to check on spread of myxomatosis through the belt lying just inland of the Divide, the presence of *An. annulipes* 'as revealed by the discovery of adults in the warrens or local breeding, (was) the one factor common to all those parts of the affected area in which we (were) able to make adequate and timely observations' (Ratcliffe *et al.* 1952). Since then, the association between epizootics and the abundance and activity of this mosquito has been observed in many outbreak areas, for example by Douglas in north-western, western, and south-eastern Victoria; by Myers in the Riverina and north-eastern Victoria; by Dyce and Poole, and Waterhouse, on the Western Slopes in northern New South Wales; and by Calaby in Western Australia.

Feeding behaviour

Anopheles annulipes is typically a dusk and night biter; and while it exploits a wide range of hosts, it has a marked predilection for feeding on rabbits. The precipitin test results point to rabbits and cattle being the two major sources of blood for *An. annulipes*, Lee *et al.* (1954) remarking on the anomaly of an indigenous mosquito coming to depend primarily on the blood of two animals introduced into Australia by the white man. It was long suspected that sheep must provide a blood source for *An. annulipes*, as in many areas where it was abundant sheep and rabbits are the only mammals reliably available; and this was confirmed by precipitin tests of blood spots collected in New England by Waterhouse.

Relevant ecology

Anopheles annulipes will be found breeding in many, if not most, of the situations favoured by *C. annulirostris* and like the latter is produced in abundance by the billabongs and swamps associated with the Murray and its tributaries. However, it exploits a number of situations not normally colonized by the *Culex*, such as the extensive surface waters present on the Riverina plains after heavy winter rain, the slowly moving water on the margins of rivers and creeks (particularly when and where a substantial growth of the filamentous alga, *Spirogyra*, has developed), and pools formed in the course of the drying up of small spring-fed hill streams, which are such an important feature of the extensive belt of undulating pastoral country in eastern Australia. These little streams may only run for a few days in the warmer parts of the year; but it is by no means unusual for the rainfall distribution to be such that pools persist at various points along their courses well into the spring and early summer, or that such *annulipes* breeding places are created by autumn falls.

Anopheles annulipes is exceptionally well qualified to be a vector of an infection of the wild rabbit in Australia. It is obviously attracted to areas carrying dense rabbit populations and it is much better equipped than *C. annulirostris* to maintain itself in the lower rainfall regions. Inland of the Divide it habitually exploits rabbit burrows for daytime harbour: the drier the conditions above ground, the deeper the insects descend into the warrens. Myers was the first to obtain evidence (in the form of freshly fed individuals in the catches of mosquitoes leaving burrows at dusk) of *An. annulipes* attacking rabbits underground. In the Riverina the much lower percentage of freshly engorged insects in the 'warren-out' catches than in those of the 'warren-in' traps indicated that most of the feeding on rabbits took place in the normal way, that is above ground during the hours of darkness (see Myers, 1956). However, in the course of surveys made later in the more inland parts of New South Wales, using warren-out traps and rabbit-baited drop traps, he concluded that underground feeding was likely to be more frequent and significant under more arid conditions.

The substantial numbers of engorged individuals appearing in the warren-mouth catches allowed the feeding activity of *An. annulipes* to be followed with considerable precision; and both Myers in the Riverina and Douglas in western Victoria were able to correlate the onset of epizootics not merely with the abundance of this vector but with the commencement of its large-scale attacks on rabbits.

This mosquito is a strong flier, and apparently quite long lived. In Western Australia, Calaby found adults resting in warrens three to four miles from the nearest breeding site; and observations made by H. J. Frith in and around the Murrumbidgee Irrigation Area pointed clearly to mass movements, assisted by southerly winds, over distances of ten miles and more. In south-western New South Wales, Fennessy found adults harbouring in warrens in considerable numbers two months after the local breeding waters had dried up.

Anopheles annulipes bites man readily enough for catches on human bait to provide a satisfactory measure of its changing abundance—a technique which proved useful in the coastal study areas where it does not harbour in warrens and was found difficult to collect in drop traps. It was much harder to obtain precise and conclusive evidence of its vector role in the coastal belt, where it is always a constituent of the local mosquito fauna, and usually an important one. While Douglas found the correlation between outbreaks and its activity clear enough at Yarram, on the Victorian coast, this was not the case in the Araluen Valley in New South Wales. In that study area *An. annulipes* was generally the most abundant mosquito, and received particular attention from Dyce. As has already been mentioned, the late summer epizootic in 1955 (which was the only major outbreak occurring there during the period of intensive study) took place when the anopheline was declining in numbers, and the virus-recovery data showed that while it may have played a vector role it was almost certainly an unimportant one. The performance of this mosquito appeared so anomalous in comparison with its role inland that Dyce tried to determine whether cattle—a readily available alternative blood source—might have affected its vector efficiency.

The collection of blood-fed adults from daytime resting places and the determination of the blood sources by precipitin tests was routine procedure at Araluen. The results obtained from unbiased samples showed a 4:1 preference for cattle over rabbits, which suggested that when the population density of *An. annulipes* was barely sufficient to raise myxomatosis transmission to an epizootic level, the damping and deflecting effect of the local herd might conceivably have been sufficient to delay or prevent the development of an outbreak dependent on the activity of this species. Another possible explanation of the rather puzzling situation is that *Anopheles annulipes* includes more than one biological entity, and that the form occurring in the coastal belt of New South Wales differs from the one ranging west of the Divide in characteristics affecting its efficiency as a vector. This is by no means impossible, nor even very unlikely.

Adults of *An. annulipes* are present at all times of the year; and although the species is active throughout the warmest months, there is usually a marked tendency for it to have spring and autumn peaks of abundance. This has a very important effect on the timetable of disease activity in many areas. For instance, along the Murray and its tributaries, where both *An. annulipes* and *C. annulirostris* are abundant in season, if conditions favour a massive spring emergence of the former species an intense epizootic will develop in November, or even earlier, and there will be no susceptible rabbit population for *C. annulirostris* to 'work on' when it appears in large numbers in December. In those regions that lie beyond the southern limit of *C. annulirostris* in Victoria, it is the autumn wave of *An. annulipes* that is generally responsible for high-grade myxomatosis outbreaks; and disease activity may continue well into the winter, presumably (though this is not proven) as a result of continuing, low-intensity mosquito transmission.

Culex pipiens australicus *Dobrotworsky and Drummond*[1]

Distribution

Co-extensive with the rabbit: found from Tasmania northward into Queensland and westward to Western Australia. Rare or virtually absent in the more arid areas.

Vector status

It was not possible to determine with certainty the overall importance of *C. p. australicus* as a vector. In the southern part of Victoria it replaces *C. annulirostris* as the dominant *Culex*, and there seems little doubt that in this region it can, and does, assume an important vector role. Virus was recovered from a small pool of *C. p. australicus* collected in November 1952 towards the end of an epizootic near Bacchus Marsh; but the main Victorian evidence of its role in transmission derives from the importance of rabbit as a blood source and the correlation of its local abundance with disease outbreaks, repeatedly observed by Douglas.

In New South Wales, A. L. Dyce (personal communication) came

---

[1] The trinomial indicates that this mosquito has been allotted only subspecific rank. However, as no other member of the cosmopolitan *Culex pipiens* complex is known to occur in Australia, *C. p australicus* has the biological attributes of a 'good' species in this country.

to the conclusion that some outbreaks he investigated in the Coolamon district (lying just north of the Murrumbidgee River and the Riverina) depended almost entirely on the activities of $C. p. australicus$ and $Ae. alboannulatus$ (see below). But Myers's observations led him to conclude that it was of negligible importance in the Riverina, where it tended to be completely overshadowed and outnumbered by either $C. annulirostris$ or $An. annulipes$ in all the local study areas. He failed to recover virus from a pool collected at Urana in December, 1952.

Feeding behaviour

$Culex pipiens australicus$ is active only at and after dusk, and is very rarely indeed attracted to man. Lee $et al.$ were in a position to say, on the results of precipitin tests carried out up to 1953, that it was a rabbit and fowl biter. In subsequent years, large numbers of rabbit-positives have been obtained from specimens collected entering and leaving warrens, mainly by Douglas in western and north-western Victoria.

While the importance of the rabbit as a blood source for $C. p. australicus$ is beyond question, most of the collections of engorged adults have been strongly biased and the significance of alternative hosts remains undetermined. Myers (1956) experimented with drop traps operated at various heights above the ground, using chicken and possum as bait in addition to rabbits. Although his catches of $C. p. australicus$ were too small to be more than suggestive, his results indicated that this mosquito might show a preference for feeding in the tree canopy (on birds and marsupials) where conditions favoured such an activity.

Relevant ecology

The breeding behaviour and seasonal timetable of $C. p. australicus$ are best summarized by comparison with the two species already dealt with. It will be found breeding in many, if not most, of the habitats favoured by $C. annulirostris$ and $An. annulipes$. Myers noted that where lagoons went through a regular seasonal cycle, becoming stabilized after winter floods, $C. p. australicus$ tended to start breeding some weeks before $C. annulirostris$ which apparently required clearer and more vegetated water. Like $An. annulipes$, $C. p. australicus$ tends to have spring and autumn peaks of abundance.

Like the anopheline also, in all but the more humid parts of its range $C. p. australicus$ has the habit of sheltering in warrens during the day; and there is evidence (Douglas, $in litt.$) that it occasionally feeds on rabbits underground. It is hardly an exaggeration to say that the burrow-harbouring habit is about the only attribute of this mosquito that prevented its study as a potential vector from being quite unprofitable; for it is a very 'shy' biter and difficult to collect in drop traps. In contrast to $An. annulipes$, in the eyes of most of the field workers $C. p. australicus$ remained to the end a rather characterless and puzzling species. Greater success in clarifying its role in transmission would probably have been achieved had it not usually been associated with one or other of the two most important vector species.

Aedes (Finlaya) alboannulatus ($Macquart$) and
Aedes (Finlaya) rubrithorax ($Macquart$)

Distribution

One or other of these two mosquitoes will be found in every region where rabbits occur except in the low-rainfall areas. Inland of the Divide their occurrence is associated with savannah-woodland and belts of timber along watercourses.

Vector status and relevant biology

These are the two commonest members of a group of closely related Finlayas, with rather slight differences in their ecology. In a sylvan habitat they will almost invariably be among the more abundant of locally occurring mosquitoes. They both exploit a variety of breeding sites associated with streams and rivers—rock pools, marginal seepages, etc.—while $Ae. alboannulatus$ will breed in temporary pools that appear after heavy rain or flooding, providing they persist long enough. Both species bite during the day and through the dusk, ceasing activity at nightfall or shortly after.

Our knowledge of the association of these two mosquitoes with rabbits and myxomatosis transmission stems almost entirely from the observations made at Colo Vale (Lee $et al.$ 1957) and in the Araluen Valley where Dyce, as has already been mentioned, recovered virus from a pool of $Ae. alboannulatus$. Confirmatory data on rabbit feeding came from Waterhouse in New England. Both species attack man readily, and rabbits 'fairly freely'; and for both marsupials are an important natural blood source. Double feeding (indicated either by blood spots being positive for two anti-sera or by insects containing partly digested blood being attracted to baits) was established in the case of *Ae. alboannulatus*.

As was mentioned earlier in this chapter, the only recoveries of virus made at Colo Vale were from two mixed batches of *alboannulatus* and *rubrithorax*, which were the dominant species in the area (the latter being the more consistently prevalent). Although both species were active when transmission was at its height, there was no obvious correlation between myxomatosis activity and the abundance of either of them—but there was no clearer association between the disease and the recorded abundance of any other species. There can be little doubt that they acted as vectors; and it is probable that one or other was the most important vector in the area. At Araluen, on the other hand, the evidence pointed clearly to *Ae. alboannulatus* being among the least important of the several species involved in transmission at the time.

Calaby, Gooding & Tomlinson (1960) concluded that *Ae. alboannulatus* was of negligible importance as a vector in Western Australia: its abundance could not be correlated with disease activity and 'more likely vectors' were always present in greater numbers in outbreak areas. Douglas has not produced any evidence that either species was associated with transmission in Victoria, though they are common in that State. It seems safe to conclude, however, that one or other of these two mosquitoes must frequently contribute to the spread of infection in timbered country. As was mentioned in the notes on *C. p. australicus*, Dyce considered, on presumptive evidence, that *Ae. alboannulatus* was jointly responsible with that species for outbreaks in the Coolamon district; and in the Australian Capital Territory and the surrounding tableland it is difficult to point to a more likely vector than *alboannulatus* in many of the myxomatosis outbreaks that have been recorded.

Aedes (Ochlerotatus) theobaldi (*Taylor*) *and*
Aedes (Ochlerotatus) sagax (*Skuse*)

Distribution

These two species are typically mosquitoes of the inland of eastern Australia, though they intrude at times and places into the coastal belt. *Aedes sagax* ranges across the continent, but is an uncommon species in Western Australia.

Biology and vector status

In the rabbit-infested areas of eastern Australia, inland of the Divide, these two mosquitoes are the commonest representatives of the subgenus *Ochlerotatus*, species of which are specially adapted to breed in transient water. This adaptive opportunism is based on resistant eggs, which are laid in the zone of damp soil around shrinking flood-water pools. In the Riverina, and other regions with a comparable amount and distribution of rainfall, *Ae. theobaldi* and *sagax* normally have a wave of abundance in early spring, but they are liable to appear at any time of the year when conditions favour their breeding.

Being day biters which attack man readily and viciously, they are the most obtrusive pest mosquitoes of the inland pastoral country. Frequently, though not invariably, their activity will continue into the period of dusk but very rarely indeed after nightfall. They are general feeders, attacking livestock, poultry and rabbits in addition to man. Observations have been inadequate to indicate whether they have any marked blood-source preferences, although Myers got the impression that they were attracted more to cattle than to rabbits. Occasional blood-fed specimens of *Ae. theobaldi* have been taken in warren-out traps in western Victoria by Douglas, and have proved rabbit positive. On the whole it would seem safe to conclude that availability is the most important factor determining the source of their blood meals.

The first (and a very good) opportunity to assess the effectiveness of *Ae. theobaldi* and *sagax* as vectors came in the Riverina in the spring of 1951. The initial spread of myxomatosis, during the preceding summer, had taken place at a time when *C. annulirostris* was the locally dominant mosquito; and the disease was virtually confined to the river frontages and the immediate neighbourhood of permanent water. After the epizootics died away, leaving very few survivors, the infection was known to be smouldering in many places, some well away from the epizootic areas. The rabbit population of the region as a whole remained almost unaffected; and conditions favoured a widespread spring emergence of *Ae. theobaldi* and *sagax*. The expected wave of these mosquitoes appeared in September, with little or no associated flaring-up of the disease (see Fig. 22). The seasonal epizootic only got under way when *Anopheles annulipes* made its appearance some months later (Ratcliffe *et al.* 1952). This sequence of events was repeated in 1953, except for the fact that the mass emergence of aedines took place earlier, in July (Myers *et al.* 1954).

The lack of correlation between myxomatosis outbreaks and waves of abundance of *Ae. theobaldi* and *sagax* was a consistent and striking feature of the field transmission picture built up in a decade of observations. With the rare exceptions discussed under *Ae. vittiger* and *camptorhynchus* below, this applies to the *Ochlerotatus* group as a whole. There can be little doubt that these mosquitoes must sometimes transmit the infection—Myers recovered virus from one out of three batches of *Ae. theobaldi* collected at Lake Urana in November 1952—and it is possible that they may assist in its maintenance and dispersal. As the field observations failed to reveal any feature of their behaviour which alone could explain the inability of these aedines to engender epizootics, one must conclude that there are a number of factors involved, each tending to reduce the likelihood of successive feeds on rabbits by the average individual. A combination of the following factors provides the most probable explanation:

(*a*) The absence of any preference for rabbits among a range of acceptable blood sources. (It is significant that rabbits are favoured by both of the main and unquestionably most effective vectors—*C. annulirostris* and *An. annulipes.*)

(*b*) The relatively short individual life span. The waves of these aedines can be observed to peter out in ten to twenty days.

(*c*) The shifting nature of their populations. This is strongly suggested by field observations: it would favour these species acting as dispersal agents, but would militate against their causing an intense local epizootic.

(*d*) Their tenacity in feeding. No mosquitoes show a greater determination to engorge fully once they have started to feed: this habit would tend to reduce the number of host animals bitten.

(*e*) The restriction of their activity to a part only of the period during which rabbits are normally above ground.

Aedes (Ochlerotatus) vittiger (*Skuse*) and
Aedes (Ochlerotatus) camptorhynchus (*Thomson*)

Distribution

Aedes vittiger is an eastern species, and an intruder from the north into the rabbit-infested regions: it is uncommon in Victoria and the southern half of New South Wales. *Aedes camptorhynchus* ranges across southern Australia (including Tasmania). Being a brackish-water breeder, it is confined mainly to the coastal belt; but it occurs well inland in Western Australia, where the ground water tends to be brackish.

Vector status

The claim of these two species to be included in the list of myxomatosis vectors rests on the correlation of outbreaks with waves of their abundance on one or two occasions only, plus the fact that they both have been shown to feed readily on rabbits.

During the first season of Waterhouse's (unpublished) studies in New England, *Ae. vittiger* enjoyed a period of exceptional local abundance. The autumn, winter and early spring of 1954 had been very dry, resulting in the reduction to a minimum of mosquito-breeding habitat, and precluding any rapid or general build up of, for example, *Anopheles annulipes*. Heavy rain in mid-October left water lying everywhere; and *Ae. vittiger* was by far the most abundant mosquito produced by the flood-water pools. Almost immediately after the appearance of adults, myxomatosis flared up and during the month of November many high-grade local epizootics were recorded. When *Culex annulirostris* emerged in numbers in December, disease activity became more or less general. By that time the numbers of *Ae. vittiger* had greatly diminished.

It must be presumed that *Ae. vittiger* was the main vector in the November outbreaks. It was observed to attack rabbits freely, and dominated the drop-trap catches. (Sheep were found to provide the other main local blood source: in Queensland the species is well known as a worrier of cattle.) Multiple feeding was indicated by a proportion of the insects attracted to human bait and to rabbits containing partly digested blood.

Most of the field observations made on *Ae. camptorhynchus* point to the fact that it should be classed with *Ae. theobaldi* and *sagax* as an ineffective vector. Thus in his study area at Yarram, on the Victorian coast, Douglas (*in litt.*) recorded the same absence of correlation between myxomatosis activity and the spring wave of *camptorhynchus* as did Myers in the Riverina with the two other aedines, and the same dependence on *An. annulipes* appearing later in the year. However, a number of intense outbreaks occurred along the Victorian coast, and on offshore islands, in the early spring of 1951, when subsequently acquired knowledge of the local mosquito fauna would suggest that *camptorhynchus* (and the related *Ae. (O.) nigrithorax*, which usually occurs with it in lesser numbers) would be the only prevalent mosquitoes.

An interesting epizootic occurred in south-western Australia in the summer 1952–53 following unseasonable rain that fell in March over an area centered on Jennacubbine, about 100 miles north-east of Perth. *Aedes camptorhynchus*—an autumn-breeding mosquito in that part of Australia—'is believed to have been the chief vector. Considerable numbers of that species were sheltering in rabbit burrows, together with much smaller numbers of *A. annulipes* and *Culex* spp.' (Calaby *et al.* 1960).

There is no doubt that *Ae. camptorhynchus* feeds readily on rabbit, although like other *Ochlerotatus* species it is catholic in its tastes. It was observed feeding on a diseased rabbit at Jennacubbine; and in Victoria, in 1955–56, G. W. Douglas (personal communication) collected it in numbers in drop traps. All engorged specimens taken during that season turned out to be rabbit positive.

With many, if not most, mosquito species that feed on rabbits but have no special predilection for them as a blood source, the line between effectiveness and ineffectiveness as a vector must depend on a rather delicate balance in the operation of several ecological and behavioural factors. When it is realized that the intensity of transmission needed to lift the infection to an epizootic level from one which merely maintains it in a rabbit population is much less than one might expect (as will be explained later in this chapter), one should not be surprised to find a normally ineffective species occasionally assuming a vector role. Environmental conditions tending to increase the longevity of the adult insects or their chances of encountering a rabbit rather than, say, a cow when seeking a second or third blood meal, might bring this about.

Aedes (Finlaya) notoscriptus (*Skuse*)

Distribution

Almost Australia-wide. Breeding readily in tanks and containers, it has adapted itself to the neighbourhood of habitations and is now best known as a domestic mosquito. As a sylvan species, breeding mostly in tree-holes, it is common in the eastern coastal belt.

Vector status and relevant biology

Our information on this species as an actual and potential vector is derived mostly from Colo Vale and Araluen (Dyce & Lee, 1962; A. L. Dyce, personal communication). At the former site it was almost as prevalent as *Ae. alboannulatus*. It is both a day and night biter, peak activity usually occurring at dusk and dawn. It enters rabbit-baited drop traps readily; and virus was recovered from one of six batches of the species collected at Araluen during the 1955 epizootic. Dyce considers *Ae. notoscriptus* likely to be quite an important vector in the scrubby forest around Sydney and other parts of the coastal belt where myxomatosis works unobtrusively but effectively; and he has pointed out that because of its breeding habits it is a mosquito whose local populations could be built up and maintained artificially if it were desired to do so.

Culiseta *spp.*

Distribution

Only those species occurring in the hill country and ranges of Victoria are considered here.

Vector status and relevant biology

Ten species of the genus *Culiseta* (which until recently was known as *Theobaldia*) occur in Victoria. Most of them have a preference for cool habitats, and are intolerant of high temperatures and low humidities (Dobrotworsky 1954). *Culiseta frenchi* (Edwards), *C. hilli* (Edwards) and *C. victoriensis* (Dob.) bite man readily, and their prevalence in season along creek and river frontages is thus obvious. Their main breeding grounds have still to be discovered; Dobrotworsky believes they will turn out to be the burrows of land crayfish.

Culiseta inconspicua (Lee), another of the more common species, breeds in the pools formed by drying creeks and in other ground water.

Douglas (*in litt.*) considers that the species just mentioned act as vectors in the hill country of southern Victoria, chiefly in the neighbourhood of creeks and rivers. He has shown that they are attracted to and feed on rabbits; and outbreaks such as those occurring at Jack River and Hiawatha in December/January 1952–53, and in foothill country in the Yarram district in 1953–54 and 1955–56, have coincided with the peaks of their seasonal activity. (*Anopheles annulipes* was also present in these areas, but was more active some distance from the streams: it was associated with foothill outbreaks later in the autumn.) With their seasonal timetable and limited ranging, Douglas regards the Culisetas as replacing *Culex annulirostris*, ecologically and as vectors, in southern and eastern Victoria.

Miscellaneous mosquito species

Those mosquitoes known to be vectors of any importance are included among the species dealt with under the preceding headings. There remain a few which deserve brief mention because they are known to feed on rabbits and have been recorded in one or other of the study areas under circumstances suggesting that they could, on occasion, play some part in myxomatosis transmission.

The rather dramatic appearance of *Ae. vittiger* in New England has already been described, because there was good presumptive evidence that for a short period it played a major vector role. There were other instances of *Ochlerotatus* species with somewhat similar breeding and feeding habits suddenly appearing in considerable numbers in the study areas—not with any indication that they contributed importantly to myxomatosis transmission, but with the probability that they made some contribution to it. *Aedes* (*Ochlerotatus*) *imperfectus* Dobrotworsky was prevalent in the Araluen Valley in the late summer of 1955, and Dyce recovered virus from the single batch (of 34 specimens) that he collected. *Aedes* (*Ochlerotatus*) *spilotus* Marks was very abundant in western Victoria (Goroke district) during the wet spring of 1955; and as Douglas (*in litt.*) showed that it was feeding on rabbits, it could well have played a minor vector role (*An. annulipes* was prevalent and active at the time). In the very wet late summer of 1956, *Ae.* (*O.*) *procax* (Skuse) appeared at Colo Vale 'in quite large numbers. They fed avidly on man and positive precipitin tests were also recorded for rabbit and possum feeding' (Lee *et al.* 1957). Myxomatosis transmission was fairly active at the time.

One of the very few surveys carried out in the arid zone was made in 1953 by Waterhouse in eight-inch rainfall country lying north of the Broken Hill railway line in South Australia, where some widespread Grade 2 outbreaks of myxomatosis had been recorded (e.g. on Koonamore station). Of the mosquitoes he collected and observed, the most likely vector was *Aedes* (*Macleaya*) *tremula* (Theobald). This species is normally a tree-hole breeder; and where it found adequate breeding places in that semi-desert region remains obscure.

Simuliidae and other small blood-sucking Diptera

Apart from mosquitoes, the only insects for which positive evidence of vector activity was obtained during the field studies were one or two species of Simuliidae (which should be referred to as 'black flies' to conform with overseas practice and to avoid the confusing usage of the term 'sandflies'). The Australian simuliids are all stream-breeders, the different forms having specific preferences within the available spectrum of habitats ranging from small trickles in some grassy paddock to the mainstreams of major rivers, such as the Murray.

Simulium melatum Wharton was identified as a vector by Mykytowycz (1957) in January 1951 near Corowa, on the River Murray. Its role in transmission during that first epizootic was established by the same technique used to incriminate *Culex annulirostris* (which has already been described); but as the simuliid almost always fed on the rabbits' ears, it was possible to mark the site of individual bites and note which of them developed primary lesions. In common with other simuliids, *S. melatum* has its peak of activity in the late afternoon, and its attacks on rabbits ceased before *C. annulirostris* started biting.

Mykytowycz had his attention drawn to *S. melatum* by the observation that many of the infected rabbits near Corowa had 'lop' ears, due to the size of the lesions on the pinnae. During the spread of myxomatosis in southern Queensland a little later in that same year, a C.S.I.R.O. veterinarian reported that a significant percentage of the diseased rabbits he saw on a trip through the outbreak area had lop ears, suggesting that simuliids (possibly the locally prevalent

Austrosimulium pestilens M. & M.) might have played some part in transmission.

Austrosimulium furiosum (Skuse) is the only other species for which positive evidence of vector activity was obtained. K. Myers (personal communication) exposed two laboratory rabbits to the biting of *A. furiosum* during an outbreak near Albury in the spring of 1952, and one of them became infected. The species is found in all six States, and periodically occurs in great abundance over an extended area (when it is liable to cause considerable worry to stock, and even be a nuisance to human beings). It feeds readily on rabbits and has been collected in drop-traps at Colo Vale (Dyce & Lee, 1962), in New England by Waterhouse (unpublished observations), and in Victoria, in considerable numbers, by Douglas (*in litt.*).

Conditions during the spring of 1952 were exceptionally favourable for simuliids; and *A. furiosum* was abundant in a belt extending from at least the south-east of South Australia into northern New South Wales. Fennessy reported it in big numbers in north-western Victoria (the Mallee), in places 40 miles from the Murray, the nearest river. In South Australia, some of the localities in which Waterhouse collected the species were even further from known or likely breeding places; and its markedly patchy distribution suggested to him that it tended to move in 'tight swarms'. There can be little doubt that *A. furiosum* has an exceptional flight range.

A clear indication of simuliids acting as vectors was obtained during the 1952 spring. Quite a number of localized outbreaks were reported, and investigated, in which simuliids were found to be active to the virtual exclusion of other possible vectors. Myers was able to locate more than half-a-dozen such outbreaks in the area around Albury (it was at one of these sites that his experimental rabbits were exposed); and others were investigated in Victoria, for example along the frontage of a small stream near Bacchus Marsh. At Karte, in the South Australian Mallee, Waterhouse recorded a grade 1 outbreak associated with *A. furiosum* in abundance.

Species other than the two named above are probably involved in transmission, for two or three are known to feed on rabbits at times. Douglas took *Austrosimulium bancrofti* (Taylor) and *Simulium nicholsoni* M. & M. with some regularity in his drop traps (in addition to *S. melatum* and *A. furiosum*). A. L. Dyce (personal communication) observed *A. bancrofti* feeding on rabbits at Moree. At Colo Vale the simuliid catches on rabbit bait included two species of the genus *Cnephia* (Dyce & Lee, 1962).

In summary, it can be said of simuliids as vectors that although of minor importance in comparison with mosquitoes a few species seem to be capable of initiating and maintaining quite intense localized outbreaks. In addition, it is not unlikely that highly mobile forms such as *A. furiosum* may at times assist the dispersal of the infection (a role which, as has already been mentioned, some mosquitoes of the aedine subgenus *Ochlerotatus* may share). The pattern of myxomatosis outbreaks during the 1952–53 summer pointed to there having been a widespread dispersal of the virus at some time during the spring, when there was abundant evidence of *furiosum* activity and mobility.

Other small Diptera had, of course, to be considered as candidate vectors, notably the Ceratopogonidae (biting midges, commonly called sandflies in Australia) and species of *Phlebotomus* (Psychodidae) —insects to which the term sandfly is applied overseas, but for which Australians have no common name. Quite a lot of attention was given to these insects, particularly by Dyce & Lee (and E. J. Reye on ceratopogonids); but the study of their ecology and behaviour in the field is beset with great technical difficulties. All that it is safe to say about them is that the possibility of their assisting in myxomatosis transmission, and even being primarily responsible for some local outbreaks for which no alternative vectors can be suggested, cannot be dismissed. Dyce (*in litt.*) has pointed out that at least two widespread species of ceratopogonids, *Leptoconops stygius* Skuse and *Lasiohelea townsvillensis* (Taylor) are confirmed rabbit feeders; while *Phlebotomus* spp., also known to feed on rabbits, are proving to be commoner insects than was once suspected, particularly in the lower-rainfall regions. Dyce found the latter to be 'quite abundant' in the mulga (*Acacia aneura*) near Cunnamulla in south-west Queensland.

In an unpublished report of his observations in the arid pastoral country of South Australia (see under Miscellaneous Mosquitoes above), Waterhouse learned from the manager of Koonamore that myxomatosis activity had coincided fairly closely with the noticeable prevalence of 'sandflies' or 'midges'. He assumed that they were ceratopogonids; but it seems more likely that they were *Phlebotomus* sp.

Fleas and other ectoparasites

The results of the early field experiments of Bull and Mules (see chapter 1) led to the expectation that the spread of myxomatosis would be materially assisted by stickfast fleas, particularly in the drier areas where *Echnidophaga* spp. are prevalent and seem to have become primarily rabbit parasites. These expectations were not fulfilled, and we had no grounds for attributing any epizootic to the activities of these insects.

Between 1951 and 1954 the Western Australian authorities operated a number of inoculation centres in different parts of the State in their efforts to assist the establishment and spread of the virus.

At some of the myxomatosis infection centres, e.g. Mingenew, Moonijin, Wongan Hills, and Katanning, the rabbits were flea-infested, and at the first three places the infestation was moderate to heavy. No epizootic developed in any of these places. At Mingenew 1400 marked infected rabbits were released over 12 months. Occasional unmarked sick rabbits were seen, the maximum in any day being three. *E. myrmecobii* was present at the year round, and its numbers were high in summer. At Moonijin the rabbits supported a fairly high *E. myrmecobii* population for a 4-month period when the area was under observation. During that time 630 diseased rabbits were released, but only an occasional case was seen, and on one occasion there was a diminution in the rabbit population of three warrens (Calaby *et al.* 1960).

K. Myers (personal communication) has been carrying out ecological studies of the rabbit in inland New South Wales and Queensland since 1961, and the populations he has been sampling have been through two or three periods of heavy infestation by stickfast fleas. No epizootics developed during these periods: they occurred at other times, during periods of mosquito abundance.

In several localized Victorian outbreaks (and at least one in New South Wales) the affected rabbit population was found to be infested by the cat flea (*Ctenocephalides felis*) with no flying vectors obviously active at the time. However, none of these outbreaks (which were picked up on regional surveys) was intensively investigated and it cannot be stated unequivocally that no other vector could have been involved.

Mykytowycz (1958b) showed that two ectoparasites commonly infesting wild rabbits in Australia, the louse *Haemodipsus ventricosus*

(Denny) and the mite *Cheyletiella parasitivorax* (Megnin), were capable of transmitting the infection; but he considered it unlikely that either 'could originate and maintain a myxomatosis outbreak of epizootic proportions'.

When the active spread of an epizootic is observed to be limited or delayed by netting fences, it must be concluded that transmission is predominantly by ectoparasite exchange. (We believe that contact infection, in the strict sense, can be ruled out in these circumstances, as we have so often recorded the failure of myxomatosis to maintain itself when the virus has been introduced into crowded rabbit populations, where frequent and close contacts between individuals must have been unavoidable.) This 'fence effect' has been noticed not infrequently in Tasmania, and was very obvious (K. Myers, personal communication) during a myxomatosis outbreak which virtually exterminated the rabbits in some recently established enclosures at Canberra. This outbreak was of a type commonly encountered in parts of the Australian Capital Territory, with the infection smouldering at a barely observable level during late summer and autumn, and flaring up to epizootic proportions with the advent of cold weather.

VECTOR MOBILITY: THE CASE OF WOODY ISLAND

Brereton (1953) investigated the spread of myxomatosis along the frontages of the Darling and some of its tributaries in the summer of 1950–51, combining first sightings of diseased rabbits by local landholders[1] with his own observations. The results suggested that the infection spread at a remarkably steady rate—about three miles per day—irrespective of whether its local incidence was high or low, leading him to advance the theory that the disease spread by 'creeping' rather than 'jumping', and as a result of the movement of rabbits rather than of vectors.

In general, those who had the task of following the initial spread of myxomatosis, particularly during the 1951–52 and 1952–53 seasons, saw a pattern of dispersal (e.g. with penetration into isolated clearings

---

[1] Some of the first-sightings on the Darling, the reliability of which Brereton considered beyond question, and which he recorded in an unpublished report, pre-dated the earliest recorded epizootics on the Murray (in mid-December near Corowa) spotlighting our incomplete knowledge of what happened to the virus in the field between the dying down of the infection at the experimental release sites and the spectacular summer flare up.

and valleys) which seemed explicable only if there was a continual, or at any rate frequent, ranging of infected insects over distances of several miles. Flight ranges of this order have been mentioned above in connexion with *Anopheles annulipes* and *Austrosimulium furiosum*. A. L. Dyce (personal communication), while working at Moree, was able to record the dispersal of *Aedes theobaldi* over a distance of some 40 miles from an isolated mass-breeding site.

While the normal movement of vectors—mainly in the course of their search for blood meals, but sometimes in response to an instinct to disperse fairly widely from their breeding grounds—seems adequate to explain the observed pattern of disease activity, longer-distance movements resulting from the passive carriage of insects on winds almost certainly take place from time to time. This was clearly indicated during the initial spread of myxomatosis, when vectors were, in effect, 'tagged' by the virus.

Shortly after the first epizootics along the Murray, myxomatosis was recorded at a number of isolated points in north-western Victoria, some well away from the river; and a most interesting series of occurrences came to light westward across South Australia. One of the earliest records of the disease in that State was the infection of a domestic rabbit in the outer suburbs of Adelaide. Small foci of the disease were then discovered (*a*) at Ardrossan, on Yorke Peninsula, (*b*) just inland of Port Pirie, on Spencer's Gulf, (*c*) at Tumby Bay on the east side and (*d*) Streaky Bay on the west side of Eyre Peninsula, and (*e*) at Koonibba, north-west of Ceduna. The last mentioned place is 400 miles from the Murray River. No explanation other than the passive carriage of insects on easterly winds can sensibly be advanced to explain these widely separated occurrences (Lines, 1952; Ratcliffe *et al.* 1952).

Another remarkable situation came to light the following year, when in February 1952 myxomatosis broke out in a population of rabbits inhabiting a small island off the Queensland coast, 200 miles north of the New South Wales border. The outbreak was reported by the State Department of Agriculture and Stock, and was subsequently investigated on behalf of C.S.I.R.O. by Dr E. J. Reye (unpublished observations). The island in question, Woody Island, lies at the northern end of Sandy Strait, which separates Fraser Island from the mainland. The only permanent inhabitants are the lighthouse keeper and his family, and visits (e.g. by fishermen) are discouraged. Rabbits were introduced to Woody Island in 1858 by a ship quarantined there, and by 1950 (after some vicissitudes) had achieved a steady high level of abundance. After the first sighting of a diseased animal, an intense epizootic developed which left very few survivors. Woody Island is separated from the nearest wild rabbit populations on the mainland by a distance of at least 200 miles. The likelihood of an infected insect, or insects, being transported from the south in a coastal vessel appeared excessively remote when the factors involved were considered; and the only reasonable explanation offering was the wind-transport of vectors from some area of disease activity near the Queensland–New South Wales border. Reye, after examining the meteorological picture for January and February, came to the conclusion that conditions which might have favoured the mass-transport of insects on the wind, in the right direction, occurred round about 12 January.

EPIZOOTIC PATTERNS AND TRANSMISSION

The early outbreaks investigated by the C.S.I.R.O. team took the form of intense epizootics which normally worked themselves out in four to six weeks after the infection was first noted in the local rabbit population. Sick rabbits were obvious during daytime inspections, and carcasses abundant. The outbreaks were associated with easily recordable vector activity. The species concerned usually built up in numbers somewhat abruptly but remained prevalent well after the epizootic had died down, indicating that their abundance was in excess of that required for effective transmission.

It was not long before we were introduced to outbreaks which followed a very different pattern. Thus Dyce and Poole, working out of Moree for a period from 1952 on, observed the results of myxomatosis operating at a barely recordable level in a river-flat area near Texas on the Queensland border. In the course of ten to twelve months, a reduction in the rabbit population equivalent to that resulting from the intense epizootics investigated in the south had been achieved. Closer to Moree, disease activity of a somewhat similar character prevailed, producing over a period of about twenty weeks an estimated 90 % mortality, with diseased animals rarely seen (Marshall *et al.* 1955). At Colo Vale, during the first year after the deliberate introduction of the virus, no clearly defined epizootic occurred; but protracted disease activity, at low intensity, effected a substantial reduction of the rabbit population—a situation broadly comparable with the cases just mentioned.

Such long-drawn-out low-intensity outbreaks, and some of the well-defined late summer epizootics observed subsequently in the coastal belt, lacked any clear association with high, or noticeably heightened, mosquito activity. It therefore seemed desirable to know the order of increase in the transmission rate required to lift disease activity from a mere maintenance level (at which an infected rabbit would pass on the disease to no more than one other individual, on the average, before it died) to a level classifiable as epizootic.

Professor P. A. P. Moran, of the Australian National University, came to our assistance with a calculation based on a numerical model, making certain reasonable assumptions regarding mosquito biting, etc. He was asked to give an approximate figure for the transmission rate, measured by the mean number of rabbits infected from one diseased animal, which would produce an outbreak in which close on 100 % of the population would become infected within six weeks. For the purpose of the calculation, it was assumed that the outbreak was observed to have started when 1 % of the population had a generalized infection, and that the virus involved was slightly attenuated, resulting in a mean survival time of twenty-one days. He found that an increase in the transmission rating from unity to three was just insufficient to produce the required result, indicating that the postulated speed of epizootic spread would be achieved by transmission at between $3\frac{1}{2}$ and 4 times the rate obtaining under 'smouldering' conditions.

The translation of Professor Moran's figure into simple terms of mosquito abundance is dangerous for many and obvious reasons, and, like others used for comparable purposes, his model was an over-simplification. However, his calculation makes it clear that in many local situations outbreaks could develop as a result of changes in vector abundance and activity which would be difficult to measure by the rather crude techniques available to the field workers; and it helps to explain the disappointing results of the virus recovery attempts. The first-studied epizootics, in the Riverina, may be said to have misled us in what to expect. In that region, the sharp seasonal differences in breeding conditions resulted in a rapidly attained abundance of the vector species representing something like 100-fold (or perhaps even greater) increase above off-season levels.

TRANSMISSION DURING INTER-EPIZOOTIC PERIODS

What happens to the myxoma virus between epizootics, and the nature of the transmission by which the infection is maintained (often at a sub-observational level) in an area, are of considerable epidemiological interest. Not much information on these questions came out of the field studies, because intensive work was confined in the main to periods of active transmission. Valuable data, however, were obtained from Colo Vale. Here the landholders on whose properties the work was carried out were exceptionally interested and cooperative; and with their help, bushmanship and local knowledge it was possible to step up the intensity of observations to a level which it was impracticable to achieve elsewhere. In this way the occurrence of diseased rabbits was recorded with sufficient frequency to demonstrate that the infection was maintained continuously in the area, almost certainly as a result of mosquito transmission. Potential vectors were shown to be available, even though in small numbers, at all times of the year (Lee *et al.* 1957; A. L. Dyce, personal communication).

After the conclusion of his investigations in the eastern Riverina, Myers (*in litt.*) made an assessment of the round-the-year availability of insect vectors, including those not effective enough to initiate epizootics but obviously capable of assisting the maintenance of the infection at a low level of incidence. He was considering a regional situation, very different from the one at Colo Vale which is a small area almost isolated by a belt of mountain forest. In the Riverina, the virus must disappear from some local rabbit populations as a result of the virtual elimination of susceptible individuals during intense epizootics, having to be re-introduced before the next season's outbreak can get under way in the new generation. He showed that winged vectors were normally present over the greater part of the year—with aedine mosquitoes such as *theobaldi* and the *alboannulatus* group, and simuliids, occurring during the off-seasons of the major vector species—and it was only during a brief period in the winter that it was necessary to postulate transmission by contact or ectoparasite exchange. Breeding is in progress at that time, with frequent combats and sexual advances to females. 'These encounters lead to wounding, and to very close contacts (licking of eyes, ears, nose and anal regions, sitting nose to nose or side by side for hours on end) between numerous individuals.' Observable disease activity was often associated with the initiation of breeding.

SUMMARY OF OBSERVATIONS AND CONCLUSIONS

It will be clear from the information presented in this chapter that the ecological framework of myxomatosis outbreaks varies widely from region to region. This variation was covered to some extent by the location of the study areas in which detailed observations were undertaken. Circumstances allowed only a limited number of epizootics to be intensively investigated entomologically: the vector situation in many outbreaks had to be interpreted in the light of knowledge of the behaviour and performance of the locally occurring insects acquired elsewhere (i.e. in the special study areas).

Two species of mosquitoes were revealed to be of outstanding importance as vectors—*Culex annulirostris* and *Anopheles annulipes*. The latter habitually rests in rabbit warrens, which provide harbour and high humidity during the day—and in addition readily available blood feeds—and in this way can remain active as a vector in places where others would be eliminated by the drying up of the environment. Outbreaks attributable to one or other of these species are the main feature of myxomatosis activity over a very substantial part of the rabbit-infested area of the Australian mainland between the coastal belt and the low-rainfall zone. In parts of southern Victoria, where *C. annulirostris* does not occur, species of the genus *Culiseta* take over its role, transmitting the infection during summer in the neighbourhood of hill streams.

On an overall rating, the widely distributed *Culex pipiens australicus* probably ranks third in importance as a vector. (It proved difficult to study and assess.) Usually acting in support, as it were, of one or other of the two major vectors, there is reason to believe that it may be of prime local importance at times, for example in Victoria.

In some regions, notably the coastal belt of New South Wales, outbreaks typically depend on the combined activity of several mosquito species (which may be categorized for convenience as minor vectors). Where this applies, the ecological situation is usually complex and the general level of mosquito abundance moderate to low in comparison with that attained, in season, in the classical habitats of the major vectors. (*Anopheles annulipes* and *C. annulirostris* are present in most coastal areas, claiming inclusion in the local 'vector groups'.) The more important of the minor vectors are probably species of the aedine subgenus *Finlaya*—*Ae. alboannulatus, rubrithorax* and *notoscriptus*.

Species of the aedine subgenus *Ochlerotatus* play rather anomalous roles. *Aedes theobaldi* and *sagax*, the most abundant casual-water breeders in the eastern plains, are ineffective vectors. The southern, predominantly coastal, *Ae. camptorhynchus* also seems incapable in general of initiating epizootics, though there is some evidence that it may on occasion do so. One or two other *Ochlerotatus* species can claim temporary and local vector status, as a result of sudden appearances in high numbers. Under such circumstances *Ae. vittiger* played a major role during one season in northern New South Wales; but the more usual situation is exemplified by *Ae. imperfectus* in one of the coastal study areas, where it merely merited consideration as a member of the local 'vector group'. Some of the commoner species of *Ochlerotatus*, which tend to be very mobile, probably assist in the dispersal of the virus.

Two species of Simiuliidae (black flies)—*Simulium melatum* and *Austrosimulium furiosum*—were identified as vectors, and others are probably involved in transmission at times. Simuliid activity gives rise, in some years, to sharp localized epizootics; and the exceptionally mobile *A. furiosum* probably assists in virus dispersal and with it the widespread maintenance of myxomatosis endemicity. In the overall picture, simuliids are of minor importance compared with mosquitoes as vectors.

Very little relevant information could be obtained on biting midges (Ceratopogonidae) and sandflies (Psychodidae); but there was enough to indicate that these small blood-sucking Diptera could not be dismissed offhand as being of no significance. It is conceivable that their unobtrusive activity might provide the explanation of some of the many outbreaks investigated in which no indication of the responsible vectors could be obtained. Species of the psychodid genus *Phlebotomus* must be regarded as of possible vector importance, particularly in some of the drier areas.

Despite the indications of early field experiments, no evidence was obtained that stickfast fleas (*Echidnophaga* spp.) were able to initiate and sustain epizootics.

Transmission other than by winged vectors must frequently occur; and it seems reasonable to conclude that it plays a part in the maintenance of the infection at low incidence. The field observations provided evidence that under certain conditions transmission by contact (in the breeding season), ectoparasite exchange, or through the agency of thistle thorns could increase myxomatosis incidence to an epizootic level.

In parts of eastern Australia, at any rate, disease activity of this kind may in the aggregate be of real significance, both from the viewpoint of rabbit mortality and the maintenance of endemicity. However the broad picture, seen in perspective with the pattern changing seasonally over the twelve years during which myxomatosis has now been under observation, demonstrates unequivocally that the infection depends primarily on winged water-breeding vectors to attain epizootic levels. Allowing for some well understood ecological complications (e.g. excessive stream flow militates against maximum production of *An. annulipes*), an exceptionally wet season is almost automatically associated with exceptional myxomatosis activity. In some regions (e.g. parts of Western Australia) effective widespread epizootics have only taken place when the normal rainfall distribution was varied by the occurrence of unseasonable heavy rain. And the identity of the (mosquito) vectors mainly responsible for the more extensive and intense disease activity that develops under such circumstances is usually obvious. Field evidence of the mobility of mosquitoes and other insects liable to pick up the infection helps to explain why the spread of myxomatosis in response to the sudden development of favourable conditions can be rapid and widespread.

CHAPTER 13

CHANGES IN THE VIRULENCE OF MYXOMA VIRUS FOR *ORYCTOLAGUS CUNICULUS*

Both 'virulence' and 'attenuation' are polarized words, with all the semantic difficulties that this implies. Yet in the consideration of myxomatosis of *Oryctolagus*, which is the main topic of this book, change in what is commonly described as virulence is by far the most important viral property. Myxomatosis of *Oryctolagus* is, indeed, one of the few natural infections in which virulence can be equated with lethality, both in the field and in the laboratory.

Prior to 1950 no population of *Oryctolagus cuniculus* had been exposed to infection with myxoma virus, apart from the experiment on the Dufeke Estate in Sweden (see chapter 17) and sporadic outbreaks in domesticated rabbits in North and South America (see chapter 15). In 1950 myxomatosis was introduced into Australia, and since 1954 the disease has been co-extensive with the rabbit population of that continent. In 1952 it was introduced into France, and within a decade had spread to virtually all populations of wild *O. cuniculus* in Europe and North Africa. In 1954 the virus was introduced into wild rabbits on Tierra del Fuego by the Chilean government. Further introductions have been made into populations of wild *Oryctolagus* elsewhere in Chile, which is the only part of the mainland of the Americas where these animals occur. Only in New Zealand and in a few isolated islands in the Americas and elsewhere do substantial populations of wild *Oryctolagus* now exist which have not been exposed to myxomatosis.

Discussion of the virulence of myxoma virus is primarily concerned with its effect on the non-immune, genetically unselected, adult *Oryctolagus cuniculus*. This is the host in which myxomatosis, either as a sporadic disease in the Americas or as a panzootic in Australia and Europe, has caused such tremendous mortalities. The most straightforward assessment of virulence which can be made is to relate it to lethality, and for this combination of virus and host animal this is a possible, and indeed the best, measure of virulence.

Since Hurst's (1937b) description of neuromyxoma it has been known that there were strains of myxoma virus of reduced lethality, that is strains existed which were of very high virulence, and of low virulence for *O. cuniculus*. When it was realized that virus strains of reduced virulence were also occurring naturally in the field in Australia (Fenner, 1953b; Mykytowycz, 1953) and in Europe (Fenner & Marshall, 1955; Hudson, Thompson & Mansi, 1955) an attempt was made to devise a reliable and simple method for evaluating the virulence of strains of myxoma virus recovered from field cases (Fenner & Marshall, 1957).

THE LABORATORY ASSESSMENT OF THE VIRULENCE OF FIELD STRAINS

Since the great majority of infections of wild rabbits in Australia are caused by mosquitoes, a method of inoculation was adopted which closely resembled mosquito bite infection, namely the intradermal inoculation of a small dose of virus (10 ID 50). Had expense, space, and labour been of no importance the inoculation of large groups of adult rabbits, and measurement of the case-mortality rate, could have provided a direct measure of lethality and thus of virulence. However, since it was necessary to discriminate between strains which killed 99 and 90 % of infected rabbits, very large groups would have been necessary—quite unmanageable numbers when hundreds of strains had to be tested. It was fortunate, therefore, to find that the survival time was correlated with the case-mortality rate (Fig. 14), and the procedure finally adopted for survey purposes was to inoculate groups of five healthy adult laboratory rabbits intradermally in the flank with doses of 10 ID 50 of virus, observe the rate of development and the type of lesions produced, and determine the mean survival time.

Before considering the results of the application of these procedures to large numbers of strains of virus recovered from the field over several years it is necessary to emphasize the importance of standardization of the procedure.

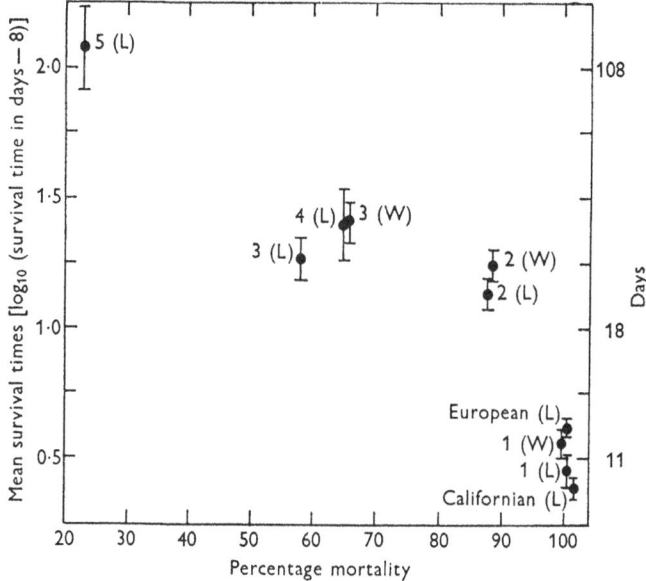

Fig. 14. The relationship between mean survival times and mortality rates with several strains of myxoma virus of widely differing virulence. Transformed means shown with ± twice the standard error. W, genetically fully susceptible wild rabbits used for test; L, laboratory rabbits used for test; Californian, composite group of all virulent Californian strains; 1, standard laboratory strain; European, composite group of highly virulent European strains; 2, Aust/Corowa/12–52/2 (KM13); 3, Aust/Uriarra/2–53/1; 4, France/Loiret/4–55/1; 5, England/Nottingham/4–55/1 (attenuated). Data from Fenner & Marshall (1957); Marshall, Regnery & Grodhaus (1963).

(a) Age of the rabbit

Young rabbits survived for much shorter times (and with strains of slightly reduced virulence the case-mortality rate was higher) than did adult animals (see page 110). No animal less than four months of age was used in the virulence tests.

(b) Route of inoculation

With a strain of reduced virulence Mykytowycz (1956) observed considerable differences in the mortality rates and the survival times, depending upon the route of inoculation, and his results have been confirmed and extended (Fenner & Marshall, 1957; F. Fenner, unpublished observations). Infection by inoculation of a small dose of virus intradermally in the flank, and by mosquito bite, gave very similar results; and the former method was routinely used. The site of intradermal inoculation may be important, since injection into the extremity of the ear is sometimes associated with slow development of the disease (Chapple & Muirhead-Thompson, 1964).

(c) Temperature of animal house

Marshall (1959) demonstrated that the environmental temperature had a profound effect on the response of rabbits inoculated in the standard fashion. High environmental temperatures diminish and low environmental temperatures accentuate the severity of the disease (see chapter 10). All animals used in the standard virulence test were kept in animal rooms maintained at a constant temperature of about 70° F.

(d) Dosage of virus

The dosage of virus inoculated intradermally could be varied over a rather wide range (1000-fold), with only minor differences in the mean survival times, although the time of the appearance of the primary lesion was much earlier when large doses were used. For instance, an increase of the dose of strains of reduced virulence ten thousand times only shortened the mean survival times from twenty-five to seventeen days (Jacotot, Vallée & Virat, 1957), and from twenty-eight to twenty days (Fenner & Marshall, 1957). Small differences in dosage were therefore not important as far as the determination of the grade of virulence of virus strains were concerned.

(e) Regularity of the results observed in successive years

Since the results reported are concerned with secular trends in virulence an assurance was required that month-by-month and year-by-year results were comparable. This was particularly important in view of the demonstration by Mykytowycz (1956) of a significantly higher mortality rate during the winter months than in the summer among rabbits infected with an attenuated strain of myxoma virus and kept in an unheated animal house. Tests of normal laboratory rabbits infected in the standard fashion with either a very virulent virus (standard laboratory strain—Fenner & Marshall, 1957) and a strain of slightly reduced virulence (KM 13—Marshall & Fenner,

1958; Marshall & Douglas, 1961), and kept in an animal house which was heated in winter, gave consistent results at different times of the year, and in different years.

(f) Comparison of the response of wild and laboratory rabbits

Non-immune wild rabbits from populations which had not been selected for resistance to myxomatosis by prior exposure responded somewhat more slowly, and lived somewhat longer, than laboratory rabbits challenged under identical conditions (Table 35). However,

TABLE 35. *Mean survival times of captured Australian wild rabbits and laboratory rabbits challenged by the intradermal inoculation of three strains of myxoma virus of differing virulence*

Tests were carried out before the development of genetic resistance in wild rabbits (data from Fenner & Marshall, 1957).

Virus strain		Mean survival time (days)	
Designation	Virulence Grade	Laboratory rabbits	Wild rabbits
Standard laboratory strain	I	$10\cdot8\pm0\cdot3$	$11\cdot6\pm0\cdot4$
Aust/Corowa/12–52/2	III	$21\cdot5\pm1\cdot9$	$25\cdot4\pm2\cdot4$
Aust/Uriarra/2–53/1	IV	$26\cdot2\pm3\cdot7$	$34\cdot3\pm7\cdot6$

it was not possible to establish a laboratory colony of Australian wild rabbits, and after about 1956 the development of genetic resistance (see chapter 14) precluded the use of wild rabbits for virulence tests. The responses of genetically unselected wild rabbits and laboratory rabbits were sufficiently similar to make it reasonable to assess virulence in terms of the survival times and mortality rates found in tests on laboratory rabbits.

THE VIRULENCE OF STRAINS RECOVERED FROM THE FIELD IN AUSTRALIA

The standardized test just described was used to assess the virulence of 672 strains of myxoma virus recovered from naturally occurring cases of myxomatosis in Australia, over the period 1951–59, as well as that of two laboratory strains, the standard laboratory strain, which was used to initiate the Australian outbreaks, and Hurst's neuro-myxoma, which was derived from the standard laboratory strain by serial intracerebral passage.

Mode of collection of field strains, and their handling prior to inoculation

Prior to 1955 lesion material was collected from infected wild rabbits by officers of the Wildlife Survey Section of C.S.I.R.O. who were concerned with myxomatosis investigations, stored in glycerol-saline, and despatched to the laboratories in Canberra. After that date collection was carried out on a much wider scale, with the help of officers of the Wildlife Survey Section of C.S.I.R.O., of the Pastures Protection Boards in New South Wales, and of the rabbit inspectors of the Department of Lands in Victoria. Bijou bottles containing 50 % glycerol-saline were widely distributed, together with tin canisters for mailing purposes and a proforma on which details of the material collected were recorded.

On arrival in the laboratory the lesion material was washed free of glycerol-saline and either stored at $-10°$, or ground with alundum in a mortar and pestle and suspended in gelatin saline before storage at $-60°$ C. This virus was assayed on the chorioallantoic membrane and a dilution used for inoculation which contained ten rabbit-infectious doses per inoculum.

Material sent to the laboratories from abroad was always passaged once and tissue from the resultant lesion used for subsequent experiments. This was not always carried out with low titre virus obtained from the field in Australia. Recent experiments (F. Fenner, unpublished observations) suggest that failure to passage low titre material could cause confusing results.

About one third of the samples of rabbit lesion material collected in 1958–59 yielded so little virus when they were tested in 1960 that the undiluted suspension had to be used to ensure infection. In 1963, replicate portions of the rabbit lesion material which had been stored in a deep-freeze and refrigerator since then were again ground up and suspended in diluent. Six rabbits were inoculated intradermally with this material, used undiluted. From one of them a tissue slice was taken from the primary lesion on the sixth day after inoculation. This was ground up and assayed on the chorioallantoic membrane. A dilution containing 10 ID_{50} (usually a 10^{-4} dilution) was inoculated into another group of rabbits. The survival times are shown in Table 36. In most cases the response to the inoculation of the low-titre undiluted lesion suspension was the same in 1960 and 1963. However, there were striking differences between these and the response to the diluted passaged virus, exemplified in the most extreme form by samples 960 and 1006.

TABLE 36. *The survival times in days of laboratory rabbits inoculated with undiluted suspensions of rabbit lesion material in 1960 and 1963, and with 10–20 ID_{50} of passaged virus in 1963*

Sample No.	Tests made with undiluted suspensions		Tests made with diluted suspension of passaged virus
	1960	1963	1963
694	18, 20, 22, S*	18, 28, 34, S, S, S	15, 16, 17, 18, 19, 20
702	24, S, S, S	21, 21, 24, 35, 38	17, 20, 20, 21, 22, 35
846	20, 29, S, S	18, S, S, S, S, S	15, 18, 18, 21, 30, 30
908	12, 15, 20, 20	19, 19, 20, 30, 30, 43	15, 17, 17, 21, 21
960	S, S, S, S	S, S, S, S	18, 19, 21, 22, 31, 32
1006	18, S, S, S	S, S, S, S	16, 17, 19, 20, 23, 26
1026	62, S, S, S	18, 19, 20, 28, 29, 32	20, 21, 21, 21, 21, 28

* S, survivor.

The mechanism of this alteration in lethality has not been elucidated. Burnet (1955), commenting on rather similar findings of other investigators with stored or dried fibroma or myxoma virus, suggested that these effects might be explicable on the basis of multiple independent 'virulence genes'. This seems most unlikely, and we are currently investigating the possibility that the large amount of inactive virus present in the undiluted preparations alters the host response to the small viable component, either by interference or an effect on the immune response.

The message is clear. Field material must be used at a high dilution (if it is sufficiently potent) or must be passaged in rabbits before it is used. Two alternative methods of avoiding the interfering effect of inactive virus are to select a single pock from the chorioallantoic membrane used for titrating the original virus suspension, or to make pin-prick or mosquito bite inoculations of a series of rabbits from the skin lesion produced by inoculation of the field material. It is always possible that such manoeuvres will select what was a minority component in the original suspension of virus, but there seems to be no way to avoid this if a passage is made.

Prototype strains of differing virulence

From the point of view of biological control of the rabbit, which was the practical consideration underlying in the use of myxomatosis in Australia, the important feature was the lethality of the disease. Each

TABLE 37. *Classification of strains of myxoma virus into grades of virulence*

Virulence Grade	Mean survival time (days)	Estimated case-mortality rate (%)	Prototype strains	
			Australia	Europe
I	$\leqslant 13$	> 99	Standard laboratory strain	Lausanne
II	13–16	95–99	—	—
III	17–28	70–95	Aust/Corowa/12–52/2 (KM13)	—
IIIA*	17–22	90–95	—	—
IIIB*	23–28	70–90	—	—
IV	29–50	50–70	Aust/Uriarra/2–53/1	France/Loiret/4–55/1
V	—	< 50	Neuromyxoma	England/Notts/4–55/1 (attenuated)

* This subdivision was introduced in 1963.

of the 672 strains tested was allocated to one of five grades of virulence; depending upon the mortality rate (for Grade V strains, the least virulent), or the mean survival time for all other strains. Representative strains belonging to the virulence Grades I, III, IV and V were designated as prototype strains, and the symptomatology of the disease caused by these strains in normal laboratory rabbits is described below.

Grade I virulence

Prototype strain; the standard laboratory strain (Pl. XII, fig. 1). Mean survival time, 10·8 days. Range, eight to fifteen days. Mortality rate, 100%. This is the strain used to initiate the disease in Australia, and it has been repeatedly re-introduced into the Australian wild rabbit population in the course of annual inoculation campaigns. It was recovered from a naturally infected *Oryctolagus* by

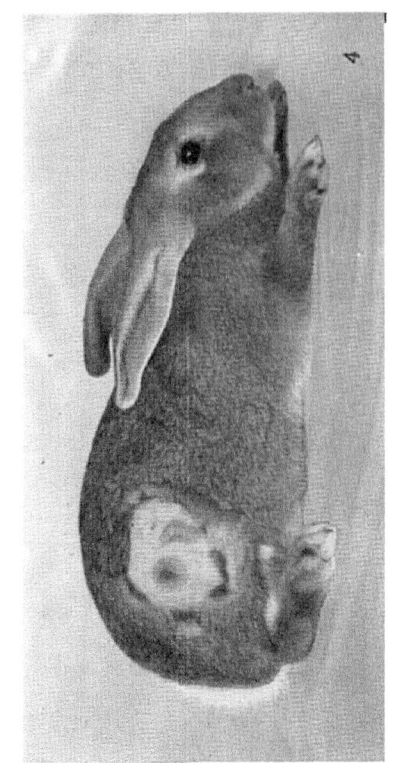

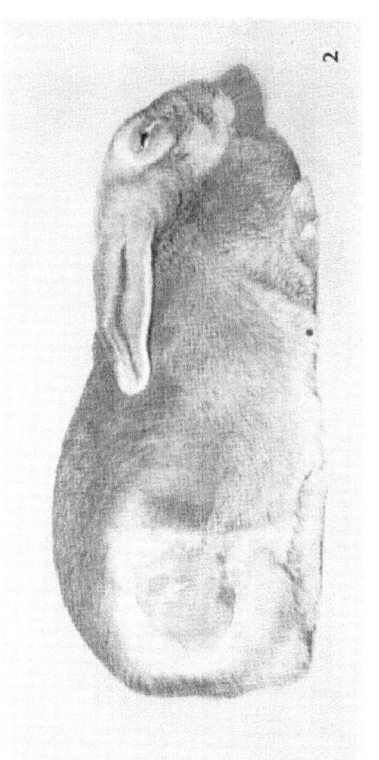

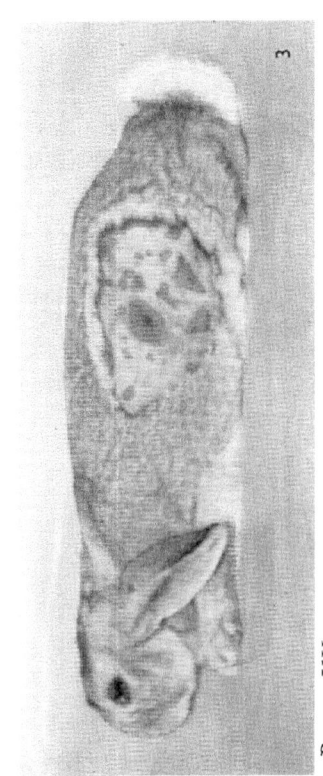

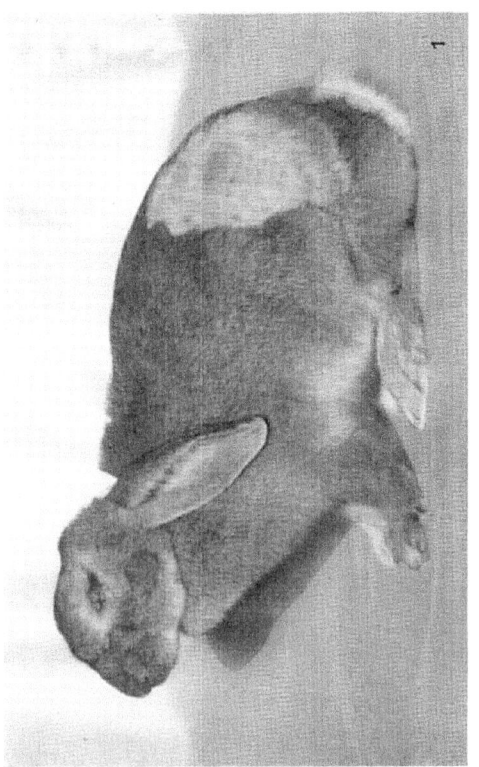

PLATE XII

Dr A. Moses of the Oswaldo Cruz Institute in Brazil (Moses, 1911), and sent by him to the Rockefeller Institute in New York in 1924 (Rivers, 1930). It was subsequently sent from New York to England, and was there designated 'strain B' by Martin (1936). Later it was taken to Australia (Bull & Mules, 1944) where it was ultimately used to initiate the widespread epizootics which broke out in 1950. The strain was maintained by serial passage in laboratory rabbits for 40 years, but since 1951 has been stored, without further passage, at -70° C.

In an extensive experience of this strain of virus in the Canberra laboratories no recoveries have been recorded in genetically unselected rabbits maintained under normal conditions, and inoculated with diluted or freshly prepared virus. Tests with 127 clones of this virus recovered from single pocks on the chorioallantoic membrane failed to reveal any clones associated with prolonged survival times.

The lump produced at the inoculation site is first seen on the third day and later becomes large, hard and convex, but not protuberant. The roughly circular margin merges gradually into the surrounding skin, there being no sharp demarcation between the tumour and normal skin. The skin over the primary lesion remains unbroken throughout the disease. Secondary skin lesions are recognizable by the sixth and seventh day, and by the ninth day they are widely distributed over the body and ears. Lumps occur on the limbs only in

---

PLATE XII. Appearance of laboratory rabbits infected by the standard method with the standard laboratory strain and three derivatives of this (from Fenner & Marshall, 1957).

Fig. 1. Standard laboratory strain, ten days after inoculation. Rabbit died later same day. Note the completely closed eyes and profuse conjunctival discharge, general oedema of the head, and lack of clear definition of the edges of the primary lesion.

Fig. 2. Aust/Corowa/12–52/2 (KM13), twenty-one days after inoculation. Rabbit died on the twenty-eighth day. Note that the eyes are not quite closed though there is considerable oedematous swelling of the head, and that the primary lesion and nearby secondary skin lesions are flat and clearly demarcated from the adjacent normal skin.

Fig. 3. Aust/Uriarra/2–53/1 (Uriarra), twenty-four days after inoculation. Rabbit recovered. The eyes are not closed although the lids are deformed with nodules, and there is little general oedema of the head. Secondary lesions are numerous and they and the primary lesions are relatively flat and clearly demarcated from the adjacent normal skin.

Fig. 4. Neuromyxoma, ten days after inoculation. The rabbit recovered and was never sick at any stage. The primary lesion is clearly demarcated from the adjacent normal skin, and is already regressing, the centre being depressed and dark purple in colour. There are very few secondary lesions in the skin, and the only abnormality on the face is one small nodule on the right upper eyelid.

cases which survive longer than usual. Thickening of the eyelids is first seen on the sixth or seventh day, and the eyes are usually completely closed by the ninth day. There is an opalescent discharge from the eyes which becomes copious and more turbid during the last two or three days of life. Oedematous swelling of the head, base of the ears, and perineum becomes pronounced in the later stages of the disease, and the slow respiration is often accompanied by a semipurulent nasal secretion. The rabbit continues to eat and drink until shortly before death, and body fat is abundant at the time of death.

The virus attains a high concentration in primary and secondary skin lesions, and mosquito transmission occurs readily.

Most of the investigations on the pathogenesis of myxomatosis (chapter 8) and on mosquito transmission (chapter 11), were carried out with this 'standard laboratory strain'.

Grade II virulence

No prototype strain has been designated for this virulence grade, because the symptomatology is identical with that just described for the standard laboratory strain. The category was introduced because some field strains, although invariably lethal in the laboratory, produced infections in which the mean survival time was a few days longer than was ever found with the standard laboratory strain. At a time when the major trend towards less virulent strains of virus was not yet apparent these strains of Grade II virulence were differentiated from the standard laboratory strain as indicating that some genetic alteration in the virus had occurred and been preserved in the field.

Grade III virulence

Prototype strain; KM 13 (Australia/Corowa/12–52/2) (Pl. XII, fig. 2). Mean survival time, 21·5 days. Range, thirteen days to recovery. Mortality rate, 88 %. This strain was recovered from a pool of *Anopheles annulipes* mosquitoes in 1952 (Myers *et al.* 1954).

A large batch of early rabbit passage material derived from this strain was prepared in 1953 and has been maintained on dry ice since then. This has maintained its titre, and has been used without further passage in tests on the genetic resistance of wild rabbits (see chapter 14).

The primary lesion at the inoculation site appears on the third or occasionally the fourth day, and becomes large, flat and soft. It sometimes becomes harder and slightly convex in the later stages of a fatal infection. Occasionally it never progresses beyond a thickening of the skin. The margin is irregular, and in the later stages becomes clearly demarcated from the surrounding skin, especially in the cases which live for longer periods than usual. The secondary skin lesions are of the same general character as the primary lesion, but may become more nodular, particularly on the legs. The first appearance of various signs coincides with the time pattern observed with the standard laboratory strain, but the subsequent development is much more gradual. The eyes are rarely closed before the fourteenth to sixteenth day, and an occasional animal dies or recovers without the eyes closing completely. The eyelids are sometimes irregularly distorted rather than generally thickened, but discrete nodules are rare. The oedematous swelling of the head is less pronounced than in infections with the standard laboratory strain, but in advanced severe cases there may be numbers of semi-confluent tumours over the head and around the mouth. Occasionally the whole course of the disease is relatively mild but the severe case which survives for much longer than twenty days presents a wretched picture in the later stages. There is a purulent nasal discharge and the breathing is laboured, the animal is emaciated, the eyes are closed and bulbous, with a purulent discharge, associated with secondary bacterial infection, and strings of mucus are passed with the faeces. There is bulbous oedematous swelling of the perineum, the ears are pendulous, and the head, legs and body are covered with small sometimes nodular, lumps.

In animals which recover, the skin over the flat tumours produced by the KM 13 strain becomes dry and scaly, but the lump itself remains fleshy for a considerable period before it dries out and sloughs. Virus has been recovered from tissue slices of such secondary tumours up to 60 days after infection, and successful mosquito transmission has been recorded 40 days after inoculation (Fenner *et al.* 1956).

Since 1955 more field isolates have been allocated to this group than to any other. In more recent surveys (since 1963) it has been subdivided into virulence Grades III A and III B (see Table 37).

Grade IV virulence

Prototype strain; Uriarra (Australia/Uriarra/2–53/1) (Pl. XII, fig. 3). Mean survival time, thirty days. Range, fifteen days to survival. Mortality rate 58 %. This was the first strain of reduced virulence recovered from the field to be fully described (Mykytowycz, 1953).

Mykytowycz (1956) has described the characteristics of infections with this strain induced in wild rabbits by contact and by mosquito bite, and Jacotot, Vallée & Virat (1955 d) have described the symptomatology in laboratory rabbits infected by injection. The following description applies to laboratory rabbits infected by the intradermal inoculation of 10 ID_{50}. The local lesion at the inoculation site appears on the third or fourth day and soon becomes hard, red and slightly convex. The colour often deepens to purple about the ninth or tenth day. The irregular margin at first merges gradually into the surrounding skin, but except in acutely fatal cases it becomes clearly demarcated by about the twelfth day. In some rabbits the centre of the flat primary lesion becomes necrotic and forms a black scab during the third week of the disease, but in other animals this process is deferred for another week or even longer. The central scab sloughs out during the fifth week in animals which recover. Secondary skin lesions are usually numerous, relatively flat, red in colour, and have clearly demarcated margins. In milder cases there are discrete nodules on the eyelids instead of a general oedematous thickening. Discharge from the eyes and nose is less than in infections with more virulent strains, but there is often extensive perineal oedema.

The virus content of secondary skin lesions was found to vary considerably from one animal to another (Fenner *et al.* 1956). It was always high between the eighth and fourteenth days, but in some animals it then fell precipitately, and in others remained high for a further two weeks (see Fig. 12).

Grade V virulence

Prototype strain; neuromyxoma (Pl. XII, fig. 4). This is the most attenuated derivative of the standard laboratory strain that is known. Virtually all rabbits survive infection with small doses of neuromyxoma, although occasionally animals die after infection with larger doses. It was recovered by Hurst (1937*b*) who passed the standard laboratory strain in series, in laboratory rabbits, by the intracerebral route. Material from the eighth passage yielded this virus.

The local lesion produced at the inoculation site is hard, red and convex. The centre becomes depressed and purple very early and the periphery is well demarcated by the fifth day. Regression and scabbing follow soon afterwards. Secondary skin lesions are scanty, and consist of small nodules usually less than 5 mm. in diameter. These may occur anywhere on the body, or on the eyelids, or perineum. Generalized oedema of the head and perineum is never seen in infections with small doses of this strain of virus, and the general health of the rabbit is hardly affected.

The virus content of the skin lesions reaches its peak about the sixth day, but at its highest remains well below the level observed with the more virulent strains (see Fig. 12). The clinical regression of the skin lesions after the seventh day is accompanied by a pronounced fall in the virus concentration in the lesions. Virus is only occasionally recovered, and then in low concentration, from the bloodstream.

Mosquitoes rarely obtain enough virus when probing through the skin lesions produced by neuromyxoma virus to transmit infection (see Table 30), and it is unlikely that a virus of this type would survive in nature. No viruses have been obtained from naturally infected rabbits which are as attenuated as neuromyxoma. Field strains have been allocated to Grade V if the inoculated rabbits suffered a mild nodular type of disease, and more than 50 % recovered.

Figure 15 summarizes the salient features of the symptomatology of infections with the prototype strains of myxoma virus; the rate or progression, the degree of swelling, and the time of death or healing.

Secular trends in virulence in Australia

Figure 16 summarizes the results of the application of these criteria to the classification of the virulence of the Australian field strains of myxoma virus. The uniformly very high mortality rate observed in the field during 1950–51, when myxomatosis first spread in Australia, and in laboratory tests involving the inoculation of several hundreds of rabbits with small doses of the standard laboratory strain, indicated that the material used to introduce myxomatosis into the

Australian wild rabbit population was relatively homogeneous, at least in its virulence. The divergence from the initially very high virulence, first observed within a year of the spread of the virus, and the subsequent appearance of still less virulent strains, must therefore be interpreted as being due to mutations of the virus which were subsequently selected for, so that it was possible to recover altered strains by the very superficial sampling that was made each year.

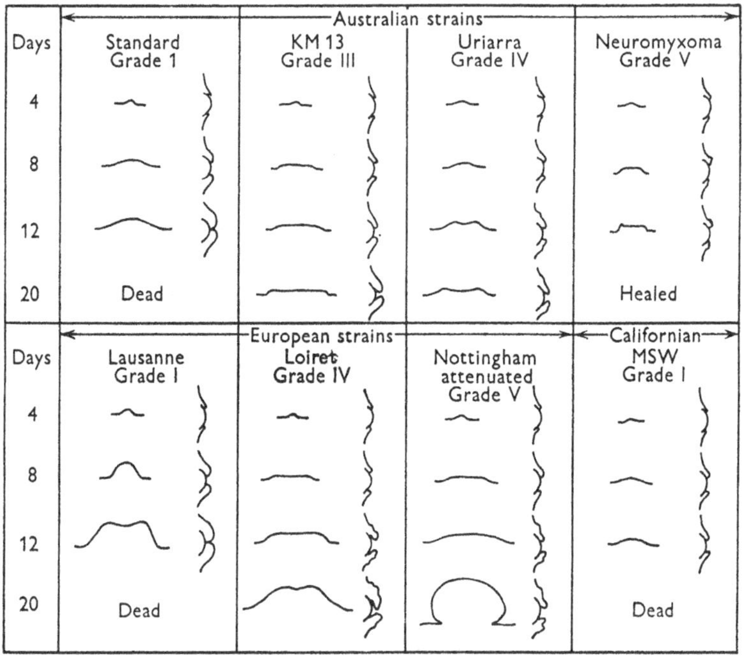

Fig. 15. Diagrammatic representation of the progressive development of the primary lesion and swelling of the eyelids in rabbits infected with the prototype strains of myxoma virus. The grades of severity follow the definitions set out in Table 37 (modified from Fenner & Marshall, 1957).

After its initial escape from the trial sites in northern Victoria late in 1950 highly virulent myxoma virus spread rapidly over an enormous area of south eastern Australia (see chapter 16). It might be thought that the subsequent appearance of variants of reduced virulence over wide geographic areas represents a similar spread of a rare but highly successful mutant over the same area. Although we believe that viruses may often be carried over areas of many square miles by infected mosquitoes, there is strong evidence that strains of reduced virulence have appeared in many widely separated areas (e.g. Tasmania, New South Wales and Western Australia) as distinct events. Sometimes these attenuated strains may over-winter and cause an epizootic during the succeeding summer, but new mutant

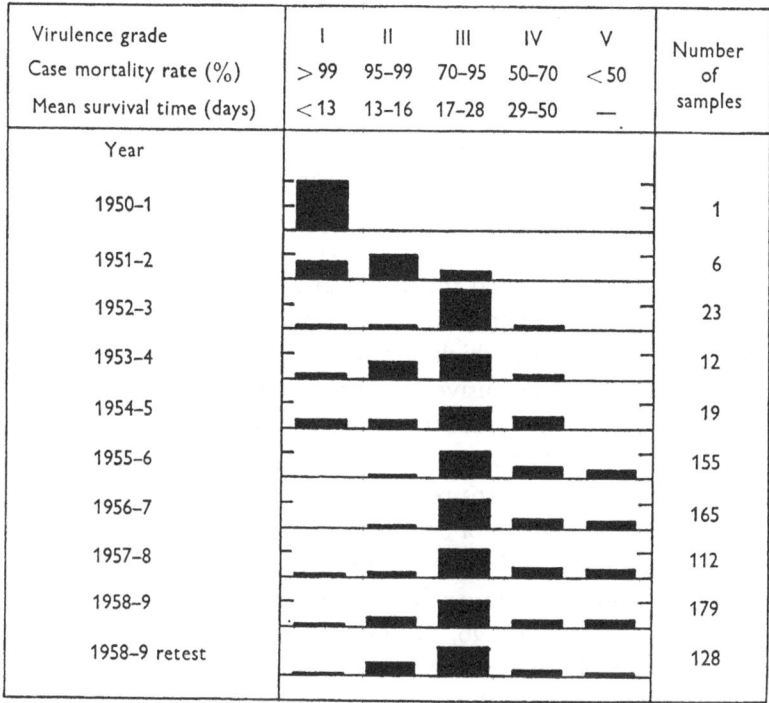

Fig. 16. Histogram showing the virulence of Australian field strains recovered between 1951 and 1959. Because of the summer epizootics the data were analysed over the year 1 July to 30 June (data from Marshall & Fenner, 1960). '1958–59 (retest)' indicates results obtained in 1963 when thirty of the strains which were originally tested only with lesion suspensions used undiluted, or diluted up to 1/5, were retested with highly diluted preparations made after one rabbit passage. These results were substituted for those obtained in 1959, and added to the 1959 results obtained with preparations then used at dilutions of 10^{-2} and higher. When passage could not be made with material originally used at concentrations greater than 1/5 the results were excluded from the 'retest' (F. Fenner, unpublished observations).

strains also emerge each year. The vast populations of wild rabbits which supported the initial spread of the virus had disappeared by 1955 and have never returned, so that the size of the susceptible host population has been greatly diminished. Furthermore, experience in both Tasmania and Western Australia, which are far distant from south-eastern mainland Australia, has been strikingly similar in that strains of Grade III virulence appeared there and have become dominant just as in south-eastern Australia. There seems no reason to doubt that variants of reduced virulence have emerged on many different occasions, and in many different places.

The virulence grades, as such, have no genetical meaning. They merely represent a convenient way of handling the mass of data. It is likely (on analogy with vaccinia virus) that the genetic material of myxoma virus contains 250–300 cistrons, alteration in many of which might lead to altered virulence of the virus for *Oryctolagus*. There may be thousands of different mutations which would affect virulence. We have no way of analysing the material available at this level.

It is clear, however, that the initial situation, with a homogeneous and very highly virulent virus strain, had altered within a year of introduction to one in which several recognizably different strains were present. More extensive sampling after 1955 revealed that circulating strains of virus were by then quite heterogeneous in respect of their virulence. For the next four years the frequency of strains allocated to the five groups remained almost unchanged.

We must enter a word of caution about the degree of attenuation and the frequency of highly attenuated strains suggested by the results of Marshall & Fenner (1960), in view of the results described on page 215 and in Table 36. The lowest row of Fig. 16 indicates what is most likely to be the correct assessment of virulence for 1958–59. It was constructed with data from which results obtained with concentrated suspensions of lesion material were excluded, or replaced by the results obtained when such material had been passaged once in rabbits and used at high dilution. It is probable that alterations of the same nature should be made to all results since (and including) those shown for 1955–56.

There is no doubt from these revised results that strains as virulent as the standard laboratory strain have disappeared, in spite of the inoculation campaigns (see chapter 16). Strains of Grade III virulence have been dominant throughout the period 1955–59, and strains of Grade IV virulence (like the prototype strain, Uriarra) have long occurred in nature, but have remained relatively uncommon. Strains of Grade V virulence do occur in Australian field material, but they are quite rare ($\sim 5\%$).

This leads to a changed assessment of the degree but not the fact of attenuation. One interpretation of these results would be that by 1955–56 something like a climax association had been reached in the mosquito-transmitted myxomatosis in genetically unselected Australian wild rabbits. This did not result in the attainment of uniformity (such as we have postulated, on admittedly inadequate grounds, in the myxoma virus—*Sylvilagus* associations described in chapter 15), but in a situation where there was a rough balance between the frequency of a number of virus strains which differed considerably in their virulence. Possible effects of increased genetic resistance of the host on this balance are discussed in chapter 18.

VIRULENCE TESTS WITH STRAINS RECOVERED FROM THE FIELD IN EUROPE

Myxomatosis was introduced into the wild *Oryctolagus* population of Europe by the inoculation of two rabbits with the highly virulent Lausanne strain of virus in June 1952. Following this introduction changes in virulence have occurred, but they have not been followed as systematically as in Australia.

The virus used to initiate the European outbreak differed in its origin and its symptomatology from that used in Australia, and many of the less virulent strains recovered in Europe have also produced a different type of lesion. For this reason certain European strains have been designated as prototype strains, which characterize grades of virulence similar to those adopted for the Australian material. A description of these prototype strains follows.

Grade I virulence—Europe

Prototype strain; Lausanne (Brazil/Campinas/1949/1) (Pl. XIII, fig. 5). Mean survival time 12·9 days. Range ten to sixteen days. Mortality rate 100 %. This is the strain which was used to initiate the European epizootics in June 1952. The symptomatology of rabbits infected with this strain differs appreciably from that observed in infections with the standard laboratory strain and all derivatives of it, and is characterized by the great proliferation of the skin lesions, so that they rapidly become large hemispherical tumours. In this respect, and in its lethality, it resembles all other strains that have been recovered from rabbits in South America (see chapter 15).

A hard protuberant convex tumour is produced at the site of inoculation. It reaches its maximum size by the tenth day, when the shining skin over it breaks down and oozes serous fluid. This is followed by ulceration of the centre of the lesion. The colour also differs from that of lesions produced by the derivatives of the standard laboratory strain, being purple from the fourth day onwards and later becoming black. The protuberance of the tumour gives the impression that it is clearly demarcated from the surrounding skin, but in fact the edges merge gradually into normal skin.

Generalization is more florid and proliferative than in infections with the standard laboratory strain, but follows the same time sequence. Secondary lesions appear on the body, legs and feet, but rarely on the ears. The head and perineal region are extremely swollen and there is a copious semipurulent discharge from eyes and nose. Death occurs before the rabbit loses much weight.

The virus content of the skin lesions follows much the same time-sequence as in infections with the standard laboratory strain, and

---

PLATE XIII. Appearance of rabbits infected by the standard method with the Lausanne strain and two derivatives of it, and with a virulent Californian strain of myxoma virus (from Fenner & Marshall, 1957).

Fig. 5. Brazil/Campinas/1949/1 (Lausanne), ten days after inoculation. The rabbit died on the twelfth day. The primary lesion is very large, hard, protuberant and deep purple in colour. The head is very oedematous and the eyes are completely closed with a profuse conjunctival discharge. Secondary lesions are raised and purple in colour and are not demarcated from the surrounding skin.

Fig. 6. France/Loiret/4-55/1 (Loiret 55), twenty-five days after inoculation. The rabbit died on the twenty-sixth day. The primary lesion is very large and exudes serum. Secondary lesions are very numerous and occur all over the body. The eyes are completely closed by deformed nodular eyelids and the ears hang down due to the numerous nodules on the pinnae.

Fig. 7. England/Nottingham/4-55/1, attenuated, twenty-three days after inoculation. The rabbit survived. The primary lesion is very large and is clearly demarcated from the adjacent normal skin. There is a moderate number of secondary lesions which resemble the primary lesion in their protuberance and the clear demarcation from the surrounding skin. In spite of the numerous nodules on the eyelids the eyes were never closed and there was no general oedema of the head.

Fig. 8. U.S.A./San Francisco/1950/1 (MSW), seven days after inoculation. Rabbit died on the eighth day. The primary lesion is small, with indefinite edges. There are a few small secondary lesions in the skin and the eyelids are slightly swollen, with a thin conjunctival discharge. External lesions were rarely more severe than shown in the photograph, but the disease was invariably rapidly fatal.

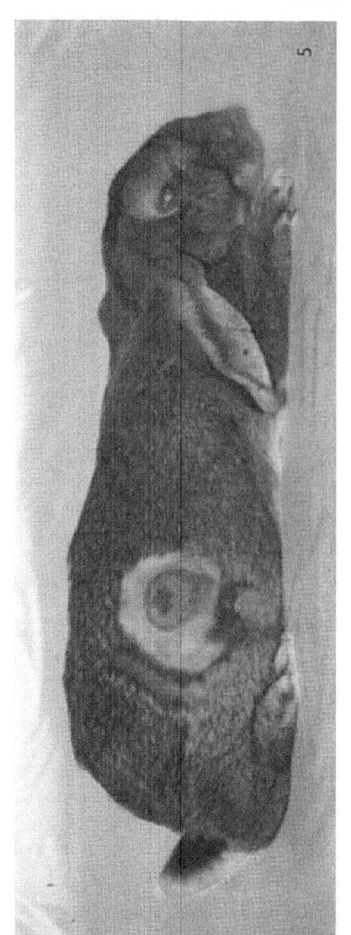

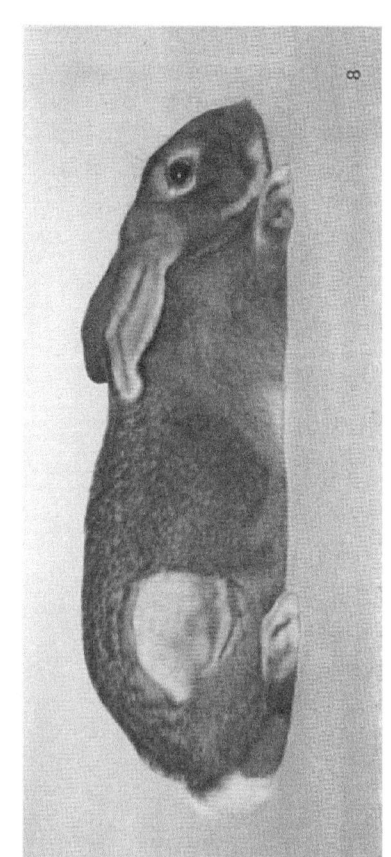

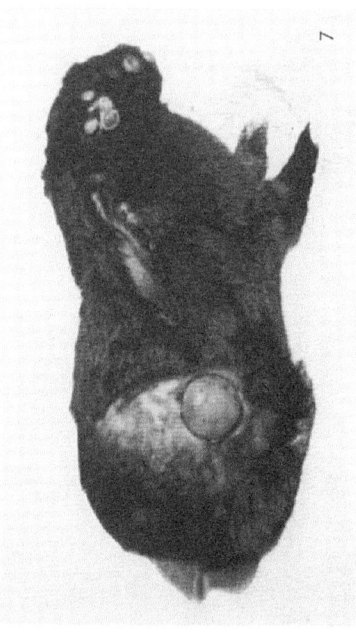

PLATE XIII

mosquitoes readily transmit after biting through skin lesions produced by the Lausanne strain (Day *et al.* 1956).

Grade III virulence—Europe

No European strains of this virulence grade have been examined in enough detail to designate one of them as a prototype strain.

Recent experiments (see page 329) have shown that many of the strains recovered from the field in Britain during 1962, and from France in 1962–63, are of Grade III virulence. Apart from slightly more raised and fleshy primary lesions, and more nodular secondary lesions, they closely resemble the Australian prototype strain of Grade III virulence, and some are indistinguishable from the latter.

Grade IV virulence—Europe

Prototype strain; Loiret 55 (France/Loiret/4–55/1) (Pl. XIII, fig. 6). Mean survival time, 33 days. Range, nineteen days to recovery. Mortality rate, 65 %. This strain was the first strain of reduced virulence to be recognized in Europe (Fenner & Marshall, 1955), although not the first to occur there.

Although attenuated, its clinical characteristics resemble those of infections due to the Lausanne strain much more closely than any of the Australian strains. Differences from the Lausanne strain reside principally in the rate of development of the symptoms. The primary lesion usually appears on the third day, but remains red in colour and relatively flat until about the eighth day. Proliferation is progressive, however, so that by the fourteenth day the primary lesion, and the very numerous secondary skin lesions, are large and prominent, but they rarely become purple in colour. The edges are more clearly demarcated from the surrounding skin than in infections with the Lausanne strain, and the eyelids are grossly deformed rather than included in a general extreme oedema of the head. In the later stages the rabbit becomes emaciated and covered with large red protuberant skin lesions, which occur all over the body, and on the ears. There is gross oedematous swelling of the perineum and genitalia, which, in bucks, frequently leads to rupture of the scrotum and consequent secondary infection, which may be fatal.

The virus content of the skin lesions rises rather more slowly than with the Lausanne strain, but attains the same high level and persists for a long time (see Fig. 13).

Grade V virulence—Europe

Prototype strain; Nottingham 55, attenuated strain (England/Nottingham/4–55/2 (attenuated)) (Pl. XIII, fig. 7). Range of survival time, fourteen days to recovery. Mortality rate 23 %. This strain is the attenuated component of what appears to be a mixture of strains recovered from a naturally infected rabbit in Nottinghamshire, England, in April 1955.

The local lesion does not appear at the inoculation site until the fifth day, and remains relatively flat and rubbery in consistency until about the twelfth day. At this stage the lesion either begins to regress and form a scab, or it rapidly proliferates into a large, hard, protuberant tumour which is purple in colour and clearly demarcated from the surrounding skin. The latter type of tumour often persists into the fifth week of the disease, before it gradually regresses and scabs. Secondary lesions appear on the ninth day and are nodular and clearly demarcated. There is severe oedema of the perineum, but elsewhere oedema is almost absent, and lesions of the eyelids consist of localized nodules. The disease develops slowly and runs a relatively benign course, rather like a generalized fibromatosis.

The virus content of the tumours rises more slowly than with other strains, but by the eighth day the titres attained are similar. After the eighth day the virus content usually diminishes fairly rapidly but in a few rabbits it remains at a high level for a prolonged period.

The occurrence in Europe of strains producing flat skin lesions

In addition to these strains, which although differing in virulence from each other can be differentiated from the Australian strains by the characteristically protuberant skin lesions, other strains have been recovered from the field in Europe which are indistinguishable from some of the Australian field strains. For example, Andrewes *et al.* (1956) noted that the strain England/Sussex/10–54/1, which was recovered from a pool of wild-caught *Anopheles maculipennis*, produced flat lesions very similar to those found in rabbits infected with strains like KM 13 or Uriarra, and Chapple & Bowen (1963) have recently commented upon the same feature in two strains recovered in England during 1962.

All strains of myxoma virus recovered from naturally infected rabbits in South America are characterized by the florid symptomatology just described for the Lausanne strain (see chapter 15). The standard laboratory strain differed from these in the flat nature of the skin lesions, and it may well be that this difference arose during its long serial passage in *Oryctolagus* in the laboratory. Natural passage of the Lausanne strain in European wild rabbits also results in the emergence of some strains characterized by flat rather than elevated skin lesions, and they were the most common type observed in a large collection of material obtained from all parts of Britain late in 1962 (Fenner and Chapple, 1965).

During recent experiments on plaque assays of myxoma virus on rabbit kidney cells monolayers, Woodroofe & Fenner (1965) recovered a plaque mutant from stock Lausanne virus. This differed from the Lausanne strain not only in virulence (mean survival time in standard tests, sixteen days) but also in symptomatology, the skin lesions at the site of inoculation and elsewhere being only slightly raised, instead of protuberant. Thus what we had earlier thought of as a characteristic feature of European myxoma strains, namely the proliferative skin lesions, can be lost in a single mutational step.

Secular trends in virulence in Europe

Although the data prior to 1962 is very limited, and tests carried out in different laboratories by somewhat different procedures are hard to compare, it is worth trying to summarize the results of virulence tests carried out with strains from the field in France and in Britain. The reason for considering these separately is that there are important differences in the mode of transmission of myxomatosis in these two countries; in Britain the flea is the only important vector, whereas in France there are major summer mosquito-borne epizootics as well as year-round flea transmission (see chapter 17).

Only those results obtained by the standard procedure described earlier are included in Table 38. In both Britain and France moderately attenuated strains, most of which produce flat rather than protuberant skin lesions, are now much more common than highly virulent strains. In spite of the absence of any deliberate re-introduction of virulent virus by inoculation, however, an appreciable number of highly virulent strains, with symptomatology of the Lausanne type, were recovered from several different counties of Britain in October and November 1962 (Fenner & Chapple, 1965).

TABLE 38. *The virulence of strains of myxoma virus obtained from the field in France and England (data from Fenner & Marshall, 1957; Fenner & Chapple, 1965)*

Virulence Grade ...	I	II	III A	III B	IV	V
Mean survival time (days) ...	< 13	13–16	17–22	23–28	29–50	—
			France			
1952	1	—	—	—	—	—
1953	2	—	—	—	—	—
1954	5	—	—	—	2	—
1960	—	—	—	1	—	—
1961	—	—	—	—	1	—
1962	—	—	—	—	1	1
1963	—	—	3	2	1	—
			Britain			
1953	1	—	—	—	—	—
1954	2	—	2	—	—	—
1955	2	—	—	—	—	—
1962	4·1%	17·6%	38·8%	24·8%	14·4%	0·5%

THE 'PURITY' OF ATTENUATED STRAINS OF MYXOMA VIRUS

All investigators have been impressed with the uniformity in the response of laboratory rabbits to highly virulent strains of myxoma virus, and by the variability in their response to infection with field strains of reduced virulence. Another result, for which no adequate explanation is yet available, is the great variation in the symptomatology of rabbits infected with attenuated strains by different methods, such as contact and intradermal inoculation (Mykytowycz, 1956). These differences could be due to either or both of two factors (a) the highly virulent strain is a pure clone of virus, in respect of its virulence, and the attenuated strains are mixtures, or (b) laboratory rabbits vary in their genetic susceptibility to myxomatosis, but the overwhelming virulence of the standard laboratory strain obscures these variations. Differences associated with route of inoculation could be due to differing growth of different components of a mixture, but are more likely to be due to ill-understood differences in the pathogenesis of the infection.

Experience subsequent to 1953, when these problems were first posed, has shown that both laboratory and wild rabbit populations do include individual animals of high and low genetic susceptibility to myxomatosis. Experiments by Fenner & Marshall (1957), described below, showed that at least one attenuated field strain was a mixture, but three others tested appeared to be homogeneous in respect of their virulence.

Table 39 summarizes the results of tests made with the original rabbit material used in high dilution, and several single pock derivatives of four attenuated strains of myxoma virus, two from Australia and two from Europe. There were no statistical differences between the different preparations of the first three viruses, but the 'strain' England/Nottingham/4–55/1 clearly appeared to contain components of very high and very low virulence. In the original tests carried out with the material received from England (in the form of a freeze-dried preparation of the first laboratory passage virus) three of the ten rabbits died with the same symptomatology as rabbits infected with the Lausanne strain of virus, five suffered a mild disease associated with very protuberant skin lesions, and two sustained a severe slowly progressive infection. Passage material from the acutely fatal cases produced an acutely lethal disease, and from the mild cases a similar mild disease. It was not unexpected, therefore, to find that single pock material derived from the original strain of virus yielded one clone of high virulence and two of very low virulence. The pock size was correlated with rabbit virulence (see Fig. 3), the virulent component producing large pocks and the attenuated component small pocks.

Hudson & Mansi (1955), using the same virus strain, claimed that there was a progressive change in the severity of the disease produced with virus reaped at various stages of the infection. At the time they did not suspect the mixed nature of the original inoculum, and they ascribed considerable epidemiological importance to this progressive change from high to low virulence. In retrospect, it is likely that this result may have been due to the inoculation of a large amount of inactive virus and a small amount of viable virus with the suspensions prepared late in the course of the disease. Repetition of this type of experiment, with the difference that no tissues were reaped later than the twenty-first day, when their virus titre was still relatively high,

TABLE 39. *Infection of laboratory rabbits with small doses of the parent strain and single pock derivatives (pure clones) of strains Aust/Corowa/12–52/2, Aust/Uriarra/2–53/1, France/Loiret/4–55/1, and England/Nottingham/4–55/1 (data from Fenner & Marshall, 1957)*

Strain of virus	Survival times (days)
Aust/Corowa/12–52/2	
Parent strain	13, 15, 17, 18, 19, 22, 26, 27, S*
Single pock: No. 1	20, 22, 27, 34, 41
No. 2	13, 18, 20, 24, S
No. 3	21, 22, 23, 26, 29
No. 4	17, 24, 40, S, S
No. 5	17, 20, 22, 22, 27
Aust/Uriarra/2–53/1	
Parent strain	17, 18, 31, 31, 34, S, S, S, S, S, S
Single pock: No. 1	22, 32, 33, S
No. 2	13, 17, 18, 49, S
No. 3	18, 25, S, S, S
No. 4	16, S, S, S, S
No. 5	32, 45, S, S, S,
France/Loiret/4–55/1	
Parent strain	20, 22, 27, 27, 29, 38, S, S, S, S
Single pock: No. 1	48, S, S, S, S
No. 2	26, 27, 31, 32, S
No. 3	28, 33, 38, 43
No. 4	19, 19, 28, 42, S
England/Nottingham/4–55/1	
Parent strain	11, 11, 13, S, S, S, S, S, S, S
Single pock: No. 1	S, S, S, S, S
No. 2	39, S, S, S, S
No. 3	11, 12, 12, 12, 13

*S, Rabbit recovered from infection.

and all inoculations were made with high dilutions of the tissue suspension, showed that no such change occurred with the KM 13 virus. A moderately severe progressive infection with England/ Nottingham/4–55/1 did in fact yield both virulent and attenuated viruses, although there was no progressive change (Fenner & Marshall, 1957). We would now ascribe this to the mixed nature of the original inoculum rather than to any reproducible change in virulence with the stage of the disease in individual rabbits.

It seems reasonable, therefore, to regard most natural infections, which are caused by small doses of virus, as being initiated by pure clones of virus. However, different viruses could sometimes multiply in nearby cells of the skin, so that the mixture could sometimes be serially passaged.

Concurrent or sequential infections with attenuated and virulent strains

Hurst (1937c) noted that concurrent infection of rabbits with fibroma and virulent myxoma virus, in the same or in different inoculation sites, sometimes resulted in non-lethal myxomatosis.

TABLE 40. *The results of prior infection of rabbits with small doses of attenuated strains on the response to infection with highly virulent strains of myxoma virus (data from R. Mykytowycz, personal communication, 1963)*

Initial infection				
Strain	Virulence Grade	Interval (hours)	Second infection (Grade I virulence)	Result (survival time in days)
Neuromyxoma	V	12	Standard laboratory strain	13, **17***, **22, S, S, S, S, S, S**
		12	Brazil/Campinas/1949/1 (Lausanne)	13, 13, 14, 14
		24	Brazil/Campinas/1949/1 (Lausanne)	17, **23, S, S**
Aust/Uriarra/	IV	12	Standard laboratory strain	12, 12, **21**
2–53/1		24	Standard laboratory strain	**21, S**
		12	Brazil/Campinas/1949/1 (Lausanne)	12, 13, 15, **21**
		24	Brazil/Campinas/1949/1 (Lausanne)	12, 13, **18, 26**
Aust/Corowa/	III	12	Brazil/Campinas/1949/1 (Lausanne)	11, 12, 13, 13, 13, 13
12–52/2 (KM13))		48	Brazil/Campinas/1949/1 (Lausanne)	13, 13, 14, 14, 14, **16, 16**

* Figures or letters in heavy type indicate a prolonged survival time.   **S**, Survivor.

Substantial protection occurred in the majority of cases in which fibroma was inoculated two or more days before the myxoma virus.

In Australian epizootics, where infected mosquitoes are relatively common and several strains of myxoma virus of differing virulence may be circulating, some rabbits must be infected at closely spaced intervals with different strains. To determine possible effects of such mixed infections R. Mykytowycz (personal communication, 1963) carried out sequential infections with attenuated and highly virulent strains of myxoma virus. Considering only infection by intradermal inoculation, prior infection with the virulent strains almost always led to death within the usual short survival time characteristic of that strain. The inoculation of an attenuated strain at intervals as short as twelve hours before the virulent strain usually modified the response to the latter, and a modified response was sometimes observed even with simultaneous infection. Table 40 summarizes the results. The effect was most pronounced with strains of low virulence, such as neuromyxoma; prior infection with the KM13 strain (grade III virulence) had no appreciable effect on subsequent infection with Lausanne. The mechanism of this sparing effect was not determined; both interference and the immune response may play a part. Neither the dose nor the site of intradermal inoculation was important, but heat-inactivated virus had no effect, even in large doses.

The stability of virulence of strains of myxoma virus: laboratory tests

It is clear from the foregoing discussion that under the novel stress of natural transmission in *Oryctolagus*, originally homogeneous clones of highly virulent myxoma virus have undergone mutation and selection, so that in both Australia and Europe there are now many strains of different virulence. It is impossible to try to match, in the laboratory, the scale upon which myxomatosis has occurred in nature; but a few experiments have been carried out, with results that warrant discussion.

Attenuation of highly virulent strains

Several experiments have been published involving the serial passage of the standard laboratory strain in abnormal hosts. Lush (1937) noted that its virulence was undiminished after 26 serial passages on the chorioallantoic membrane, and tests after 75 passages (F. Fenner, unpublished observations) show no alteration in rabbit virulence. Haagen & Du (1938), on the other hand, reported that serial passage on the chorioallantoic membrane of fourteen-day-old chick embryos led to a profound fall in rabbit virulence, such that later passage material produced only localized tumours in *Oryctolagus*, which were then immune to challenge infection with the virulent virus.

Two experiments have been reported on the serial transmission of virulent myxoma virus by the intracerebral inoculation of rabbits. It was from the eighth such serial passage that Hurst (1937b) obtained his greatly attenuated 'neuromyxoma' strain. Repeating the procedure Rhodes (1938) recorded a progressive change in the histological picture of infected rabbits, similar to that reported by Hurst, but there was no change in rabbit virulence.

Jacotot, Vallée & Virat (1955 b) observed no change in the virulence of Lausanne type of myxoma virus which had been exposed to two unusual host situations; (a) virus from tumours persisting in a fibroma-vaccinated rabbit for up to four months, and (b) virus recovered from the inoculation site in hares (*Lepus europaeus*) nine to fifteen days after inoculation. They did not determine whether the virulence could be altered by serial passage in either of these hosts.

Using a Californian strain of myxoma virus, Saito & McKercher (personal communication, 1963) found no change in the virulence of the MSD strain after 110 passages on the chorioallantoic membrane. When the latter material was passed in a rabbit kidney cell line, incubated at $33 \cdot 6°$ C., however, the virulence had decreased greatly after 32 passages, so that only an occasional rabbit died, and after 40 passages intradermal inoculation caused only a trivial lesion at the inoculation site and a transient rise in temperature, with solid protection against challenge infection with virulent myxoma virus (McKercher & Saito, 1964).

Increase in the virulence of attenuated strains

Jacotot, Vallée & Virat (1956) found that four successive passages by the intratesticular route did not change the virulence of a French strain of Grade III virulence when comparative tests were made by intradermal inoculation. The same workers (1955 d) failed to demonstrate any change in the virulence of the strain Australia/Uriarra/2–53/1 after six serial passages on the chorioallantoic membrane, or after three alternate egg-rabbit passages.

Increase in the virulence of highly virulent strains

Naturally occurring strains of myxoma virus from various parts of South America all appear to have the same high virulence for laboratory rabbits as the viruses originally introduced into Australia and Europe; they are invariably lethal within fourteen days (see Table 47). The survival time of rabbits inoculated with Californian strains is shorter, deaths often occurring with very slight external evidence of disease in about eight days. Since the cause of death in myxomatosis of *Oryctolagus* due to Californian virus is probably different from that operating in South American myxomatosis it is not possible to say which is the more 'virulent'. Similarly, it is not possible from tests made on laboratory rabbits to distinguish differences in the virulence of strains from various parts of South America, or in different highly virulent field isolates from Australia or Europe. However, by using as host animals rabbits which have been selected for their resistance to a moderately virulent Australian strain of virus (see chapter 14) it should be possible to distinguish such differences, if they exist.

TABLE 41. *'Hypervirulent' strains of myxoma virus, recognized as such by the inoculation of rabbits with some degree of genetic resistance to myxomatosis (data from Douglas, 1962; W. R. Sobey, personal communication, 1963)*

Source of rabbits	Conditions of test	Virus strain	Mortality rate (%)
Wild rabbits, captured at Ouyen, Spring 1961	Field enclosure	Standard laboratory strain	71
		Glenfield strain	97
Wild rabbits, captured at Karnak, Spring 1961	Animal house	Standard laboratory strain	75
		Glenfield strain	97
Laboratory rabbits selected for resistance	Animal house	Standard laboratory strain	85
		Glenfield strain	100
		U.S.A./San Francisco/1949/1	100

Using wild rabbits captured in areas where previous tests had shown that there was a substantial increase in their genetic resistance, Douglas (1962) confirmed the earlier suggestion of Sobey (1960) that the virus known as the Glenfield strain was more virulent than the standard laboratory strain (Table 41). The Glenfield strain is the name given to virus prepared by serial passage in laboratory rabbits of the virus Australia/Dubbo/2–51/1 (Fenner & Marshall, 1957), by the Glenfield Veterinary Research Station. It has been used in inoculation campaigns in the State of New South Wales since 1951, and in Victoria also since 1961.

W. R. Sobey (personal communication, 1963) has carried out tests on his genetically selected rabbits with the Glenfield strain, the standard laboratory strain, and a virulent Californian strain (U.S.A./San Francisco/1950/1(MSW)), with the results shown in Table 41.

Clearly, in genetically more resistant rabbits it is possible to discriminate between highly virulent strains of myxoma virus. The virulence of the standard laboratory strain is not the maximum attainable.

CORRELATES OF VIRULENCE

Although inoculation of laboratory rabbits by a method and with a dose of virus which simulates natural infection provides a direct measure of the virulence of myxoma virus for wild *Oryctolagus*, it is an expensive and time-consuming procedure, even when a test relying upon survival time rather than mortality rate is used. For this reason, and as a step towards the understanding of the mechanisms which underlie virulence, attempts have been made to find some other property of the virus, or of virus-host or virus-host cell reaction, which may be correlated with virulence. Three possibilities have been explored, all except the first in a preliminary fashion only. They are the recognition of the simultaneous presence of antigen and antibody in the serum of tissues of infected rabbits, production of and sensitivity to interferon, and the temperature sensitivity of viral development. The results are of considerable interest, but for the determination of virus virulence there is as yet no satisfactory alternative to the inoculation of groups of laboratory rabbits under carefully controlled conditions, and the observations of their response.

Serological recognition of attenuated strains

Mansi & Thomas (1958) made the interesting observation that in some rabbits infected with a strain of myxoma virus of reduced virulence it was possible to recognize the simultaneous presence of antigen and antibody in certain tissues by means of gel-diffusion precipitation in agar. This result presumably occurs because of the different time sequence of the appearance of different soluble antigens of myxoma virus, and of the corresponding antibodies. In laboratory rabbits infected with highly virulent strains of myxoma virus only the antigens are found in the serum and tissues, because the rabbits usually die before much antibody is produced.

An attempt has been made to classify field strains of virus as 'typical' (i.e. fully virulent) or 'atypical' (i.e. somewhat attenuated) on the basis of this test, which is performed with pieces of tissue and

237

the blood of the wild rabbit and standard myxoma antiserum and antigen. The occurrence of antigen only in the serum was taken as indicative of infection with a 'typical' strain; if antibody was found, either alone or in association with antigen, in the serum of a rabbit suffering from myxomatosis the virus was regarded as 'atypical'.

Since one is measuring the progress of an interaction between virus and host it is obvious that the test cannot provide an absolute measure of virus virulence. Apart from specific immunity, the virus-host interactions may be affected by changes in the genetic resistance of the host or by such environmental effects as ambient temperature as well as by genetic differences in the virus.

TABLE 42. *Correlation between survival times and presence of antibody and/or antigen in the serum of rabbits 11 or 12 days after they were infected with 10 ID50 of some sixty European field isolates (F. Fenner, unpublished observations)*

| | Week of disease during which rabbit died | | | | | |
Serological findings	Second	Third	Fourth	Fifth	Survivors	Total
Antigen only	31	127	28	10	1	197
Antibody only	0	5	10	1	3	19
Antigen and antibody	6	20	13	4	4	47
Neither antigen nor antibody	1	16	20	14	18	69
Totals	38	168	71	29	26	332

Even in laboratory rabbits maintained under ordinary laboratory conditions, Chapple, Bowen & Lewis (1963) found antibody as well as antigen in the serum of three out of fifteen rabbits infected with a highly virulent strain, and antigen only in 35 of 57 rabbits tested in the third or fourth week of the disease caused by strains of somewhat reduced virulence. They concluded that the gel-diffusion precipitation test could be used with confidence for the diagnosis of myxomatosis, but not for a decision on the virulence of the virus involved.

During a survey of myxoma strains recovered from wild rabbits in England in 1962 the serum of the test rabbits was examined for soluble antigen and for antibody on the eleventh or twelfth day after inoculation. Table 42 summarizes the results obtained with 332 rabbits, inoculated in groups of six with 60 strains of virus. At that stage of the disease, it was much more usual to find antigen than antibody in the serum (244 compared with 66, 69 rabbits containing neither). In almost all rabbits infected with viruses of Grade I or Grade II virulence, antigen and only antigen occurred in the serum, whereas in those which ultimately recovered antigen only was present in only one of the twenty-six animals, neither antigen nor antibody being found in eighteen.

Thus there was a general correlation between high virulence and the early presence of antigen in the serum, but all patterns were found with the common strains of Grade III virulence, and with individual rabbits which died between the third and fifth weeks. Some animals whose serum was negative when examined on the eleventh day yielded serum containing antibody on the twelfth or fourteenth day, but serial tests were carried out on very few rabbits.

Interferon production and sensitivity to interferon

Enders (1960) suggested that the virulence of poliovirus (as judged by its capacity to produce neural lesions in primates) might be correlated with the greater production of interferon by attenuated strains, and Ruiz-Gomez & Isaacs (1963 *a*, *b*) found that with several unrelated viruses there was a general correlation between 'virulence' and low sensitivity to and production of interferon.

G. M. Woodroofe (personal communication) has carried out some preliminary experiments on interferon production and sensitivity to interferon with three pairs of myxoma strains, of which the first was highly virulent for *Oryctolagus* and the second attenuated; standard laboratory strain and neuromyxoma, Lausanne and Nottingham attenuated, and MSD and MSD attenuated of McKercher & Saito (1964).

The virulent strains tended to produce less interferon than attenuated strains on the chorioallantoic membrane, but the results were irregular. All strains of myxoma virus were highly sensitive to interferon produced in the allantoic sac by influenza virus infection, when assayed on chick embryo fibroblasts; and there was a consistent suggestion that the attenuated strains were more sensitive than the virulent strains. When tested on rabbit embryo fibroblasts with an interferon prepared in rabbit kidney cells infected with influenza virus only the Nottingham attentuated strain showed a moderate sensitivity.

No satisfactory technique has been worked out for the assay of interferon production in the infected rabbit, although some of the results reported earlier on the effects of prolonged storage on 'virulence' may well be due to interferon production.

Temperature sensitivity of viral development

The temperature sensitivity of viral development has been correlated with the virulence of poliovirus. Commenting upon the general implications of this finding, Lwoff & Lwoff (1960) suggested that not only may the high virulence of the original strains of myxoma virus be correlated with their high 'rt' (i.e. high optimum temperature for viral development), but exposure during overwintering at low temperatures may have been an important mechanism in the emergence of less virulent strains of myxoma virus (see chapter 10).

This interesting suggestion has not been examined adequately. In early experiments Thompson & Coates (1942) showed that the virulent myxoma virus grew to very much the same extent in cultured cells incubated at 32, 35 and 37°, but grew poorly at 40° C.

Kilham (1959) reported that fibroma virus grew poorly in rabbit kidney cells at 38°, and not at all at 40°, whereas virulent myxoma virus grew as well at 38 as at 36°, and multiplied to some extent even at 40°. However, when tested for their capacity to produce plaques in rabbit embryo fibroblasts, in experiments performed in water baths with accurate temperature control, none of the strains of myxoma virus used in the interferon experiments reported in the previous section produced plaques at 38°, and all produced less plaques at 37 than 35·5° (G. M. Woodroofe, unpublished observations).

No experiments have been carried out on the possible differential growth rate of virulent and attenuated strains at lower temperatures (30°) which could occur in the rabbit skin in winter. Slightly more rapid growth of the less virulent strains under these conditions, analogous to the better growth of 'cold' than 'hot' poliovirus at lower temperatures, could have important epidemiological consequences, but might be hard to detect.

SUMMARY

From the time of its recognition in 1898 until Hurst reported the neuromyxoma strain in 1937 myxoma virus was believed to be of uniform and very high virulence for European rabbits. Soon after the introduction and natural spread of highly virulent myxoma virus in wild populations of European rabbits in Australia and Europe it was recognized that attenuated strains were occurring in nature.

A standardized test for virulence was devised which utilized the laboratory rabbit as a test animal, specified the dose and route of inoculation so that these were comparable to natural transmission by insects, and ensured that inoculated rabbits were kept at a uniform temperature. Under these conditions the survival times of inoculated rabbits were correlated with the mortality rate obtained in large groups of animals inoculated with a few representative strains.

Several hundred strains of virus obtained from the field in Australia and Europe have been tested and classified into one of five 'grades' of virulence depending upon the mean survival time of the inoculated rabbits. In both continents there is now a wide spectrum of viruses which vary greatly in virulence, the most common being those which kill about 90 % of inoculated laboratory rabbits, with a mean survival time of about three weeks. Several properties of the virus-host interaction which might be correlated with virulence were examined, namely interferon production and sensitivity, antigen and antibody production, and temperature sensitivity of viral development; but none could be substituted for the animal inoculation test as a means of measuring virulence.

Prototype strains of virus deriving from the Australian and European outbreaks have been designated which characterize the various virulence grades, and the clinical features of infection of laboratory rabbits with these has been described and illustrated.

Infection of rabbits in succession with an attenuated and a highly virulent strain usually produced a mild form of the disease. One example of a natural mixed infection was described, but the variable response of rabbits to the attenuated strains is ascribed primarily to differences in their innate resistance.

Tests on European rabbits with some genetic resistance to myxomatosis have led to the recognition of 'hypervirulent' strains of myxoma virus.

CHAPTER 14

CHANGES IN THE GENETIC RESISTANCE OF *ORYCTOLAGUS CUNICULUS* TO MYXOMATOSIS

In the Americas, where they evolved, myxoma and rabbit fibroma viruses appear to have reached a stable equilibrium with their mammalian hosts (chapter 15). This is characterized by the production of localized tumours of the skin (fibromas) which are transmitted by biting arthropods. In closely related host animals (e.g. other species of *Sylvilagus*) inoculation with myxoma or fibroma virus either produces no lesions, a trivial non-transmissible lump, or very occasionally a more severe generalized infection (see Table 10).

In sharp contrast, all specimens of myxoma virus recovered from 'spontaneous' infections of European laboratory or domesticated rabbits (*Oryctolagus cuniculus*) in north and south America (and hence at one remove from their natural *Sylvilagus* hosts) produce very severe generalized infections in this host animal. In laboratory rabbits myxomatosis had the reputation of being almost invariably lethal, until a variety of strains of reduced virulence were recovered from wild rabbits infected in the Australian and European epizootics. However, the early impression that the European rabbits were uniformly susceptible to myxomatosis was due to the overwhelming virulence of the commonly used laboratory strains of myxoma virus. When slightly less lethal strains were isolated, after the disease had become established among wild *Oryctolagus* in Australia, much more variation was found in the host response. It was shown in the previous chapter that this variation was only occasionally due to a mixed viral population in the inoculum, and it was concluded that the variation in host response seen after a standard type of inoculation with a small dose of virus was due primarily to differences in resistance in the rabbits, differences which were presumably determined by genetic factors.

242

After myxomatosis had become established as a natural infection in wild *Oryctolagus* in Australia, the question of possible changes in the genetic resistance of this host was widely discussed, and Fenner (1953b) forecast that 'the drastic selection imposed by myxomatosis will within a few years alter the genetic resistance of the Australian wild rabbit'.

Had strains of myxoma virus of reduced virulence not appeared when myxomatosis became established in wild *Oryctolagus* in Australia the disease would in some areas have eradicated the rabbit, and the virus itself would probably have disappeared from areas where eradication did not occur; for in the absence of rabbit fleas over-wintering of highly virulent strains would present great difficulties. However, when strains of virus became common which allowed 10 % of infected rabbits to recover, there were enough rabbits left for this prolific animal to build up substantial populations again after the end of each annual summer epizootic. If the reasonable assumption is made that a major proportion of the animals which survived did so because of the genetically determined greater resistance, then the stage was set for a progressive increase in genetic resistance, for each annual exposure to what was still a very lethal disease would select for the genetically more resistant animals.

Two types of experiments were set up to examine this problem. One group of workers (Marshall & Fenner, 1958; Marshall & Douglas, 1961; Douglas & Tighe, 1965) tested the resistance of non-immune wild rabbits from areas where records were available of previous exposure to myxomatosis, and carried out such tests at yearly or longer intervals, for several years. Sobey (*in litt.*) studied the heritability of resistance in the laboratory, by breeding and progeny testing. Both groups obtained results which are in close agreement, and it is clear that the genetic resistance of many of the wild rabbits left in Australia ten years after the virus was introduced is considerably higher than that of the rabbits of the pre-myxomatosis era.

FIELD EXPERIMENTS

The procedure used by Marshall and his collaborators was, in essence, the converse of that used in testing the virulence of strains of virus recovered from the field. A large batch of virus, of known titre, was stored in many separate ampoules in a dry-ice cabinet, and this was used each year to test groups of non-immune wild rabbits obtained from areas with a known epizootic history of myxomatosis.

Selection of the virus to be used for these experiments posed some problems. The standard laboratory strain had the advantage that in genetically unselected rabbits it was invariably lethal, after a survival time which was very constant. In 1953, when the decision as to what virus strain should be used over the next decade had to be made, it was thought that the development of increased resistance would be a slow process. It was then thought that the very high virulence of the standard laboratory strain would obscure minor changes in host resistance, and Marshall & Fenner chose the strain Australia/Corowa/12–52/2 (KM 13), which is the prototype strain of Grade III virulence and with which it is possible to detect the genetic differences in resistance which are present in normal laboratory rabbits.

For the rest, the procedure was simple in conception, although not so easy in execution, even with the generous help of many officers of the Wildlife Survey Section of C.S.I.R.O. During early spring, at the height of the rabbit breeding season and before natural myxomatosis was common, large numbers of 4–6 weeks old rabbits were captured alive, from a number of sites in New South Wales and Victoria where the past history of myxomatosis was known. They were brought into an insect-proofed animal house in Canberra, and raised until they reached the age of at least four months, by which time they could be expected to respond to myxomatosis as adults and to have lost antibody which may have been transmitted *in utero* from their mothers (Fenner & Marshall, 1954). Careful observation, and serological testing prior to challenge infection, served to exclude spontaneous infection with myxoma virus. These rabbits, and groups of normal laboratory rabbits, were then infected by the intradermal inoculation of a small dose (5 ID 50) of the special batch of KM 13 virus. Observations were made of the symptomatology, mortality rate, and survival time. After the first two years extensive testing of rabbits from many sites was replaced by tests, at intervals of one or two years, on rabbits obtained from two sites where there were severe epizootics each summer, Lake Urana in New South Wales and Maryvale in Victoria.

The results are summarized in Fig. 17. Over a period of seven years the response of wild rabbits to infection with a standard dose of virus of Grade III virulence (which produced a disease with a mortality rate of 90 % in genetically unselected wild rabbits) underwent a dramatic change. Not only did the mortality rate fall from 90 to 25 %, but the symptomatology was greatly altered (Table 43). From a situation in which all cases were severe, and 90 % were fatal, there was a progressive change in the severity of the disease, until by the seventh year only half the cases, in the large sample of 140, were

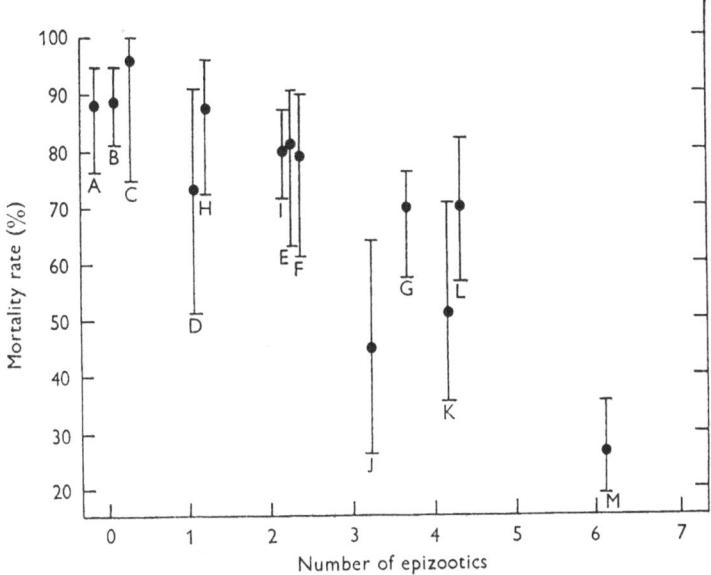

Fig. 17. The relation between mortality rates of Australian wild rabbits after challenge infection with small doses of Aust/Corowa/12–52/2 (KM 13) strain myxoma virus, and the selection pressure for increased resistance exerted by exposure of the parents of the tested rabbits to severe epizootics of myxomatosis. Data from Marshall & Fenner, 1958 (A to K): and Marshall & Douglas, 1961 (L & M). Ordinates: mortality rates (%). Abscissae: numbers of annual epizootics which occurred in areas from which rabbits were obtained, weighted for the immune rates in survivors of each epizootic. Results from Lake Urana represented by H (1953), I (1954), J (1955), K (1956) and M (1958).

classed as severe, and 30 % were mild. The possible future significance of this increased genetic resistance on the virulence of the dominant strains of myxoma virus is discussed in chapter 18.

The possibility remained that the careful attention which these wild rabbits received in the animal house in Canberra might contribute to their survival. Under natural conditions infected wild rabbits would be exposed to greater stresses such as predator activity

245

and difficulties in obtaining food. Direct comparison was not possible, but an experiment was carried out in which replicate groups of adult non-immune wild rabbits, raised under natural conditions, were tested with the same stock of virus (a derivative of KM 13, of slightly lower virulence). One group was housed in the animal house in Canberra, and the other in a field enclosure in places in the Victorian Mallee where they had been captured (Marshall & Douglas, 1961). In both situations the wild-caught rabbits were more

TABLE 43. *The severity of myxomatosis in groups of Australian wild rabbits inoculated with 5 ID50 strain Aust/Corowa/12–52/2 (KM13) (data from Marshall & Fenner, 1958; I. D. Marshall, unpublished observations)*

		Symptomatology		
Group	No. of epizootics to which population had been exposed	Severe (including fatal)	Moderate	Mild
Wild rabbits before myxomatosis	0	93%	5%	2%
Lake Urana, 1953	2	95	5	0
Lake Urana, 1954	3	93	5	2
Lake Urana, 1955	4	61	26	13
Lake Urana, 1956	5	75	14	11
Lake Urana, 1958	7	54	16	30

resistant than control animals, but the mortality rate was consistently higher in the animals maintained in the animal house (Table 44). The experiment was carried out in early summer, and the maximum afternoon temperature rose as high as 100° F. in the field enclosure, but not in the animal house. It is probable that this elevated temperature contributed to the reduced mortality (see chapter 10).

The conclusion can be drawn from this experiment that the increased genetic resistance indicated in Fig. 17 is not a laboratory artefact, but represents a real change in the rabbit population which would influence the survival of rabbits in nature.

Douglas & Tighe (1965) have continued tests on groups of wild rabbits obtained from different parts of Victoria, using both highly virulent and attenuated strains of virus. Their results, also summarized in Table 44, confirm the earlier reports and show that increased genetic resistance is now widespread amongst Australian wild rabbits.

TABLE 44. *Mortality rates in wild rabbits collected from three districts in Victoria, and held in the animal house (temperature maintained at about 70° F.), or in field enclosures (temperatures up to 100° F.) after inoculation with small doses of either virulent (standard laboratory strain) or attenuated (Aust/Corowa/12–52/2A) strains of myxoma virus (data from Marshall & Douglas, 1961; Douglas & Tighe, 1965)*

Group	Type of locality and history of myxomatosis	Time of collection	Conditions	Strain of virus	Percentage mortality
Domestic rabbits	Never exposed	—	Animal house	Aust/Corowa/12–52/2A	67
Ouyen	Semi-arid, with hot summers	Summer 1957	Animal house	Aust/Corowa/12–52/2A	21
	Annual summer epizootics since 1951	Summer 1957	Field enclosure	Aust/Corowa/12–52/2A	6
		Spring 1959	Animal house	Aust/Corowa/12–52/2A	17
		Summer 1957	Field enclosure	Standard laboratory strain	87
		Spring 1959	Animal house	Standard laboratory strain	93
		Spring 1961	Field enclosure	Standard laboratory strain	71
Maryvale	Rolling hills, with hot summers	Summer 1957	Animal house	Aust/Corowa/12–52/2A	38
	Annual summer epizootics since 1951	Summer 1957	Field enclosure	Aust/Corowa/12–52/2A	7
Penshurst	Cooler, 'stony rises' country. No epizootics. Myxomatosis has occurred at a low level through each year since 1952	Spring 1959	Animal house	Aust/Corowa/12–52/2A	27
		Summer 1959	Animal house	Standard laboratory strain	90
		Summer 1961	Animal house	Standard laboratory strain	85

They also show that the resistance in wild rabbits has now risen to an extent which allows the survival of appreciable numbers of rabbits inoculated with the standard laboratory strain of myxoma virus, which is uniformly lethal in unselected laboratory rabbits, and in wild rabbits of the pre-myxomatosis era. Tests with another virulent strain of virus (Aust/Dubbo/2–51/1—the Glenfield strain) showed that although this strain could not be distinguished from the standard laboratory strain in its virulence for laboratory rabbits, it was consistently more lethal for these partially resistant wild rabbits (see also p. 236).

LABORATORY EXPERIMENTS

W. R. Sobey (*in litt.*) has embarked on a long-term project to determine the heritability of genetic resistance to myxomatosis, with the ultimate objective of determining its physiological basis.

Initially no attempt was made to select rabbits on a generation basis. Because of the rarity of recovered rabbits when the programme was started, recovered does were retained for breeding for as long as possible. Although recovered bucks are often infertile (Sobey & Turnbull, 1956), they could be replaced more frequently, and each fertile male animal could produce many more progeny, resulting in a greater degree of selection amongst bucks. The degree of selection attained by individual rabbits was graded by a scoring system based on the ability of the rabbit to survive infection with the same strain of virus as that used by Marshall and his colleagues, namely Aust/Corowa/2–52/2 (KM 13).

Overall results obtained between 1954 and 1961 are shown in Table 45. There is clearly a trend to higher resistance, which closely resembles that seen in the experiments with wild rabbits. It is not possible to make an exact comparison, but in a rough way each grade unit of 0·5 may be equated with the selective effect of one epizootic in the field experience. On this basis the trend in mortality rates is very similar to that shown in Fig. 17.

A limited series of tests was also made with the virulent standard laboratory strain. These showed, as did the field experiments (Table 44) that resistance to the slightly attenuated strain KM 13 also implies resistance to the more virulent strain.

Since the basis of selection was quantal (death or recovery) heritability (h^2) could not be estimated by a direct parent–offspring comparison. Recourse was therefore made to intra-sire correlations (Lerner, 1950) using survival time as an index of genetic resistance. Each selected buck was mated to five does selected at random and the survival times of four of her offspring selected at random were recorded for h^2 analysis. The data obtained between 1957 and 1961

TABLE 45. *Changes in the mortality rate of rabbits challenged with Aust/ Corowa/12–52/2 (KM13) and the standard laboratory strain, after selection based on recovery from infection with strain KM13 (data from Sobey, in litt.)*

Selection grade *	Challenged with strain KM13			Challenged with standard laboratory strain		
	No. of progeny tested	No. of recoveries	Mortality rate (%)	No. of progeny tested	No. of recoveries	Mortality rate (%)
0	454	21	95·4	56	1	98
0·5	500	34	93·2	—	—	—
1·0	795	93	88·3	—	—	—
1·5	755	174	76·9	18	2	89
2·0	804	351	56·3	64	9	86
2·5	381	242	36·5	64	12	81
3·0	267	149	44·2	168	19	89

* 0 = control rabbit. If rabbit of grade 0 (or any other grade) recovered from infection it was allotted a grade of 1 (or its previous grade plus 1). Offspring of a 0 × 1 mating were graded 0·5; those which recovered after challenge infection became 1·5; and similarly for other grades.

were grouped into three Lots, each consisting of the results of tests on the 80 offspring of sixteen bucks, with the results shown in Table 46. The resistance grading (see footnote to Table 45) rose with the Lot number, from an average of 0·3 for females and 1·0 for males for Lot 1 to averages of 2·1 and 2·3 in Lot 3.

The heritabilities based on sire correlations, although limited statistically, were all of the same order, and were consistent with the type of results obtained in calculations of the mortality rates in the field and the laboratory.

The heritabilities based on dam correlations, however, showed a progressive rise with Lot number, with apparent complete heritability in the third Lot. This cannot have a genetic basis, and explanations for this anomalous result are still being sought. Since breeding had been made from recovered does an obvious maternal effect could have been passive immunity (see chapter 9), but this was excluded since the same trend was found with breedings made from uninfected does (W. R. Sobey, personal communication, 1963). A possible explanation which is being actively explored is the congenital infection of the progeny of certain does with a virus (which may be quite unrelated to myxoma virus) which interferes with myxoma virus infection in the grown progeny.

TABLE 46. *Estimates of the heritability (h^2) of resistance to myxomatosis*

Intrasire correlations based on survival time after infection with Aust/Corowa/ 12–52/2 (KM 13) as an index of genetic resistance (data from Sobey, *in litt.*)

Lot no.		h^2	95 % limits	Degree of freedom
1	Sire	0·41	(0–1)	15
	Dam	0·37	(0·19–0·53)	64
2	Sire	0·22	(0–1)	15
	Dam	0·83	(0·75–0·88)	64
3	Sire	0·38	(0–1)	15
	Dam	1·23	(0·99–1·00)	64
Totals, Lots 1–3	Sire	0·34	(0·06–0·53)	45
	Dam	0·94	(0·91–0·96)	192

In an experimental selection programme in which challenge infection of adult animals is a necessary prelude to further breeding, it is impossible to achieve more than one generation per year. Experiments have therefore to be planned on a long-term basis. The preliminary results obtained by Sobey accord with the results of the field experiments and indicate that selection by exposure to infection can lead fairly rapidly to an increased level of genetic resistance to myxomatosis.

SUMMARY

Under the selective effect of exposure for several generations to strains of myxoma virus which were lethal to 90 % of the original wild and laboratory populations of *O. cuniculus* substantial changes were observed in their genetic resistance to myxomatosis. This was first observed in groups of wild rabbits obtained from areas where there were annual epizootics of myxomatosis, and tested at yearly intervals with aliquots of a single batch of slightly attenuated myxoma virus, under conditions which excluded possible effects of passive or active immunity. Over a period of seven years the mortality rate fell from 90 to 25 %, and the proportion of severe cases from 93 to 54 %. Samples of resistant rabbits tested in the field were found to be resistant, to a lesser extent, to the highly virulent standard laboratory strain also.

Breeding and progeny testing experiments carried out in the laboratory with recovered wild and laboratory rabbits showed the same trend to increased genetic resistance, the heritability based on sire correlations being about 0·3. An unexplained maternal effect was observed in the laboratory, but there is no means of determining whether this also operates in the field.

CHAPTER 15

MYXOMATOSIS IN THE AMERICAS

Myxoma virus evolved in the Americas. Had it not been for its extreme lethality for European rabbits which were raised commercially in South America and California, and which were occasionally accidentally infected, it might well have remained an unknown infection of certain American wild rabbits. Indeed, although cases of myxomatosis in native American rabbits have been deliberately sought, only two naturally infected animals have been found, one in Brazil (Aragão, 1943), and one in California (Marshall & Regnery, 1960).

'Spontaneous' infection of European rabbits in the Americas provides evidence of the occurrence of the natural disease among wild $Sylvilagus$, and domesticated $O.$ $cuniculus$ may be regarded as highly sensitive sentinel animals, as far as the detection of enzootic myxoma virus infection among wild $Sylvilagus$ is concerned. Using the occurrence of myxomatosis in domestic rabbits as a criterion, Marshall (1961) produced a map on which Figs. 18 and 19 have been based. From these it may be postulated that South American enzootic myxomatosis is co-extensive with $S.$ $brasiliensis$ (except for southern Argentina and Chile, where the virus probably survives in wild $Oryctolagus$), and Californian enzootic myxomatosis is co-extensive with $S.$ $bachmani$.

It has been possible to examine twelve samples of South American myxoma virus, each recovered from a different outbreak in $Oryctolagus$, and ten samples of Californian myxoma virus. Only one of these (no. 15 of Table 47) was obtained directly from its natural host. The others were recovered from domesticated or laboratory $Oryctolagus$ which had been naturally infected, presumably from the $Sylvilagus$ reservoir.

The results of the inoculation of small doses of American strains of myxoma virus into laboratory rabbits, under the standardized conditions described in chapter 13, are recorded in Table 47. All strains

Fig. 18. Map of South and Central America, showing the distribution of *Sylvilagus brasiliensis* and *Oryctolagus cuniculus*, and the places where strains of myxoma virus have been recovered from naturally infected *Oryctolagus* (modified from Marshall, 1961).

253

presumed to have come directly from *Sylvilagus*, without extensive serial passage in *Oryctolagus*, were highly virulent for *Oryctolagus cuniculus*, but the symptomatology produced by South and Central American strains differed from that seen in most domestic rabbits infected with Californian strains.

TABLE 47. *The survival times of laboratory rabbits inoculated intradermally with small doses of South American and Californian strains of myxoma virus (data from Fenner & Marshall, 1957; Marshall, Regnery & Grodhaus, 1963; I. D. Marshall & F. Fenner, unpublished observations)*

No.	Designation	Survival time (days)		Symptomatology
		Mean	Range	
1	Brazil/Campinas/1949/1 (Lausanne) prototype South American type	12·9	11–16	Florid
2	Brazil/Campo Grande, Rio de Janeiro/12–53/1	10·0	9–12	Florid
3	Brazil/Jacarépaguá, Rio de Janeiro/4–54/1	12·5	12–14	Florid
4	Uruguay/Colonia/1947/1	10·2	9–12	Florid
5	Panama/Panama/1959/1	11·7	11–13	Florid
6	Panama/Panama/1959/2	11·3	11–12	Florid
7	Argentina/Buenos Aires/1958/1	11·1	10–13	Florid
8	Colombia/Tolima/1961/1	12·6	12–14	Florid
9	Colombia/Santander/1961/1	11·2	10–13	Florid
10	Colombia/Valle/1961/1	12·2	11–13	Florid
11	Colombia/Cundinamarca/1961/1	12·3	11–13	Florid
12	Colombia/Cauca/1961/1	10·6	10–11	Florid
13	USA/San Francisco/1950/1 (MSW) prototype Californian type	9·2	7–14	Slight
14	USA/San Diego/1949/1 (MSD)	12·6	10–22	Slight
15	USA/Searsville Lake/9–59/3 (from *Sylvilagus bachmani*)	8·2	7–9	Slight
16	USA/San Luis Rey/9–59/1	10·4	10–12	Slight
17	USA/San Luis Rey/9–59/2	11·2	10–12	Slight
18	USA/San Luis Rey/9–59/3	9·6	8–10	Slight
19	USA/Vista/9–59/1	10·0	9–12	Slight
20	USA/Vista/9–59/2	9·6	8–11	Slight
21	USA/Searsville Lake/9–59/1	10·2	6–12	Slight
22	USA/Searsville Lake/9–59/2	11·6	8–15	Slight

SYMPTOMATOLOGY IN *ORYCTOLAGUS* INFECTED WITH SOUTH AMERICAN STRAINS

All strains presumed to have come from the *S. brasiliensis* reservoir (nos. 1–12 of Table 47) kill rabbits in about twelve days, and all produce a very florid disease in domestic rabbits. Skin lesions, especially at the site of inoculation, are large and protuberant, and become purple in colour; and the swelling of the eyelids is extreme, as is the degree of swelling in the anogenital region. The 'Lausanne' strain (no. 1 of Table 47) typifies these viruses. The type of lesion produced by this strain is shown in Pl. XIII, fig. 5, and the characteristics of the disease produced in laboratory rabbits by this virus are fully described on page 225.

Each of the twelve strains of myxoma virus recovered from South America (including the two from Panama) produced essentially the same symptoms in *Oryctolagus* as the Lausanne strain, and each of them killed all rabbits inoculated with small doses of virus with mean survival times of less than thirteen days. The symptomatology was slightly less florid and less rapidly progressive with one of the Colombian strains, but the survival time was not prolonged. Certainly no strain recovered from South America showed reduced virulence for *Oryctolagus* of the type commonly found in any collection of strains of a similar size now made in either Australia or Europe (see chapter 13).

SYMPTOMATOLOGY IN *ORYCTOLAGUS* INFECTED WITH THE CALIFORNIAN STRAINS

The survival times of laboratory rabbits infected with small doses of Californian strains are often very short, and the symptomatology is much less florid than in rabbits infected with the South American strains (see Pl. XIII, fig. 8). In the occasional animal that survives for as long as twelve days the symptoms are more like those usually associated with myxomatosis, but the skin lesions are never so protuberant and the swelling of the face and the anogenital region is much less pronounced than in infections with South American strains.

Since the Californian virus has not been used to initiate continuing epizootics of myxomatosis in wild *Oryctolagus*, there was no occasion to describe the symptomatology of the infection in *Oryctolagus* in chapter 13. We will therefore describe the disease picture produced in laboratory rabbits by the Californian strains of myxoma virus in this chapter.

255

Fenner & Marshall (1957) chose the strain California-MSW (no. 13 of Table 47) as the prototype virulent Californian type myxoma virus. This strain had been passed eight times in domestic rabbits before it was used, but comparisons subsequently made with the virus recovered from *S. bachmani* by Marshall & Regnery (1960) (no. 15 of Table 47) showed that the two were indistinguishable. We have therefore retained California-MSW as the prototype strain of Californian myxoma virus.

Virulent Californian myxoma virus

Prototype strain: California-MSW (U.S.A./San Francisco/1950/1, no. 13 of Table 47) (Pl. XIII, fig. 8). Mean survival time 9·2 days. Range seven to fourteen days. Mortality rate 100 %. The MSW strain was derived from a domestic rabbit naturally infected in 1950.

It is highly lethal for laboratory rabbits, but the progress of the infection differs greatly from that seen in rabbits infected with South American strains or their derivatives. The primary lesion appears on the third day and slowly increases in size, but never becomes prominent. It is rubbery in consistency and the edges merge gradually into the normal skin. Swelling of the eyelids appears on the seventh day, but never progresses to closure of the eyes. It is rarely more severe than shown in Pl. XIII, fig. 8. The conjunctival discharge is slight in amount and thin and milky in nature. Skin secondaries and swelling in the anogenital region appear about the ninth day, but are often absent at the time of death. Neither progresses to an advanced stage. Manifestations of involvement of the central nervous system occur more commonly in rabbits infected with Californian strains than with any others studied, and consist of a very rapid tremor or convulsions; and appreciable amounts of virus are found in the brain (Table 15).

In the dead animal there is often a bloody discharge from the nose or the anus, and post mortem examinations show haemorrhages in the skin, stomach, intestines, lungs and other viscera to be more frequent than in fatal cases of myxomatosis due to other strains of the virus.

During our initial investigations, so many rabbits died before symptoms of advanced myxomatosis appeared that contamination

256

of the myxoma virus with another pathogen was suspected. However, no bacteria could be isolated from suspensions of the virus, and previous infection of rabbits with fibroma, or with other strains of myxoma virus, protected them from generalized myxomatosis due to the MSW strain. Large amounts of the specific soluble antigens were present in the serum at death, even when the external symptoms were slight.

The multiplication curve of virus in the skin after the intradermal inoculation of moderate doses resembled those of the standard laboratory and Lausanne strains, usually rising to an even higher level (see Fig. 13), but the local lesions were never large and secondary skin lesions were usually very slightly developed at the time of death.

As would be expected from the high titre of virus in the skin, mosquito transfer from the primary skin lesion was effective from the fifth day until the rabbit died. In the very occasional case which survived for longer than twelve days there were extensive skin lesions which were highly infectious for mosquitoes.

Seven strains of myxoma virus were recovered from naturally infected *Oryctolagus* in three different outbreaks in California in September 1959 (nos. 16–22 of Table 47). All produced a disease in laboratory rabbits very similar to that described above, and quite different from the syndrome produced by the infection of laboratory rabbits with South American strains.

It is probable that in Californian rabbitries many sporadic deaths of domestic rabbits are caused by unrecognized myxomatosis. Even as experienced an observer as Marshall has commented upon the difficulty of diagnosing Californian myxomatosis in these circumstances, especially, as so often happens, if the rabbit is dead when first seen.

In their detailed description of domestic rabbits infected with myxoma virus in California in the 1930's, Kessel *et al.* (1934) reproduce several illustrations showing rabbits with much more pronounced skin lesions than those just described. In comparing their Californian strains with myxomatosis of *Oryctolagus* caused by the classical South American strain, they reported that the progression of symptoms was somewhat slower, and the survival time longer, in rabbits infected with the Californian strains, but they did not comment upon the diminished severity of the skin lesions. The disease produced in *Oryctolagus* by the viruses prevalent in California in the 1930's was different from that produced by viruses obtained since 1949 and the relationship of these two episodes of myxomatosis in California remains conjectural. Throughout this chapter we have concentrated our attention on the Californian strains now extant, that is those studied by Marshall & Regnery.

FIBROMA IN *SYLVILAGUS* IN NORTH AMERICA

Shope's fibroma, which is caused by a virus related to myxoma virus, occurs naturally in several of the eastern and mid-western states of U.S.A., and in the neighbouring province of Ontario, Canada. Unlike myxomatosis, it has been recognized only as an infection of the common eastern cottontail (*S. floridanus*) and many recoveries of the virus have been made, notably in New Jersey, New York, Maryland, Michigan, Missouri, Illinois, Wisconsin and Ontario (Canada).

Fibroma has been recognized as a naturally occurring infection of domestic rabbits only in animals deliberately exposed in an enzootic area by R. E. Shope (personal communication, 1961), for it causes only a localized tumour of short duration in these animals. Infections of domestic rabbits are probably rare, for there is little commercial rabbit breeding in eastern U.S.A. and because they do not produce generalized signs any infections that may occur remain unnoticed.

Cottontail rabbits bearing fibromas are seen mainly in the autumn and early winter (Table 48), which may be partly due to the more extensive hunting carried out then, as well as the probable greater frequency of recent infections. During the summer and autumn there is both a larger population of susceptible rabbits, and a larger vector population, so that infection then is relatively common. Naturally and artificially produced lesions are localized to the site of inoculation or mosquito bite, and they may persist for prolonged periods. Kilham & Dalmat (1955), for example, have recorded the existence of a tumour which persisted, and was infectious for mosquitoes, for as long as ten months.

Details of the natural history of fibroma have proved difficult to elucidate. Dalmat (1958b) showed that circulating antibodies dis-

258

appeared within five months of infection, so that serological surveys have only a limited value in assessing past infection. In the laboratory, mosquitoes are able to obtain virus from cottontail fibromas for prolonged periods, and infected mosquitoes may remain infective for a long time. These features may explain the epidemiology. Overwintering is probably achieved by persistence of the infectivity of at

TABLE 48. *The seasonal incidence of fibromas in cottontails (data from Herman et al. 1956)*

Month	Patuxent Research Refuge, Maryland		Rose Lake Wildlife Experiment Station Michigan			
	1947–50	1951	1941	1942	1943	1944
January	0/2*	0/1 ⎫	—	0/51	—	—
February	0/3	0/1 ⎭				
March	0/11	3/7 ⎫	—	0/32	—	—
April	0/13	1/1 ⎭				
May	0/15	0/1 ⎫	—	—	0/13	0/30
June	0/13	0/4 ⎭				
July	0/14	0/11 ⎫	3/95	—	—	—
August	0/7	0/8 ⎭			5/55	0/55
September	0/14	0/2 ⎫	10/147	—	—	—
October	0/35	5/13 ⎭				
November	0/77	3/17 ⎫	21/224	—	7/112	1/158
December	3/19†	0/4 ⎭			—	—

* Numerator = cottontails with fibroma; denominator = total number of cottontails examined.

† All in December 1950, when 16 cottontails were examined for fibroma.

least some of the many fibromas seen during the early winter months, but there is a puzzling period during the spring and early summer when fibromas are very rarely seen, although mosquitoes are common.

A COMPARISON OF FIBROMA, CALIFORNIAN MYXOMA, AND SOUTH AMERICAN MYXOMA VIRUSES

Not only do these viruses differ in their symptom production when they infect domestic rabbits; they also cause distinctive lesions in cultured cells and on the chorioallantoic membrane, and they differ in their antigenic structure. These features, which have been discussed in more detail in other chapters, are compared in Table 49.

It is reasonable to ascribe these differences to prolonged separate evolution of the three viruses, a conclusion supported by recent experimental work with Californian species of *Sylvilagus*, which is described in the latter part of this chapter.

Preliminary studies suggest that strains from different parts of South America, although they show 'family' resemblances, may be

TABLE 49. *Comparison of South American myxoma virus, Californian myxoma virus, and Shope's fibroma virus*

Biological character	South American myxoma virus	Californian myxoma virus	Fibroma virus
Lesion in *Sylvilagus brasiliensis*	Localized fibroma, transmissible*	...†	...
Lesion in *S. bachmani*	Localized fibroma, non-transmissible	Localized fibroma, transmissible	Nil
Lesion in *S. floridanus*	Localized fibroma, non-transmissible	Localized fibroma, non-transmissible	Localized fibroma, transmissible
Disease in *Oryctolagus cuniculus*	Severe generalized; florid symptoms, transmissible	Severe generalized; slight symptoms, transmissible	Localized fibroma, non-transmissible
Pocks on chorioallantoic membrane	Moderate size	Very small	Nil
Plaques on rabbit kidney monolayers	Moderate size, hazy	Small	Heaped-up foci
Soluble antigens	Distinctive for each virus, with some cross reactions		

* Transmissible or non-transmissible refers to suitability of skin lesions as source for mechanical transmission by arthropod vectors.

† No information.

more heterogeneous than the Californian viruses. In particular, strains from Panama and Colombia show differences from other South American strains in their soluble antigens and plaque type on rabbit cells, but not in the symptomatology of the disease they produce in *Oryctolagus* (Fenner, 1965).

EPIDEMIOLOGY OF MYXOMATOSIS IN SOUTH AMERICA

The only publication dealing with the natural history of myxomatosis in South America is the classical paper of Aragão (1943), who first showed that the tapeti (*S. brasiliensis*) was a reservoir host.

Aragão showed that in this animal a tumour, very like a cottontail fibroma, appeared at the site of inoculation of the virus, and developed slowly until it reached a diameter of about 1 cm., regressing after 10–40 days. Recovered animals were immune to re-infection. Rarely, slight unilateral blepharitis was seen, and infected rabbits sometimes lost weight and very rarely, especially with very young rabbits (H. de B. Aragão, personal communication, 1953), one died.

Aragão showed that mosquitoes could mechanically transfer the virus from the skin lesions of *S. brasiliensis* or *O. cuniculus* to other wild or domestic rabbits, with the production of the sort of disease characteristic of the species of rabbit infected. He pointed out that mosquito transfer explained very well the summer incidence of outbreaks amongst commercially raised domestic rabbits, and the failure of routine quarantine to control its spread. The only aspect of the natural history of myxomatosis in South America not investigated by Aragão was how the virus survived through the winter; a problem that has resisted final solution in very many mosquito-borne viral diseases.

In the survey of myxomatosis in South America in 1961, Marshall brought to light some interesting facts about myxomatosis in different South American countries. The following details are extracted from his report (Marshall, 1961).

Brazil

Recurrent outbreaks of myxomatosis have occurred for the last half century in the vicinity of Rio de Janeiro, and Saõ Paulo. Myxomatosis has been reported at Belém, at the mouth of the Para Branch of the River Amazon, but it does not appear to occur in Rio Grande do Sul, the southernmost province of Brazil. It seems likely that *S. brasiliensis* is enzootically infected wherever it occurs in Brazil (see Fig. 18).

Three specimens of virus obtained from domestic rabbits which were presumably infected directly from the *Sylvilagus* host have been tested; nos. 1, 2 and 3 of Table 47. In laboratory rabbits each of these produced a rapidly lethal disease characterized by florid signs of generalization and large protuberant lesions at the site of intradermal inoculation.

Uruguay

Sanarelli's original outbreak occurred in the Hygiene Institute at Montevideo, in 1896, but the only subsequent recorded outbreak was in 1947. The virus from the latter outbreak (no. 4 of Table 47) had been passed 44 times in laboratory rabbits before it could be examined in our laboratories. The symptomatology in *Oryctolagus* conformed closely to the disease pattern of the other South American strains.

No species of *Sylvilagus* occurs in Uruguay. The existence of hutches of domestic rabbits in rural areas of Uruguay, in which myxomatosis has never been recognized, makes it rather unlikely that there is any other natural reservoir host of the virus there.

Only one colony of domestic rabbits was involved in the 1947 outbreak and these were brought to Colonia from El Tigre, Argentina, the first signs of disease developing shortly after their arrival. It has been impossible, at this late date, to determine the source of Sanarelli's rabbits, but it is not improbable that in 1896, as in 1947, the virus was brought in with infected domestic rabbits, from either Brazil or Argentina.

Colombia

Both *S. brasiliensis* and *S. floridanus* occur in Colombia, but myxomatosis was unknown there until very recently. Since 1958 the Government of Colombia has encouraged the Indians to rear domestic rabbits as a source of animal protein. Myxomatosis is a serious threat to the success of this project.

Strains of myxoma virus were obtained from 'spontaneous' outbreaks of myxomatosis in domestic rabbits in five different provinces of Colombia. When examined in Canberra, they were found to be similar in their virulence and symptomatology in *Oryctolagus* to other South American strains of the virus (nos. 8–12 of Table 47), but their soluble antigens show closer resemblances to the Californian than to Brazilian myxoma virus (Fenner, 1965).

Panama

For a number of years myxomatosis has occurred during August and September among the laboratory rabbits of the Gorgas Memorial Laboratory, if they are kept in unscreened outdoor hutches (E. McConnell, personal communication). *Sylvilagus brasiliensis* occurs in Panama, and two strains of virus obtained from naturally infected laboratory rabbits were found to be of the South American type in their effect on *O. cuniculus* (nos. 5 and 6 of Table 47). However, they differ from the Brazilian strains in their soluble antigens, and in the type of plaque produced on rabbit embryo fibroblasts (Fenner, 1965).

Argentina

The situation in Argentina offers unique opportunities for investigations on the natural history of myxomatosis, but these have not yet been exploited. In northern Argentina domesticated European rabbits are bred on a large scale, both for the carcass trade and for the production of Angora fur. In Tierra del Fuego, far to the south, wild European rabbits constitute a serious threat to the pastoral industry.

Myxomatosis now occurs in both areas, with Central Argentina constituting an ecological barrier to the spread of both wild rabbits and myxomatosis. In the north there must be a reservoir in some wild animal, for myxomatosis has been recorded amongst the domestic rabbits at irregular intervals since 1919 (Rosenbusch, 1919), and a strain of virus recovered from domestic rabbits proved to be of the highly virulent South American type (no. 7 of Table 47).

Sylvilagus brasiliensis occurs in forested areas of subtropical Argentina, near the Paraguayan and Brazilian borders, but it probably does not occur on the pampas. Although no evidence exists to substantiate the proposition, it is reasonable to follow Marshall (1961) and suppose that myxomatosis, enzootic in *S. brasiliensis*, escapes from time to time into the domestic rabbit colonies of northern Argentina. From here it may spread down a fairly continuous chain of rabbitries to the main rabbit breeding centres in Buenos Aires province. It is highly unlikely that this highly virulent virus is maintained from year to year in domestic rabbits.

Part of Tierra del Fuego is administered by Argentina and part by Chile. Although Argentina finally withdrew from its agreement with Chile jointly to introduce myxomatosis into the wild *Oryctolagus* population of the island, the Chilean authorities made introductions in 1954 and have carried out inoculations annually since then. The disease spread across the political border and now appears to be enzootic amongst the wild *Oryctolagus* of Argentinian Tierra del Fuego. There is no suggestion that this introduced myxomatosis has penetrated into the northern domestic rabbit breeding areas of

Argentina. Local scientists believe that ceratopogonids are the vectors, and there are no insects of this genus common to south and north Argentina.

In Argentina, therefore, there is myxomatosis enzootic in wild *Oryctolagus* in the extreme south, and myxomatosis spilling over from a wild reservoir host (probably *S. brasiliensis*) in the far north, and extending to the south by infection of domestic rabbits. Virulence tests with strains recovered annually from the two areas would give a most interesting comparison of the evolution of the virus in two different host animals, but myxomatosis does not loom very large amongst the host of urgent problems confronting veterinary authorities in Argentina.

Chile

Sylvilagus brasiliensis does not occur in Chile. This fact, and the consequent absence of enzootic myxomatosis in Chile, may be partly responsible for the fact that wild *Oryctolagus* introduced into Tierra del Fuego have spread up the western side of the Andes to localities at least 200 miles north of Santiago. There are also said to be dense populations in some of the high valleys of the Andes.

Regular inoculation campaigns of wild *Oryctolagus* with myxoma virus are carried out in many parts of Chile, but there does not seem to be any adequate assessment available of its efficacy as a method of rabbit control. No virulence tests have been carried out with strains of myxoma virus recovered from wild *Oryctolagus* in Chile.

Other countries of South and Central America

No information is available concerning myxomatosis in other countries of South and Central America. From the recent evidence of its occurrence in Colombia and Panama it is reasonable to assume that the disease is enzootic in *S. brasiliensis* over the whole of its range (Fig. 18), and outbreaks will occur among *Oryctolagus* wherever they are adequately exposed to infectious insect vectors.

EPIDEMIOLOGY OF MYXOMATOSIS IN NORTH AMERICA

Myxomatosis has been recorded in California, U.S.A., and probably occurs in Mexico also.

The occurrence of myxomatosis in California was first reported by Kessel *et al.* (1931) who diagnosed the disease as the cause of lethal infections among domesticated rabbits near San Diego. Subsequently Kessel, in a letter to Simmons (1934) wrote:

It is impossible to state whether the infection has been endemic in California for a long time, or whether it has been introduced rather recently from South or Central America. I know of no one in California who had any of the South American strains of virus prior to the 1930 outbreak. There have been outbreaks apparently of natural infections since the 1930 outbreak, but we have every reason to presume that these are Californian rather than South American.

Concerning the epidemiology of the disease in California, Kessel further commented:

In the rabbitries in which the epidemic was encountered, the incidence of morbidity was about 60% while the mortality rate of those which became infected was 100%. The outbreaks were usually sporadic in some ten rabbitries each spring season, and we usually heard nothing from breeders except during the months of May, June and July. It is impossible to do anything but speculate upon the probable source of the infection. Two possibilities are of interest: (1) That the infection is chronic the year round in a few animals and only occasionally breaks out in a few instances. (2) Some wild animal may serve as a reservoir host and the infection may be transmitted by some insect such as a mosquito.

We pointed out earlier in this chapter that there was some doubt as to whether the virus recovered by Kessel was of the same type as those recovered in California since 1949. However, during the last three years investigations by Marshall, Regnery & Grodhaus have established the validity of two of Kessel's suggestions: (*a*) the reservoir host is a wild animal, *Sylvilagus bachmani* and (*b*) the infection is transmitted by mosquitoes.

The reservoir host of myxomatosis in California

On the basis of the correlation between the geographical distribution of myxomatosis and different species of wild rabbits in western U.S.A., Calaby (1954) suggested that the brush rabbit (*Sylvilagus bachmani*) was the animal most likely to be the reservoir host of myxomatosis. Subsequent investigations by Marshall & Regnery (1960, 1963) have positively incriminated *S. bachmani* as a reservoir host of myxomatosis in California, and there is suggestive evidence

265

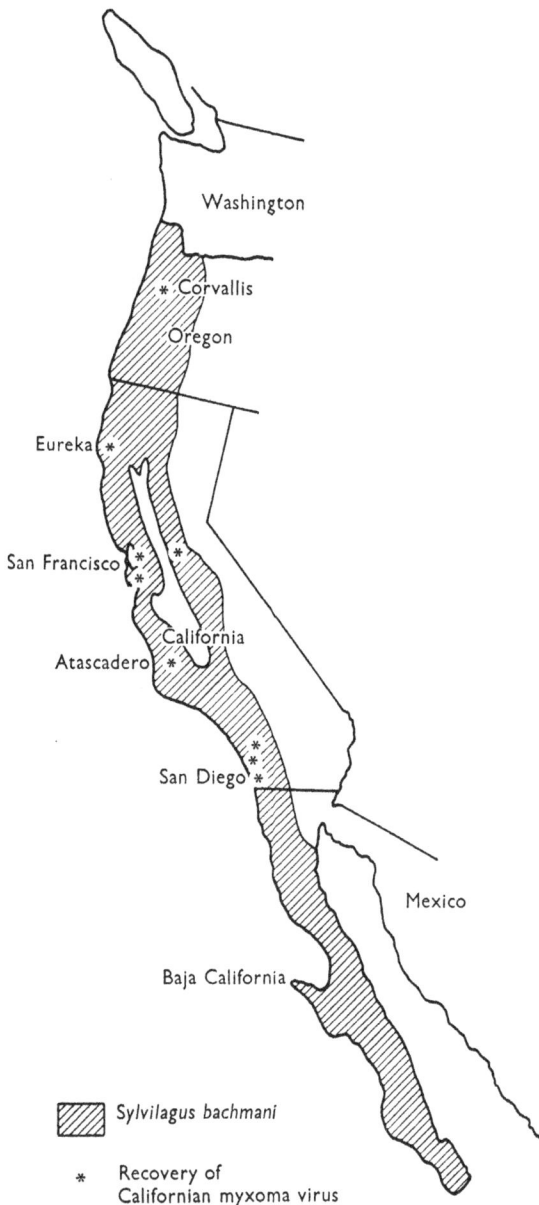

Washington

* Corvallis

Oregon

Eureka *

San Francisco * *
*

California

Atascadero *

San Diego *

Mexico

Baja California

Sylvilagus bachmani

* Recovery of
Californian myxoma virus

Fig. 19. Map of western U.S.A., showing the distribution of *Sylvilagus bachmani* and the places where strains of myxoma virus have been recovered from naturally infected *Oryctolagus* (modified from Marshall, 1961).

that it may be the only one (D. C. Regnery & I. D. Marshall, personal communication, 1962). Figure 19 shows the distribution of myxomatosis and *S. bachmani* in western U.S.A.

Difficulties experienced in raising pet rabbits at Palo Alto, California, and a retrospective diagnosis of myxomatosis as the probable cause of this difficulty, led to the recognition of myxomatosis among two groups of pet rabbits in Palo Alto in August–September 1959. The natural cases were positively diagnosed as myxomatosis by clinical signs and animal inoculation tests. Since no domestic rabbits had been moved into the area for many months the source of the virus must have been some wild animal. The only species of rabbit recovered by trapping was *S. bachmani*, of which three specimens were caught in the immediate vicinity of the colony of diseased domestic rabbits and a further twenty-six from the same general area. All three *S. bachmani* trapped near the outbreak area showed antibodies to myxoma virus, and a strain of myxoma virus was recovered from a small rather dry lesion on the foreleg of one of these animals. Subsequent experiments (Grodhaus *et al.* 1963) showed that similar tumours could be produced in *S. bachmani* by mosquito bite infection. Virus was recovered from two pools of wild-caught *Anopheles freeborni*, and the same species readily became infective after feeding through the tumours produced by Californian myxoma virus in *S. bachmani*, thus providing conclusive evidence both of the vector role of this mosquito, and the role of *S. bachmani* as a reservoir host of the infection.

Californian myxoma virus in *S. bachmani*, like the South American virus in *S. brasiliensis*, produces only a localized tumour at the site of inoculation or mosquito bite. These tumours are usually infectious for mosquitoes for about twenty days, occasionally for much longer.

In myxomatosis of *S. bachmani*, as in fibroma infection of *S. floridanus*, circulating antibodies disappear rather rapidly, so that serological surveys are not of much assistance as a epidemiological tool.

Mosquito transmission of Californian myxomatosis

Myxomatosis in Californian domestic rabbits has long been recognized as a disease of the summer months, and mosquitoes have been suspected as the vectors of the disease (Vail & McKenny, 1943). In the Palo Alto outbreak just described a total of 435 mosquitoes, belonging to fourteen different species, was collected. The majority were *Anopheles freeborni*, and Californian strain myxoma virus was recovered from two separate pools of these.

Subsequently Grodhaus *et al.* (1963) showed that each of the five species of mosquito adequately tested would transmit Californian myxomatosis in series in *S. bachmani*, if they bit through the skin tumours to acquire infectivity.

The antiquity of myxomatosis in California

There is no science of palaeo-virology, so that conclusions concerning the evolutionary history and the antiquity of viral diseases must be based on inference. The evidence summarized in Table 49 shows that the virus which causes natural myxomatosis in California is rather different in several properties from strains recovered in different parts of South America; and strains recovered from different parts of California, over a decade, are virtually identical. The known natural host of myxomatosis in California, *S. bachmani*, does not occur in South America, and *S. brasiliensis*, the natural host of myxomatosis in South America, does not occur in California.

Myxoma virus produces a benign, localized, but persistent tumour in the skin of *S. bachmani*, readily transmissible from one *S. bachmani* to another; a type of response which is compatible with a long established host-parasite relationship. Fibroma virus, it will be recalled, produces a benign localized tumour in *Oryctolagus*, which is a novel host; but these are not serially transmissible by mosquitoes, whereas the similar localized tumours which this virus produces in *S. floridanus* are infectious for mosquitoes.

Serial transmission of Californian myxoma virus by mosquitoes can occur in *Oryctolagus* (chapter 11) but the very short survival of most infected domestic rabbits rapidly removes this source of virus. Grodhaus *et al.* (1963) reported that the skin lesions of *S. bachmani* were usually infectious for mosquitoes for about twenty days and one such lesion was infectious for as long as 80 days. It is possible that in Californian myxomatosis in *S. bachmani*, just as in fibroma in the eastern cottontail, an occasional rabbit may remain infectious for many months, thus ensuring the over-wintering of the virus.

Experiments with three different strains of myxoma virus in *S. bachmani*, summarized in chapter 11, showed that only the

Californian strain of virus was adapted for natural survival in this host animal. Other experiments (D. C. Regnery & I. D. Marshall, personal communication, 1962) showed that of five species of North American leporid infected with Californian myxoma virus by mosquito bite, in only *S. bachmani* were the tumours produced themselves infective for mosquitoes (Table 12).

From all this evidence it is arguable (*a*) that the Californian strains of myxoma virus are different from strains found elsewhere in the world, but form a rather homogeneous group, and (*b*) that the Californian type of myxoma virus, and only this type, can be maintained in nature in a Californian rabbit, *S. bachmani*. All this suggests that Californian myxoma virus and *S. bachmani* form a climax association of virus and host, which is the end result of a long period of that mutual adjustment which constitutes the evolution of an infectious disease.

Experiments of the type just described have yet to be carried out with South American myxoma virus and fibroma virus, but what scattered evidence there is of mosquito transmission experiments, and tests of species specificity, is consistent with the hypothesis that each of these diseases also represents the end-product of a long process of evolution, the climax associations being *S. brasiliensis*—South American myxoma virus and *S. floridanus*—fibroma virus.

MYXOMATOSIS AND THE COLONIZATION OF THE AMERICAS BY *ORYCTOLAGUS CUNICULUS*

Enzootic myxomatosis in the Americas can be compared with trypanosomiasis in Africa. Each of these diseases causes negligible mortality in the indigenous species, but is highly lethal for European rabbits and bovines respectively.

In chapter 3 we suggested that wild *Oryctolagus* were much more likely to become successfully established on a continental scale than were domesticated European rabbits. The lack of information on attempted introductions of wild European rabbits into the Americas makes it difficult to argue with conviction that enzootic myxomatosis in California and in South America was the major factor in preventing the establishment of European rabbits there, but several pieces of evidence point that way.

The European hare, which was never domesticated, was taken to both North and South America and has become a pest in parts of South America (e.g. Uruguay). During the era of naturalization societies, it is quite likely that the wild rabbit also was taken to America and released. Certainly the climate was suitable for its pullulation, for domesticated European rabbits have become established as dense populations of feral animals in many islands off the shores of Argentina and California, and domestic rabbit breeding is a large industry in both these countries, both as a small-scale collateral activity and on large rabbit ranches.

The predator population of the Americas is vastly greater than that of Australia or New Zealand, but predators have failed to destroy more vulnerable species like *Sylvilagus* spp. and *Lepus* spp. Further, the European rabbit has become established on the mainland of South America in Chile, where it has undergone a colonizing spread not unlike that seen in Australia. It is a noteworthy fact that there is no *Sylvilagus*, and there is therefore no enzootic myxomatosis, in Chile.

Although from the nature of the case there can be no certain evidence, it is reasonable to conclude that enzootic myxomatosis has been the major factor in preventing the successful establishment of *Oryctolagus cuniculus* as a wild animal in many parts of the mainland of North and South America.

SUMMARY

We can recognize, in geographically separated parts of the Americas, three related poxviruses which have evolved in association with three different species of *Sylvilagus*. They are Californian myxoma virus in *S. bachmani*, South American myxoma virus in *S. brasiliensis*, and Shope's fibroma virus in *S. floridanus*. Each virus produces a localized benign fibroma in the host in which it has evolved; and produces either slightly more severe, or more commonly quite trivial, lesions in related species of *Sylvilagus*. In *Oryctolagus cuniculus*, the European rabbit, two of these viruses cause a lethal generalized disease, whereas the virus of *S. floridanus* (Shope's fibroma virus) produces a benign localized fibroma in *Oryctolagus* also.

The epidemiology of these benign fibromas of *Sylvilagus* is not fully understood, but extensive spread of the disease during the summer and autumn months is certainly due to the activity of arthropod vectors which acquire virus when they probe through the fibromas and transmit it mechanically to other susceptible rabbits. The mechanism of over-wintering may be similar in many of the virus-induced skin tumours of wild animals, and involve the persistence for long periods of tumours infectious for biting arthropods. Other as yet unrecognized mechanisms, or perhaps other reservoir hosts, may also be involved.

With the Californian type virus, which is the only one on which such investigations have been made, several species of *Sylvilagus* other than the natural host appear to be highly susceptible to infection, in that localized tumours are produced after mosquito-bite infections, but the lesions produced by the bites of infected mosquitoes do not themselves become infective; and such cases are therefore dead-end infections.

Owing to their great susceptibility to the myxoma viruses, and the lethal outcome of such infections, domesticated European rabbits act as 'sentinel animals' which reflect the presence of the benign disease in wild *Sylvilagus*.

It is possible that enzootic myxomatosis of *Sylvilagus* spp. was an important factor in preventing the colonization of the major part of mainland America by *Oryctolagus cuniculus*.

CHAPTER 16

MYXOMATOSIS IN AUSTRALIA
1950–63

In chapter 1 we described the events which preceded the widespread establishment of myxomatosis in south-eastern Australia in 1950. Here we will deal in somewhat greater detail with the 1950 field trials, and with the escape and dramatic spread of the infection. A generalized account will then be given of the subsequent history of myxomatosis in Australia. This will involve some consideration of the importance of topics discussed at length in other chapters, such as vector ecology (chapter 12), and changes in virus virulence (chapter 13) and in host resistance (chapter 14).

PASTEUR'S INTEREST IN THE BIOLOGICAL CONTROL OF THE RABBIT

Within twenty years of the introduction of wild rabbits into Australia in 1859, they had become the country's major pest (see chapter 3). In 1887 the Government of New South Wales offered a prize of £25,000 for the eradication of rabbits from the colony; and it is interesting to recall that this advertisement prompted the first serious thought of the biological control of a mammal pest. No less a man than Pasteur was interested in the problem, and he sent his nephew and assistant Loir to Sydney with a culture of the chicken cholera bacillus (*Pasteurella septica*), which Pasteur had previously used with success on a very small scale on the estate of Mme Pommery in France (Pasteur, 1888). The Australian quarantine authorities were as reluctant then as they were almost half a century later to allow a pathogenic micro-organism to be brought into the country, and Loir was unable to import his culture, though he did stay for eight years and set up the first laboratory of microbiology in the continent.

272

Fig. 1

Fig. 2

PLATE XIV. Scientists who have been concerned with the investigation of myxomatosis in Australia and America.

Fig. 1. From left to right: Professor F. Fenner (Australian National University), Dr I. D. Marshall (Australian National University), Dr M. F. Day (Division of Entomology, C.S.I.R.O.), Mr G. W. Douglas (Department of Crown Lands and Survey, Victoria), Professor D. C. Regnery (Stanford University), Dr G. M. Woodroofe (Australian National University) and Dr W. R. Sobey (Division of Animal Genetics, C.S.I.R.O.) Photograph V. Paral.

Fig. 2. Officers of the Wildlife Survey Section of C.S.I.R.O. From left to right: Mr F. N. Ratcliffe, Mr B. V. Fennessy, Mr J. H. Calaby, Dr R. Mykytowycz, Mr A. L. Dyce and Mr K. Myers. (Mr W. E. Poole was unavoidably omitted from the group.) Photograph E. Slater.

FIELD TRIALS OF 1950, AND THE FIRST
SPREAD OF MYXOMATOSIS

With the establishment of the C.S.I.R.O. Wildlife Survey Section in 1949, an intensive effort was made to put rabbit control on a scientific basis, and attention was again turned to the possible use of the myxoma virus as an agent of biological control. Remembering the discouraging results of the earlier field trials, carried out by force of circumstances in a low-rainfall area in South Australia, and Bull & Mules's (1944) pointer to native species of mosquitoes being likely to assist transmission, it was decided to work in the Murray Valley; and extensive field trials were carried out at several sites in this region during 1950.

The first experimental liberation of the virus was made on an irrigated dairy farm at Gunbower, in Victoria, some 150 miles downstream from Albury. The property consisted of about 500 acres of river-frontage land carrying an exceptionally dense population of rabbits, mostly living in large and well-established warrens. After an earlier failure, due to faulty virus material, the infection was successfully introduced into a number of warren colonies in different parts of the experimental area in May. Gin traps, modified to puncture the skin of a rabbit springing them and introduce virus into the wounds (Anonymous, 1942), were used for the purpose. During the weeks that followed, a total of 77 diseased rabbits were seen (in a population of over 4000) well distributed over the area; but the sighting of sick animals became less and less frequent, and by the end of July some time had passed without one being seen. Observations were not extended beyond the boundaries of the study area as the land in its immediate neighbourhood contained very few rabbits. Except for the tail end of an aedine wave which had developed just prior to the virus liberation, mosquitoes were not in evidence.

It was presumed that the infection had died out at Gunbower; and in August the field team moved upstream to start a series of experiments in the Albury area, where the River Murray leaves the foothills of the Dividing Range to flow westward as the boundary between Victoria and New South Wales. (The rabbits on the Gunbower site were involved in the great myxomatosis epizootic that swept down the Murray frontage six or seven months later, but the results were not spectacular: the main kill took place in the following season.)

The four release sites selected were near (a) Wymah, in hilly country about twenty miles north-east of Albury, (b) Rutherglen in Victoria, (c) Coreen, about twenty miles north of Corowa in New South Wales, and (d) Balldale, about fifteen miles north-east of Corowa. With the exception of the one near Rutherglen, which will be described below, these sites were all in grazing country typical of the region, none on the actual river frontage. At Balldale, the virus was liberated in a rabbit population inhabiting a belt of cypress pine (*Callitris*) on light soil, always a much-favoured rabbit habitat. Rabbits were caught in various ways, for example with ferrets, by digging out or netting at night, and were inoculated with highly virulent virus—the standard laboratory strain, as it was later to be termed.

The results at Wymah (Fennessy, unpublished observations) differed from those obtained elsewhere. After the release of inoculated rabbits, myxomatosis had a period of unobtrusive activity which resulted in a noticeable reduction in rabbit density in one part of the property. In other words, what we would later have called a localized, low-intensity, grade 2 outbreak was experienced before the infection appeared to die out.

Myers (1954) has described the procedure followed and the results obtained in the other three areas. The most detailed observations were made at the Rutherglen site, which was a fenced area of thirteen acres surrounding some 'mullock heaps' (hillocks of crushed rock) beside a worked-out gold mine. Particular attention was paid to factors of potential epidemiological importance, such as rabbit density, predation, the behaviour of rabbits infected with myxomatosis, and the presence and activities of possible insect vectors. Counts of diseased rabbits were made daily, and carcasses were counted and removed. Population counts were made at intervals. These were carried out in a carefully standardized way, and although they did not provide a measure of the total number of rabbits on the site, they can be accepted as a reliable index of population density and its changes.

Twenty-seven rabbits were inoculated with virus on 7 September, when the population count was about 700. The disease failed to become established, and a further inoculation of 66 rabbits was made on 27 October, when the population count had risen to 900 as a result of breeding. Between then and 16 December, when observations ceased, four generations of natural infection could be distinguished, each smaller than the preceding one (see Fig. 20).

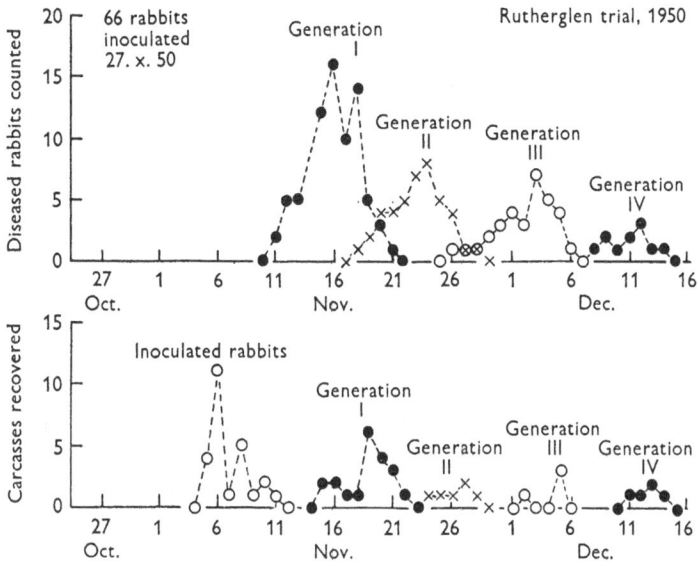

Fig. 20. Daily counts of diseased rabbits and the collected carcases of rabbits dying on the surface of the ground during the second trial at Rutherglen (from Myers, 1954).

The population count at the end of the experiment had risen to 1000. During the fourteen weeks from 7 September to 16 December, known predation (by a feral cat and a pair of little eagles) and deaths attributed to coccidiosis, accounted for about 400 kittens: naturally acquired myxomatosis was known to have killed about 70 rabbits. Insect vector activity was slight: a few aedine mosquitoes were present between August and October, and an occasional *Anopheles annulipes* was caught in November. Conditions, however, would have been most favourable for epizootic transmission by contact, or by oral or respiratory routes, namely an extremely high rabbit population density, no selective predation of diseased animals, and a high proportion of infected rabbits remaining within their warren colonies until they died. In spite of this, the ratio of infection from one generation to the next was only 0·6.

275   18-2

By the beginning of December, myxomatosis had apparently died out everywhere except at Rutherglen, where the infection was fading towards extinction. The Murray Valley trials had been thorough: they had been carried out at a variety of sites where the rabbit population was dense, and sometimes very dense indeed. Aedine mosquitoes had been reasonably abundant at Coreen; and at Balldale, where they were common, one of the three species collected—*Ae. alboannulatus*—had earlier been shown by Bull & Mules (1944) to be capable of transmitting myxomatosis under laboratory conditions. At that juncture, therefore, the 1950 trials provided what seemed a clear confirmation of the conclusion reached by Bull & Mules on the results of the South Australian trials carried out seven years earlier, i.e. that the myxoma virus held little promise of being of practical value in rabbit control.

THE EPIZOOTIC OF DECEMBER 1950–FEBRUARY 1951

Before any decision could be taken on what should be done to follow up the apparently abortive Murray Valley releases, we received the first indication of the development of a spectacular epizootic which for scale and speed of spread must be almost without parallel in the history of infections. A telephone call from the owner of the Balldale property reported sick rabbits being seen in numbers when his employees went to fumigate the warrens on the experimental site. The control operation was proceeded with, so nothing of much interest could be seen when the area was examined a few days later. There was not long to wait, however, before the significance of the flare-up of the disease at Balldale became apparent and we knew that a widespread dispersal of the virus was taking place. Within a week or less an epizootic of myxomatosis was reported on the frontage of the River Murray, ten miles or so to the south of the Balldale site; and during the latter half of December reports came in with increasing frequency revealing a spread of the disease downstream and epizootics in progress at various points along the course of the northern tributaries—the Murrumbidgee, Lachlan and Darling.

In January a start was made on the vector studies (described in chapter 12) near Corowa, and the mapping of the spread of the infection was hurriedly organized. This involved the co-option into the Wildlife team of colleagues from other C.S.I.R.O. Divisions and the co-operation of numerous field officers in State Departments. The developments brought to light by the surveys were reported by Ratcliffe *et al.* (1952).

After the New Year, the epizootic gathered momentum rapidly. By mid-February 1951, myxomatosis had appeared in practically all parts of a huge area that was to be affected before the advent of winter, and the peak of disease activity had passed in most districts. Observers were impressed by the speed with which the disease swept through local rabbit populations. When high mortalities (of the order of 90 % or better) occurred, they were often achieved within three to four weeks of the date on which the infection was first noted in the area. By the end of March, a state of general quiescence had been reached, and it was possible to collate the information that had been accumulated.

The extent of the spread of myxomatosis during the 1950–51 summer is shown in Fig. 21. No claim is made for the detailed accuracy of this map in the sparsely settled regions from which it was impracticable to obtain adequate data. Two things stand out. The first is the sheer size of the event recorded. The area over which the virus was dispersed—albeit patchily in some regions—measured nearly 1000 miles from south to north and 1100 miles from east to west. The occurrences of the disease at points along the coast of Victoria and New South Wales were the result of local successes of inoculations organized by the State authorities; attempts to establish myxomatosis during that summer in other regions, for example the northern tableland of New South Wales, were unsuccessful. Apart from these outliers, the map is a record of the natural spread of the infection.

The second thing that stands out is the dominating importance of the Murray–Darling river system in the distribution of myxomatosis activity. As we realized when we had learned something of the insect vectors, this was due to the fact that epizootic transmission had depended on summer mosquitoes that breed in persistent water, notably *Culex annulirostris*. The unusual rainfall distribution during 1950 was reflected in the great difference in the extent of disease activity in the southern and northern portions of the river system. In Victoria and southern New South Wales, 1950 was a year of

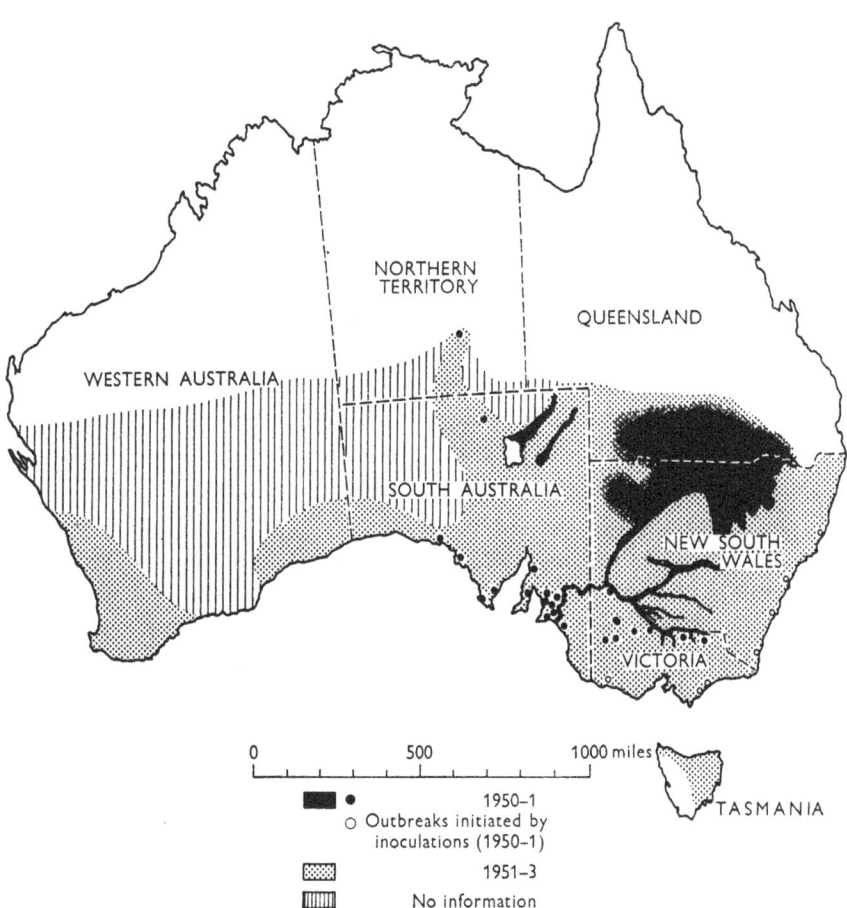

Fig. 21. Spread of myxomatosis in Australia during the three years following its escape, in 1950, from a test site near Albury. The isolated occurrences in 1950–51 ranged from single cases to outbreaks covering a few square miles. The two in central Australia, north and west of Lake Eyre, are included in the first season's dispersal although they were not reported until August and September 1951.

average or sub-average rainfall, and by mid-summer the countryside in general was very dry. In marked contrast, the northern parts of New South Wales and most of Queensland received record-breaking precipitations, resulting in flooding on an almost unprecedented scale in the northern part of the epizootic area. The last flood rains had occurred in November; but their effects in the form of abundant surface water lingered for at least a couple of months.

278

Along the main stream of the Murray, the epizootic was restricted to a very narrow strip, often only a few hundred yards wide (and usually corresponding closely to the *Eucalyptus camaldulensis* zone) except where irrigation complicated the picture. The same applied in general along the Murrumbidgee. On the Lachlan, further north, epizootics of high intensity occurred in a belt that extended two or three miles back from the river and its associated lagoons. Further north still, and particularly in the region lying east and north-east of Bourke on the Darling, which is watered by a complex system of tributary streams tapping the ranges of New England and south-eastern Queensland, disease activity lost its obvious association with the rivers and the epizootic became widespread and general.

Two other features of the spread of myxomatosis during the 1950–51 summer deserve mention. In general, the foothills of the Divide set its eastern limit, the disease rarely moving up the rivers beyond the points where they emerged on to the plains. Then there was evidence that the autumn decline in vector activity terminated the epizootic at a stage when chance effects were still apparent. Thus some eminently favourable areas, including substantial stretches of the Murray frontage, were by-passed or only feebly invaded by the disease, the active advance of which was rather suddenly curtailed in places.

The remarkable spread of myxomatosis that took place in the first three months of epizootic activity is quite inexplicable except on the assumption that the movement of infected insects was continuous and often extended. Evidence on the degree of mobility of mosquitoes and other potential myxomatosis vectors was to come later. During that first summer all one could point to was the indication of long-distance wind-carriage of infected insects provided by the isolated disease foci that developed across South Australia (see chapter 12). It became apparent that *Culex annulirostris* was the main vector associated with the intense epizootics along the frontage of the River Murray, and also that its ranging was strictly limited under the conditions ruling there at the time. However, the occurrence of several isolated short-lived outbreaks in places many miles away from the river suggested that some widely ranging insects had picked up the infection. In the recently flooded northern part of the epizootic area, it would be safe to say that *C. annulirostris* would not only

279

have been prevalent but would have been widely ranging (see notes on this species in chapter 12). One can assume also that the wet conditions would have favoured the breeding up of a variety of mosquitoes, which would have included one or two of the strongly flying rabbit-biting aedines such as *Ae. vittiger*. Simuliids would also have been prevalent.

We assumed at first that the main dispersal of the infection started from an area of intense epizootic activity on the frontage of the Murray near Balldale. However, no large-scale epizootic on the Murray was reported early enough to account satisfactorily for some of the outbreaks reported on the northern tributaries. The fact is that we do not know what was happening to the virus just prior to the first observed flare-up of the disease at Balldale: it seems almost certain that some unreported—perhaps even unobserved—myxomatosis activity was occurring somewhere. We do know, however, why disease activity in the Albury region did not become obvious until December: 1950 was a year in which conditions did not favour a spring emergence of *Anopheles annulipes*, and therefore transmission at an epizootic level had to wait until *C. annulirostris* appeared on the scene—a situation that has been observed in some subsequent seasons. Had our field trials in this region been carried out in either 1951 or 1952 the results would have been very different, and widespread myxomatosis activity would almost certainly have been evident by mid-November, and perhaps earlier.

MURRAY VALLEY ENCEPHALITIS AND MYXOMATOSIS

Early in February 1951 cases of severe human encephalitis occurred in Mildura, a fruit-growing town on the Murray River, and also in the Shepparton district of Victoria. The superficial coincidence in space and time of these cases of human encephalitis, which had not been recognized in these areas since 1925, and the first epizootics of myxomatosis in Australian wild rabbits, led to much speculation amongst medical men and the public concerning a possible relationship between the two diseases—a speculation reflected in comments in both the popular press and medical periodicals.

Careful analysis of the distribution of cases of encephalitis and of myxomatosis showed that these did not coincide, and the matter was greatly clarified by the recovery of Murray Valley encephalitis virus from the brain of a fatal human case in Victoria by French (Anonymous, 1951) and in South Australia by Miles, Fowler & Howes (1951). This virus was shown to be a member of the group B arboviruses. Tests were carried out, and their results widely promulgated, to show that it was quite a different agent from myxoma virus, although both were transmitted by mosquitoes.

In spite of these reassurances, however, public concern at the human pathogenic potential of myxomatosis was not settled until Mr R. G. (now Lord) Casey, then Minister in Charge of C.S.I.R.O., announced in the House of Representatives on 8 March 1951 that three Australian scientists, Sir Macfarlane Burnet, Sir Ian Clunies Ross (Chairman of C.S.I.R.O.) and Professor F. Fenner, had been inoculated with myxoma virus some months earlier, and had not been affected (*Hansard*, 1951). A dose of about 1000 ID50, considered to be the maximum dose likely to be acquired from a mosquito bite, had been inoculated intradermally. No local lesion or general signs developed, and no antibody to myxoma virus could be found three weeks later.

THE OVER-WINTERING AND SUBSEQUENT SPREAD OF MYXOMATOSIS

Epizootic activity had ceased by the end of March 1951. Within two or three months reports of the sightings of sick rabbits were being received, indicating that the infection had not died out. In a number of places localized outbreaks developed which, though of low intensity, produced fairly high mortalities. The records of isolated diseased rabbits and the local outbreaks practically all occurred outside the areas in which the summer epizootics had taken place, often well away from them. (There were, for instance, very few winter sightings of sick rabbits in Queensland.) The outbreaks tended to take one of two forms. Some were localized around swamps, hillside seepages, etc., suggesting that they had been initiated by infected mosquitoes 'homing' to water. Others, more diffuse as a rule, often showed a marked association with patches of thistles—mainly the saffron thistle (*Carthamus lanatus*) which forms large and dense stands in the grazing country west of the Divide.

In the late winter and early spring (August–September) a series of intense epizootics developed at points along the coast, and on offshore islands, of eastern Victoria. These would have coincided with the emergence of the brackish-water breeding *Ae. camptorhynchus* (see chapter 12), and incidentally have not been a feature of the seasonal myxomatosis picture in subsequent years.

While the events of the 1951 winter cannot be claimed to have revealed myxomatosis as an established enzootic disease in Australia, they certainly indicated that this outcome was likely. The pattern of winter outbreaks also indicated that, in addition to vectors capable of producing intense epizootics (with a local infection rate approaching 100%), mobile insects were likely to pick up the virus and establish disease foci many miles from the outbreak areas.

As a result of the spread of myxomatosis in the second and third summers or seasons (it is convenient to speak of the period from spring to autumn inclusive, when most observable activity takes place, as the myxomatosis 'season') the distribution of the virus became virtually coextensive with that of the rabbit in Australia. This does not mean that rabbit populations everywhere had experienced epizootics by mid-1953: in many regions widespread outbreaks did not occur until some years later. The natural spread of myxomatosis was assisted by State-organized inoculation campaigns, which started in 1951 and will be discussed later in this chapter.

No attempt has been made in Fig. 21 to distinguish between the second and third seasons' spread of myxomatosis, for in a small-scale map this would be impracticable and even misleading. In 1951–52 there was a greater increase in the area in which the disease was reported; but during 1952–53, when conditions favoured transmission over practically the whole of south-eastern Australia, the increase in areas experiencing intense disease activity was very much greater.

The extent and nature of the spread during the second and third seasons are exemplified by the observations made by the field team, working from Albury as a base. In 1950–51, as has already been mentioned, epizootics along the Murray and Murrumbidgee Rivers were confined to a very narrow belt, often only a few hundreds of yards wide, and the disease did not spread into the foothills. When epizootic activity started in 1951–52 it was much less restricted, extending many miles back from the rivers; and soon afterwards a huge area of undulating and hilly country became involved, extending right across the centre of Victoria and then northward, in New South Wales, over the foothills and western slopes of the Dividing Range, with many localized outbreaks beyond the margin of the epizootic zone.

The occurrences of the disease during the 1952 winter were similar to those observed in the previous year, but more numerous and extensive, preparing the way for the widespread and intense epizootic activity during the spring and 1952–53 summer. The differences between 1950–51 and the two succeeding seasons are readily explained in terms of vectors (see chapter 12). The activities of the adaptable, widely ranging *Anopheles annulipes* were suppressed in 1950–51, at any rate in the southern part of the epizootic area; but the rainfall distribution in the following years favoured this species, at times and in some places, to an exceptional degree. In addition, in the spring of 1952 highly mobile vectors such as *Austrosimulium furiosum* were very much in evidence, and probably assisted the widespread dispersal of the virus which occurred at that time and in the months that followed.

At the end of the third season, the authorities in New South Wales could report that myxomatosis had reached every part of the State (though occurrences on the tablelands had been patchy), and that the Western Division had been virtually cleared of rabbits. In Victoria the disease had been recorded over most of the State, and the first widespread epizootics had occurred in the Gippsland hills, the Mallee (the low-rainfall north-western portion of the State) and in the Western District—the last-mentioned comprising mostly late-autumn and winter outbreaks.

South Australia benefited almost as much as Victoria and New South Wales, though the epizootics in the low-rainfall regions were often associated with only moderate mortalities. It is interesting to recall that the areas in which the early field trials were carried out (Bull & Mules, 1944), and where the infection had shown no tendency to spread, were swept by myxomatosis in 1952, nine years later.

In Queensland, the main spread had occurred in 1950–51; and a westward extension of this early, widespread northern activity carried myxomatosis into the desert around and beyond Lake Eyre. This most interesting and unexpected feature of the spread of the virus took place in very sparsely settled country, and the pattern it followed remains unknown. As an outbreak of the disease occurred in the Northern Territory, south of Alice Springs, as early as September 1951, it seems reasonable to assume that long-distance vector movements were involved.

LAKE URANA: THE APPEARANCE OF ATTENUATED STRAINS

Lake Urana was one of the three Riverina study areas originally selected by Myers for intensive work on vectors. The advantages of the site for epidemiological investigations, and the study of the interaction between the myxoma virus and its rabbit host, were very soon recognized. The vector situation virtually guaranteed the occurrence of an epizootic every year; and the rabbit population, which was fairly well isolated, was large enough to ensure that there would be a significant number of survivors from any myxomatosis outbreak. Some of the most important and comprehensive field data bearing on the epidemiology of myxomatosis were to come from Lake Urana; and to insure continuity of the observations an area of sandhills towards the south end of the lake was leased and enclosed, and maintained under surveillance.

Urana is situated in the plains, about 70 miles north-east of Albury. The lake itself (see Pl. X, fig. 2) is a flat depression, some 18,000 acres in extent, in an area of internal drainage. It is flooded in wet years, and at other times provides high-quality grazing. The lake bed has a rim of sandhills, best developed at the southern end (chosen for the study area) where they carry stands of *Callitris* pine. As K. Myers (personal communication) was later to discover, Lake Urana is typical of the foci of rabbit infestation that occur throughout the inland plains. These foci are generally associated with local conditions preventing the run-off after rain getting away into the tributaries of the main rivers, thus favouring plant growth and persistence. They are also almost invariably associated with patches of sandhills, in which the permanent warrens are concentrated. Places of this kind provide refuge for rabbits during a run of dry years; and when numbers have built up in good seasons many will move out to recolonize the surrounding land.

The first intensive study at Lake Urana was undertaken between August 1951 and March 1953, and has been reported by Myers *et al.* (1954). The essential findings are presented graphically in Fig. 22. The rabbit population, harbouring mainly in the sandhills, was very

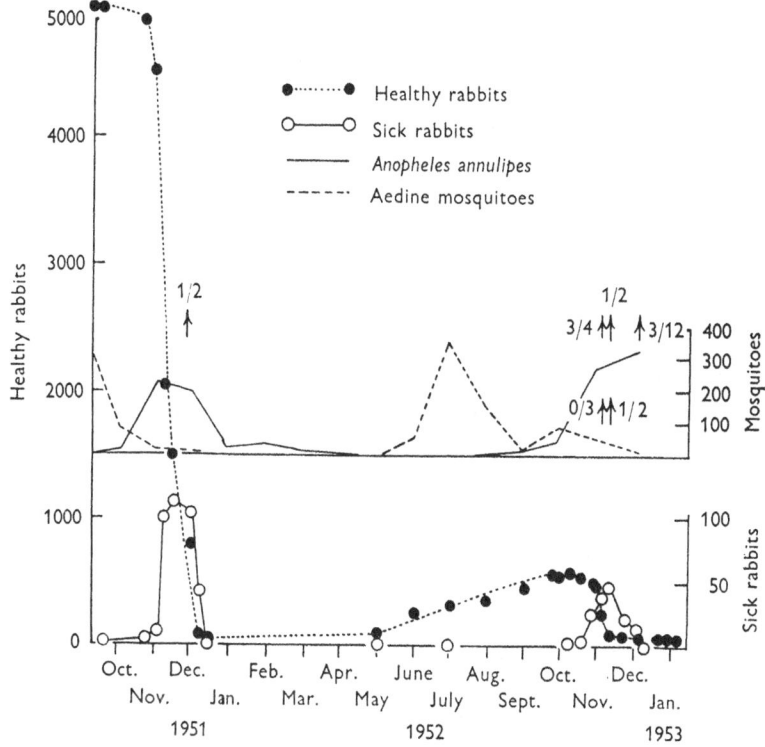

Fig. 22. Population counts of rabbits and adult mosquitoes at Lake Urana. The arrows indicate the occasions on which virus was isolated from batches of mosquitoes (*An. annulipes* in upper line, aedine mosquitoes in lower) and the proportion of isolations to batches tested (from Myers, Marshall & Fenner, 1954).

large as it had not suffered an epizootic during the explosive dispersal of myxomatosis in the previous summer. Two sections of the sandy lake margin were selected for routine counts which provided an index of the rabbit population and its changes. The counts were made just before dusk, on standard transects under uniform weather conditions, and we believe they were sufficiently accurate for the calculation of the case-mortality rates during the two epizootics of myxomatosis that were studied. Only the figures from the larger

285

of the two sections are presented in Fig. 22. Those from the other section were closely comparable for the first epizootic period, but a local landholder killed off the survivors, and thus rendered further counts and observations in that section impossible.

The population count during September 1951 averaged 5000. During that month and the next, some hundreds of rabbits were captured, inoculated with virulent myxoma virus (the standard laboratory strain) and released. The first natural cases of the disease were observed in October, and the epizootic rose to a peak in November, when there was a catastrophic fall in the rabbit population index from 5000 to 50 within a period of four weeks. Serological tests showed that most of these survivors (76 %) had escaped infection, only 24 % being recoveries. The case-mortality rate was therefore 99·8 %.

The greatly reduced population in the study area was augmented in June 1952 by an influx of rabbits from the lake bed, parts of which became flooded. Young rabbits started emerging from the warrens in July, and their numbers rose steeply in September and October. The population count stood at 550 in October, when 50 rabbits were caught, inoculated and released. Before the infection had time to spread from these animals myxomatosis, which had been observed to persist at low intensity in the area throughout the winter, flared up in a sharp epizootic. This second outbreak reduced the population count from 550 to 60; and it remained at that low level for the rest of the summer, despite the presence of large numbers of infected mosquitoes (see Fig. 22). Serological tests showed that all the survivors had been infected and recovered. After making allowance for the persistence in the population of some old immune rabbits from the epizootic of the year before, the case-mortality in 1952 was calculated to be 90 %.

The results of this study at Lake Urana provided the stimulus for much of the Australian work on myxomatosis that has been described in chapters 13 (virulence of the virus), 14 (genetic resistance of the host), and 10 (environmental effects). There is no doubt that the dramatic fall in case-mortality rate, from 2 survivors per 1000 to 100 survivors per 1000, was due to the presence of somewhat attenuated strains of virus which dominated the second outbreak, in spite of the liberation of some 50 rabbits which had been inoculated with the standard laboratory strain. The presence of this attenuated virus in turn permitted the survival of enough rabbits for effective selection for host resistance to occur, with results that became evident in the Urana rabbit population some years later (see Fig. 17).

INOCULATION CAMPAIGNS IN AUSTRALIA

It was apparent early in 1951 that myxomatosis would spread effectively in Australia; and as soon as the fears of the public that it would cause disease in man and his domestic animals had been dispelled, farmers' and graziers' organizations began pressing for the full exploitation of this new weapon of rabbit control. They urged the not unwilling State and Federal authorities to disseminate myxoma virus widely in the wild rabbit population. Some officially sponsored inoculations were carried out before the end of the 1950–51 summer, in regions to which the disease showed no signs of spreading naturally. Full-scale inoculation campaigns did not get under way, however, until the following spring; and their pattern varied according to the administrative organization of rabbit control in the different States.

Except for some material obtained from the Veterinary Research Institute in Melbourne for use in the earliest releases, and a few batches prepared at research institutions in Adelaide and Brisbane for use in South Australia and Queensland respectively, all the virus material used for inoculation in the field was prepared either by the Commonwealth Serum Laboratories in Melbourne or the Veterinary Research Station at Glenfield, New South Wales. The former is administered by the Commonwealth Department of Health; the latter is part of the New South Wales Department of Agriculture. Once standardized production had got under way at C.S.L. and Glenfield, failures in the field due to faulty material, which had marked one or two of the early inoculations, were never encountered.

The virus used for the Glenfield preparation was originally recovered from a naturally infected rabbit sent in from Dubbo in February 1951 for identification as a case of myxomatosis: it became widely known as the Glenfield strain. For the C.S.L. preparation, which was freeze-dried and sent out in sealed ampoules, the standard laboratory strain (see chapter 13) was used for ten years. In the spring of 1961 the Glenfield strain was substituted, at the request of the Victorian State authorities, because there was evidence that it was rather more virulent and more readily transmitted (see chapter 13). The distribution of dried virus from Glenfield was restricted to New South Wales, being made available to landholders direct on request, or through the various Pastures Protection Boards. The C.S.L. virus was distributed to all parts of Australia, as a result of bulk purchases by the authorities in the various States (other than New South Wales) sponsoring inoculation campaigns and orders received from individual landholders.

The degree to which their inoculation campaigns were organized, and the persistence with which they were carried on after the first year or two, varied greatly from State to State. The Government of Tasmania at first legislated against the introduction and dissemination of the myxoma virus. An illegal introduction was made, however, and some myxomatosis activity was recorded in the northern part of the State during the 1951–52 season. The official attitude was reversed shortly after and an inoculation campaign was organized during the summer of 1952–53. The Department of Agriculture set up a number of centres, well distributed over the State, to which landholders were invited to bring rabbits for inoculation. In some districts, where rural employees made good money trapping rabbits in the winter, pressure was put on landholders not to have rabbits inoculated. Probably in part because of this, the total number of animals treated at the centres was not large; but the campaign (which was not repeated in subsequent years) resulted in myxomatosis becoming widely established throughout the settled parts of the island.

South Australia—a State which, at that time, lacked any effective centralized rabbit control authority—also staged only one organized inoculation campaign. It has been described by Lines (1952). In the spring of 1951, the South Australian Department of Agriculture advertised dates on which landholders could bring rabbits in to be inoculated at some 50 centres distributed across the agricultural belt, from Eyre Peninsula in the west to Mount Gambier in the south-east. Some trouble was experienced with the first batches of (locally prepared) virus material, and the western inoculation centres were re-opened in December and January. Altogether about 20,000 rabbits were infected, some being brought in from 200 miles away.

The campaigns in Queensland have been reported on by Brebner (Anonymous, 1959). The initiative came from the Co-ordinating Board, a body responsible to the Minister for Public Lands for vermin and noxious weeds control. For the first (1951–52) season, virus material was prepared at the Sherwood Laboratory of the Lands Department's Biological Section; thereafter ampoules were obtained from the Commonwealth Serum Laboratories. The Department of Agriculture and Stock made its field staff available to inoculate rabbits brought in to advertised points throughout the infested area. When the never very satisfactory landholder response fell away after three or four years, Stock Inspectors only obtained virus (from Sherwood, where a supply was held) if and when they received requests to inoculate. In recent years less than a score of ampoules have been sent out annually.

In New South Wales, rabbit control is the responsibility of local bodies known as Pastures Protection Boards, which differ greatly in the energy and efficacy with which they tackle the problem. The distribution of virus from Glenfield, largely through the P.P. Boards, is still proceeding at the time of writing. Some of the recipient boards have merely acted as retailers of the virus, passing vials on to landholders who have requested them. Some have inoculated rabbits brought in by landholders on advertised dates; and at any rate one Board took the initiative of sending a rabbit inspector round the district in a specially equipped vehicle, not only inoculating rabbits for landholders but catching, infecting, and releasing rabbits at strategic points. Up to and including the 1958–59 season, Glenfield was sending out between 3400 and 7800 vials of virus material annually, each vial containing twenty-five doses, in response to requests from P.P. Boards and landholders. Since then the amount of virus distributed has dropped considerably (though an abnormally large number of requests were received in 1961–62), but Glenfield is still meeting some hundreds of requests and sending out between 1500 and 2000 vials annually. There is no means of ascertaining the average number of rabbits infected and liberated in the field per vial sent out, although we know it is less, and suspect it is much less, than the number of dose units contained. All one can safely say is that the distribution of virus results in the liberation of some thousands of inoculated rabbits each year, fairly well distributed over the State and in a pattern bearing some relation to population density, and that in the early and mid-1950's the rabbits inoculated and released annually in New South Wales were probably numbered in tens of thousands.

The two States in which inoculation campaigns have been organized most intensively and on the largest scale are Western Australia and Victoria. In Western Australia, responsibility for rabbit control is divided between local Vermin Boards and a body known as the Agriculture Protection Board, which has the authority to formulate control policy. The Chief Vermin Control Officer in the Department of Agriculture is, *ex officio*, the Board's Chief Executive Officer. The organization of the inoculation campaign in Western Australia has been described by Calaby *et al.* (1960). In the early spring of 1951, and during the next two years, a total of twenty-one infection centres were set up forming a strategic network over that portion of the State in which the rabbit is an economic problem. The sites for the centres were carefully selected with regard to rabbit abundance and potential vector breeding grounds. Most of the centres were near streams, a few being in dry areas where the rabbits were known to be regularly infested with stickfast fleas (*Echidnophaga* spp.). Each centre was in charge of an attendant who, until ampouled virus was readily obtainable, maintained infection in caged animals, and caught and inoculated rabbits. At first the infected rabbits were exposed in cages on the banks of the streams; later they were released in warrens. The centres were usually closed when it became apparent that there was little chance of a local outbreak of myxomatosis developing, or when the infection was judged to be well established in the surrounding area. All the centres were closed down by the middle of 1954, a total of over 27,000 rabbits having been infected and released during the period of their operations.

In addition to the output of the inoculation centres, and mainly after their closure, many thousands of rabbits were inoculated by mobile units and in drives conducted by the more active local Vermin Boards. From 1952 on ampoules of C.S.L. virus were in adequate supply to retail to individual landholders; but the number distributed fell away sharply after two or three years. Both the infection centres and the mobile units operated the year round, but the latter endeavoured to take advantage of any unseasonable wet weather.

No personnel were employed full time on myxomatosis dissemination after 1958, when the Agriculture Protection Board's main effort was concentrated on poisoning programmes. Inoculations at a reduced level continued, however, and are still continuing. They have been handled in the main by the A.P.B. field staff—the Vermin Control Officers. In some areas 'inoculation days' have been advertised; but this has only resulted in worthwhile numbers of rabbits being treated in the Geraldton district, which has benefited more from myxomatosis than any other part of the State (Tomlinson & Gooding, in litt.). The Glenfield strain of the virus, as prepared and marketed by the Commonwealth Serum Laboratories, has been used since 1961.

The Chief Vermin Control Officer, Mr A. R. Tomlinson, who was largely responsible for planning the inoculation campaign, knowing his State and its climate did not expect to see speedy results on a large scale. His intention, as he put it, was to get the virus 'seeded' throughout the rabbit-infested south-west, so that the infection would be present in any region when conditions turned up favouring the development of an epizootic. In this he can claim to have been successful. Naturally enough, the inoculation campaign had the moral and political support of landholders at its inception; but when widespread and intense outbreaks failed to develop, and particularly when men who had obtained virus for use on their properties saw no benefits from their inoculations, criticism began to mount and many expressed the opinion that the continuation of the campaign was a waste of public money. As we shall see, some useful outbreaks of myxomatosis occurred during the period when inoculations were taking place, but they were obviously associated with exceptional and unseasonable rain rather than with the efforts of the infection centres and the mobile units.

The credit for having given the most enthusiastic and prolonged attention to field inoculations must go to the authorities in Victoria, and particularly to Mr G. W. Douglas of the Vermin and Noxious Weeds Destruction Board in the Department of Crown Lands and Survey, which has the responsibility of dealing with vermin control problems in that State. Douglas maintains close liaison with the Commonwealth Serum Laboratories, and during the four years from 1957–61 he and his staff were responsible for providing C.S.L. with myxoma lesion material from which the freeze-dried and ampouled product was prepared. Prior to 1957 the Walter and Eliza Hall Institute for Medical Research had performed this task; and in 1961 the C.S.L. acquired adequately secure facilities for maintaining infected rabbits on their premises. Demonstration experiments are carried out regularly in the field to assess the mortality caused by the virus in samples of regional rabbit populations, and public interest in inoculations is stimulated by the issue of pamphlets and other forms of publicity (see Douglas, 1958, 1962).

TABLE 50. *Number of rabbits inoculated each season by Lands Inspectors in Victoria, to nearest thousand (from Douglas, in litt.)*

1950–51	(Estimated 5,000)
1951–52	40,000
1952–53	12,000
1953–54	30,000
1954–55	25,000
1955–56	15,000
1956–57	26,000
1957–58	30,000
1958–59	15,000
1959–60	14,000
1960–61	8,000
1961–62	28,000*
1962–63	18,000

* Well-publicized switch from the Standard to the Glenfield strain of the virus took place in spring of 1961.

Most of the Victorian inoculations are carried out by the Lands Department's 140 inspectors, each of whom has up to half a dozen centres in his district which he attends to inoculate rabbits on advertised dates. Records have been kept of the rabbits inoculated each year by inspectors, and the totals (in round figures) are set out in Table 50. In addition to the efforts of the Lands Inspectors, some hundreds of ampoules of virus are purchased each year direct from C.S.L. by Victorian landholders.

The annual campaign is planned to fit in with the seasonal pattern of epizootics in the different parts of the State and the indications of probable vector activity. In the north, along the Murray and its tributaries, inoculations are arranged in the spring. In the central and southern parts of Victoria inoculations are usually carried out twice, first before Christmas (November or December) with the later effort, which is regarded as more important, being made in January or February. The seasonal conditions are watched, and arrangements made for the modification of the normal timetable if this seems desirable. Thus in a dry year, when *Anopheles annulipes* activity will be suppressed, inoculations in the North will be postponed until late November or December, when *Culex annulirostris* can be expected to appear in numbers. On the other hand in a wet year, inoculations may be advanced in this region to the beginning of September or even August. To exploit conditions that promised to be exceptionally favourable for transmission, a mobile unit has been put into the field on several occasions. This unit works through a district, inoculating and releasing rabbits which are caught by spotlighting at night.

At the time of writing, inoculation campaigns have been carried out in Victoria annually for twelve years, and it is the intention of the authorities to continue them—although they are paying increasing attention to regional poisoning campaigns—for as long as their tests show that the strains of virus available will produce worthwhile mortalities in samples of the wild rabbit population (G. W. Douglas, personal communication).

THE VALUE OF INOCULATION CAMPAIGNS

The natural spread of myxomatosis during the first three summers after the 'escape' of the virus demonstrated its potentiality for dispersal under conditions that favoured the activity of mobile vectors, and with adequate numbers of susceptible rabbits. Surveillance was unavoidably so superficial, with the disease dispersed over nearly a million square miles, that only on rare occasions was there even circumstantial evidence of the relationship between an inoculation campaign and an outbreak of myxomatosis. Some of the early localized outbreaks in regions where myxomatosis had not occurred previously, for example in the coastal belt of New South Wales and Victoria in 1951, were undoubtedly initiated by inoculation; but after the disease became enzootic over the greater part of the rabbit-infested area of Australia it was impossible to distinguish with certainty between outbreaks due to enzootic and introduced strains. Tests of virus strains recovered from the field after 1955, when the first reasonably large samples were obtained, showed that very few were as virulent as the Glenfield and standard laboratory strains which were used in inoculation campaigns.

An attempt was made to assess the value of inoculation in the summer of 1954–55, at Lake Urana and Merricumbene (the C.S.I.R.O. study area in the Araluen Valley), both in New South Wales (Fenner, Poole, Marshall & Dyce, 1957). The experiments were originally prompted by what was thought at that time to be the greater stability of high virulence exhibited by the Lausanne strain,

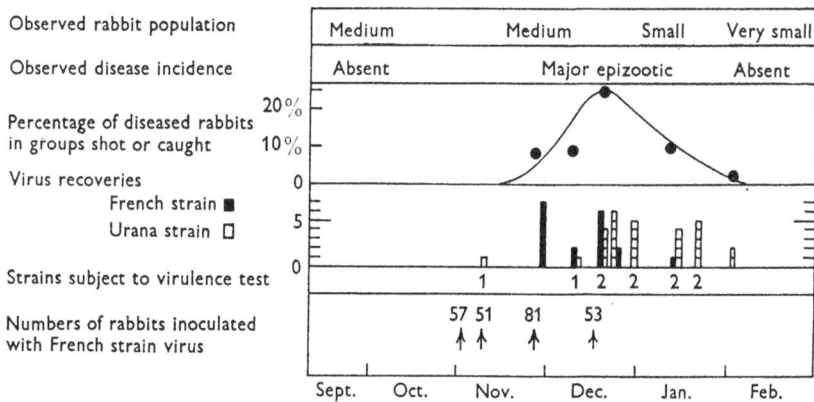

Fig. 23. The epizootic which followed the introduction of the French strain of myxoma virus into the wild rabbit population at Lake Urana—September 1954 to February 1955. Graphs show the incidence of the disease, the size of the rabbit population, the type of virus as determined by the intradermal screening test, and the schedule of inoculation of the French strain of myxoma virus (from Fenner, Poole, Marshall & Dyce, 1957).

which had been introduced into France by Delille in 1952; subsequent experience of myxomatosis in Europe showed that many attenuated variants of the Lausanne strain appeared after 1954. The results of the two experiments are illustrated in Figs. 23 and 24.

Inoculation with a French strain of Grade I virulence was carried out on a large scale at Lake Urana, and on a smaller but still fairly substantial scale at Merricumbene. The capture and inoculation of rabbits were done at a time when vector activity was expected to produce an epizootic; but as no outbreak developed at Merricumbene until late in the summer, the period covered by inoculations at that site was prolonged to over four months. In both places cases of myxomatosis were found amongst rabbits shot or captured during the period of the inoculations, and these were shown to be due to an

294

enzootic Australian strain of Grade III virulence. This finding altered the character of the experiments, and allowed them to be used as a comparison of the relative performance of a Grade I and a Grade III strain under epizootic conditions.

The flat and raised skin lesions produced by the enzootic Australian and the French strain respectively provided the basis of screening tests, and the distribution of cases due to each strain of virus was determined. The early part of the epizootic at Lake Urana was

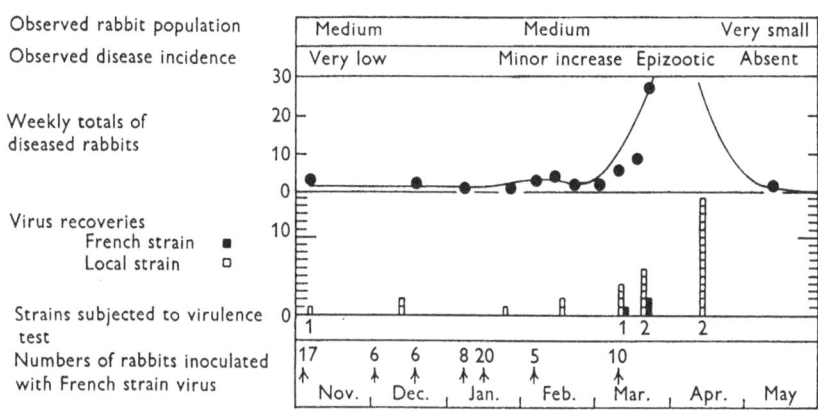

Fig. 24. The epizootic which followed the introduction of the French strain of myxoma virus into the wild rabbit population at Merricumbene—November 1954 to May 1955. Graphs show the incidence of the disease, the size of the rabbit population, the type of virus as determined by the intradermal screening test, and the schedule of inoculation of the French strain of myxoma virus (from Fenner *et al.* 1957).

dominated by the French strain, which probably killed about 70 % of the initial population of susceptible rabbits. In the later stages of the epizootic it was completely replaced by the attenuated strain, which was the only virus surviving through the winter of 1955 to give rise to an epizootic in the succeeding spring. The pattern at Merricumbene was similar in that the French virus was only re-covered in the early stages of the sharp epizootic that developed in March–April 1955. It did not achieve dominance even at this time, however, and there is no reason to believe that it could have been responsible for more than a small minority of the total number of deaths.

It is clear from the records of these experiments that under con-ditions of epizootic transmission the slightly attenuated strains, such as the strains of Grade III virulence that emerged in the field in Australia, have a selective advantage over the fully virulent virus, which can be postulated on the basis of the survival times of infective host animals (Fenner *et al.* 1956). For any area in which slightly attenuated enzootic strains of the virus are present, the practical implications of the results of the Lake Urana and Merricumbene experiments are unequivocal. The scale on which rabbits were captured and inoculated at Merricumbene would be within the capacity of an ordinarily energetic landholder (even though the inoculations had to be extended over a longer period than most men would be willing to contemplate) and the resulting benefit in the form of an increased kill was slight, and probably quite insignificant. At Lake Urana the indications from observed vector activity of the approaching epizootic were much more definite, and it was possible to time the inoculations with some precision in relation to the developing outbreak. The liberation of the 242 rabbits inoculated with the French strain started three to four weeks before the epizootic was observed to be well under way, with the final batch released just prior to its peak. The effectiveness of the timing of the inoculations, and the large number of rabbits treated, outweighed the advantage possessed by the enzootic strain occurring naturally in the area and resulted in a substantial majority of the local rabbit population being infected by the French virus. The result, of course, was an increase in the total mortality. This increase although not large in itself (it could hardly have exceeded 10 %, and was probably less than that) would represent a reduction of a much higher order in the number of potential survivors with some degree of innate resistance. No ordinary landholder could be expected to make an effort comparable with the one staged at Lake Urana, nor would he be armed with the entomological information required for such accurate timing of the inoculations.

What overall assessment can be made of the value of inoculations in Australia, with enzootic myxomatosis likely to be present in any local population of rabbits? In 1955 or 1956 we would have been prepared to make the unpopular statement that the organizational effort involved was not justified. The implication of the Merricumbene trial was clear; and we knew that the great majority of landholders would inoculate far fewer rabbits than were released in that area. The testing of virus samples recovered in the field had shown that the virulent virus used in the inoculation campaigns could rarely have established itself or spread to any extent. During 1954–55, and again in the following season, many effective outbreaks of myxomatosis occurred, for example in New South Wales, with so little relation to the inoculation effort that even the field inspectors (who would naturally have liked to claim success from their inoculations) reported them as having developed from locally persisting infection. There seemed little doubt that the great majority of inoculations represented wasted effort; and it seemed sensible to conclude that the time of those involved in the field campaigns would have been spent more profitably on other aspects of rabbit control, especially the organizing of poisoning drives in those areas where there were a substantial number of survivors from myxomatosis outbreaks.

In recent years we have been forced to reconsider this early verdict. The cost and effort involved in the capture and inoculation of a score or so of rabbits by a landholder are small. Even though the effort might be wasted in a season when vector activity was generally at a high level, resulting in widespread epizootics, in other seasons there would always be the chance of the infection being able to exploit conditions that were only locally favourable. No one really knows the pattern to which the infection shrinks, so to speak, in an unfavourable year. However, we do know that a substantial part of the total benefit derived from myxomatosis is due to the summation of many small, localized and often seemingly insignificant outbreaks. Furthermore, as Sobey (1960) has emphasized, introductions of virulent virus represent the only way in which man can hope to affect the epizootic situation and do something to defer the development of genetic resistance in the wild rabbit population. Then there is the possibility discussed in chapter 18, that is that dependence on mosquito transmission might lead to the selection of more virulent strains in the field to balance the progressive increase in genetic resistance in the rabbit population, at least until no further changes in the virus were possible. To facilitate this evolutionary process, it would obviously be desirable to ensure that virulent strains were always present in the field, even though most of the virulent virus would tend to be short-lived under today's conditions.

MYXOMATOSIS IN AUSTRALIA DURING THE DECADE 1954-63

As we have seen, by the end of the 1952-53 season myxomatosis, by natural spread and with man's assistance, had reached virtually every part of the continent in which rabbits are found. Although it probably took another season or two for the infection to become established in the last few 'clear' areas, the year 1954 marks the beginning of what might be termed the stabilized situation in which overt myxomatosis activity has waxed and waned in response to changes in vector activity and to the fluctuating numbers of susceptible individuals in the wild rabbit population. In any Land Inspector's district in Victoria, or P.P. Board district in New South Wales—indeed in any rabbit-infested region in Australia—the aggregate effects of myxomatosis are now derived from a patchwork pattern of localized outbreaks, varying in intensity and spatial distribution from year to year. Periodically, at intervals varying widely from one region to another, conditions exceptionally favourable for transmission will occur, and with them a widespread epizootic bringing rabbit numbers down throughout the area.

For five or six years after the experimental releases of the myxoma virus in the Murray Valley, and its escape from one of the trial sites, there were a substantial number of trained observers in the field involved in vector studies and extensive surveys, and they were the source of much valuable information on myxomatosis activity. During those early years also meetings of Federal and State authorities were convened annually, and reports of the season's myxomatosis performance in each State were prepared for these conferences. After 1956, when most of the intensive field studies had been terminated and the novelty of myxomatosis had worn off, the collection and collation of field data on an Australia-wide basis virtually ceased. Thus the effects of myxomatosis in recent years has to be assessed on information that is far from adequate, consisting in the main of observations and reports of a superficial nature made by men who were rarely in a position to do more than record the presence or absence of disease activity and gross changes in local rabbit numbers. It will be appreciated, therefore, that we can only describe the recent history of myxomatosis in Australia in very general terms.

It is hardly necessary to state that myxomatosis has brought, and is still bringing, almost incalculable benefit to Australia. It has made a substantial contribution to the solution of the country's rabbit problem; and although the problem that remains is still challenging, it now appears to be manageable. A natural first step in defining the effects of myxomatosis is to attempt an estimate of the average overall reduction in rabbit numbers. Such an estimate is misleading, to the extent that rabbits in 1950 and during the three or four preceding years reached something like their greatest abundance over most of southern Australia, and they would presumably have suffered some decline in numbers without the help of the virus. The situation between the two World Wars would provide a more satisfactory basis for comparison, but it is impossible for the present generation to conjure up an accurate picture of rabbit infestations during that period.

Douglas has been in a position to observe the rabbit situation in Victoria both in detail and in perspective, and he estimates (personal communication) that the population over the State as a whole is now fluctuating around 20 % of the pre-myxomatosis level. The history and performance of the disease vary greatly from region to region. In the Mallee (the north-west corner of the State) and the Wimmera (adjacent and to the south), outbreaks occur in most years; but intense, widespread epizootics can only be expected at intervals of about four years and organized poisoning on a substantial scale has been necessary to deal with the infestations that have built up. The frequent outbreaks on the plains of the Western District, lying between the Wimmera and the coast, are usually of low intensity, presumably because suitable vectors are deficient. Over the rest of the State useful outbreaks can be relied on to occur in most years—say in three or four years out of five. The most consistently benefited regions are the hill country of south Gippsland (east of Melbourne), the Central Highlands, and parts of the hilly North-East. Even in these areas, however, poisoning has to be carried out from time to time to achieve the degree of control required by the authorities.

Broadly speaking, and making allowance for differences in climate, the Victorian picture is repeated in the other States. The overall level of rabbit infestation has been very greatly reduced, and the once dreaded phenomenon of the 'rabbit drought', which so often occurred as the final phase of a sequence of good seasons, now appears to be a thing of the past. There are regions in every State where vector activity of the kind needed for the development of a widespread epizootic can only be expected during seasons of exceptionally high rainfall such as might occur one year in five, or even less frequently; and in such areas rabbits may breed up to quite a high density when the rainfall is adequate to promote plant growth, but not sufficient to cause local flooding and the creation of mosquito breeding grounds. Every State can point to certain areas within its borders in which rabbits have never managed to stage a recovery after the first wholesale myxomatosis kills; but most States also have persistent 'trouble spots' in which, despite periodic myxomatosis activity, rabbits constantly threaten to attain dangerous levels, and sometimes succeed in doing so. It is often difficult to advance a good ecological explanation for these extreme situations (that is those areas in which rabbits have virtually ceased to be a problem, and those in which myxomatosis has not prevented the frequent attainment of high densities) but local variations in the intensity of predation are probably involved.

Apart from the ability of predators to keep rabbit numbers down once they have been reduced to a low level by other factors (see chapter 4), A. L. Dyce and G. W. Douglas have both pointed out (personal communication) that predation can, in effect, augment the virulence of locally prevalent attenuated virus strains. The great majority of rabbits which would normally recover from infection by strains of Grade III virulence pass through a stage of comparative helplessness when they fall easy victims to predators, which congregate in areas of myxomatosis activity. At times and in some places myiasis—fly strike—may also intervene, to reduce further the chances of recovery. When the trend of increasing host resistance and the prevalence of somewhat attenuated strains of the virus first became apparent, it was commonly believed that their combined effects would become strikingly obvious within a few years. This has not proved to be the case; and while the arrest of the trend of progressive attentuation (see chapter 13) provides a partial explanation, there can be no doubt that predation has been a buffering factor of real significance.

An overall reduction in the myxomatosis case-mortality rate must have occurred since the first drop recorded at Lake Urana, mentioned earlier in this chapter, which was the result of virus attenuation before host resistance could have contributed its effect. However, without data from field studies, which it was impracticable to carry out on an adequate scale, no clear evidence for it can be produced. A reduction in vector activity resulting in lower infection rates would have the same effect on rabbit population levels, and seasonal conditions since 1956 have on the whole been less favourable for vector breeding and activity than they were in the preceding years. Some recent epizootics, for example in Victoria, have appeared to be nearly as effective as those which marked the early years of myxomatosis, making it clear that the virus is far from being a spent force as an agent of rabbit control (Douglas, *in litt.*). Had a thorough check of the post-epizootic situation been carried out, however, there is little doubt that it would have revealed more recovered animals than had survived the intense early outbreaks.

The catastrophic effect of the first three myxomatosis seasons on the Australian rabbit population has already been mentioned. The 1954–55 season was if anything more favourable for vector activity than was 1952–53, and at its conclusion rabbit numbers were at the lowest level ever recorded. Since then they have increased significantly, though in a varying and often rather insidious manner. Three factors have been involved in the changing situation: (*a*) an enhanced recovery rate, which has been mitigated by (*b*) predation, and probably augmented by (*c*) a generally lower infection rate. On the data available, it is impossible to attempt a quantitative expression of any of them.

There remain some points in the myxomatosis picture in States other than Victoria which deserve special mention. Western Australia is on the whole a dry State. Although the main rabbit problem is concentrated in the climatically favoured south-western portion, the rainfall there has a strongly marked winter incidence, which militates against intensive vector activity. As a result, the level of myxomatosis activity in Western Australia has been, in general, lower than in the eastern States. While the disease has helped materially to keep rabbit numbers down, it has left a problem which causes the authorities some anxiety and has necessitated the regular organization of poisoning drives.

The infection appears to be enzootic throughout the rabbit population of Western Australia. It builds up to epizootic proportions in a sporadic and unpredictable manner, most of the outbreaks being highly localized and developing in response to unusual weather conditions. Widespread and intense epizootics of the type that have occurred commonly in the east are almost unknown. One occurred in the autumn of 1955 (when rabbits were already reduced to low numbers, as a result of a couple of poor seasons and an intense poisoning campaign) over an area about 120 miles long and up to 60 miles wide, following a swathe of heavy cyclonic rain that crossed the south-west in mid-February.

An exception to this general picture is provided by the area around Geraldton, about 250 miles north of Perth, which was once among the worst rabbit infested regions of the State. One of the earliest intense outbreaks was recorded there in 1951–52, covering a substantial area. Myxomatosis has been 'working' there on and off ever since, assisted by re-introductions of the virus, and the rabbit population has never managed to stage a significant recovery. As was mentioned earlier in this chapter, because of the obvious benefit that has accrued from myxomatosis, landholders in the Geraldton area have had an enthusiasm for inoculation which has not been shown elsewhere.

An interesting situation has been reported (but not investigated at first hand) on the Nullarbor Plain, an unsettled tract extending towards the South Australian border and crossed by the transcontinental railway. The Nullarbor has become an important commercial trapping area since rabbit numbers in more accessible and settled country have been reduced by myxomatosis and other factors. The infection is enzootic in the area and it flares up into a patchwork of outbreaks following storm rains. Disease activity forces the trappers to shift their operations from one part of the Plain to another, but the rabbit population is apparently also continually re-adjusting itself, and trappers claim they can get a paying catch if they return to a trapped area after an interval of two or three months.

Very little can be said about myxomatosis in South Australia, as reports from that State have been very meagre. A feature of the situation during the years when observers were in the field was the frequency with which local rabbit populations were reported to recover within twelve months from the effects of high-grade epizootics, suggesting that reproduction might often be augmented by immigration of rabbits into the affected areas. (The probability that a seasonal movement of rabbits occurred to a significant degree in South Australia, and in adjacent portions of Victoria with a similar climate, was discussed in chapter 4.) Since the advent of myxomatosis, rabbit numbers have remained relatively very low in the drier parts of the State, and this must be attributed primarily to the virus.

It is also impossible to say much about Queensland, owing to the paucity of available information. K. Myers, whose current programme involves some work in that State, reports (personal communication) that good summer rains are by no means always followed by epizootics, the indications being that myxomatosis activity is especially favoured by the much less reliable early (spring) rains. Rabbits have built up, albeit patchily, to quite high numbers in parts of southern Queensland.

New South Wales has experienced a significant overall increase in rabbit numbers since 1955, the trough year; but as conditions affecting the existence of rabbits and of water-breeding insects vary enormously from one part of the State to another, a rather complex situation prevails. Despite the occurrence, beyond the River Darling, of some concentrations of rabbits sufficient to attract the attention of trappers operating mobile freezers, the reduction from pre-myxomatosis levels has been as marked and as persistent in the dry far west as in almost any part of the State. K. Myers (personal communication), who has been sampling rabbit populations in the Western Division during the past few years, attributes the failure of the rabbit to stage a general come-back there in no small part to the heavy incidence of predation, particularly by foxes—and this would probably apply to the drier parts of South Australia. He reports that in the Riverina, where so much of his early field work was carried out, the overall reduction in rabbits seems to be significantly greater than the 80 % estimated by Douglas for Victoria. The once heavily infested river-frontage zone, which was swept by the intense early epizootics, remains virtually free of rabbits; and they have not re-infested, to any significant extent, the open plains away from their specially favoured habitats and refuges. In his experience, the rabbit population least affected by myxomatosis is to be found in the alpine grasslands of the Snowy Mountains area.

It has been mentioned that in every State there are regions in which widespread and effective myxomatosis outbreaks can only be expected to occur at fairly long intervals. The northern tableland of New South Wales (New England) is an example. Despite the fact that it is an area enjoying good rainfall, by Australian standards, it needs an exceptional amount and distribution of rain to ensure adequate vector activity over the region as a whole. E. J. Waterhouse happened to start his New England vector studies in one of these exceptional years, and was thus able to pinpoint their essential features. After analysing the local meteorological records, he concluded (personal communication) that a season of general and intense myxomatosis activity could only be expected to occur once in ten to fifteen years in that region. In more normal seasons outbreaks would be localized and mostly of low intensity.

A broad picture of myxomatosis fluctuations in New South Wales can be derived from the special reports prepared each month by Pastures Protection Boards. These reports were requested by a committee set up by the Minister for Agriculture in 1958 (to advise on rabbit control policy) when its members realized that there were no arrangements for recording the incidence of outbreaks. The only information which the committee considered it realistic to expect all boards to supply was the number of properties in their districts on which myxomatosis was known to have occurred during the month in question. There are in all 59 P.P. Boards in New South Wales. Most board districts contain between 1000 and 2000 properties, the western boards having a few hundred only, and some of the coastal ones well over 2000.

Mr B. V. Fennessy, a member of the Rabbit Control Advisory Committee, has collated the information given in the P.P. Board reports and has made the results available to us. The consolidated figures for 1961–62 (favourable for myxomatosis) and 1962–63 (unfavourable) are presented graphically in Fig. 25. 'More than 80 properties' and '20–80 properties' were selected somewhat arbitrarily as indices of widespread (probably intense) and moderate disease activity. Three points emerge from the P.P. Board data: (a) myxomatosis is active somewhere in the State during every month of every year, but (*b*) the majority of outbreaks occur in mid to late summer; (*c*) the aggregated area of outbreaks, in most years, represents a very small fraction indeed of the total area of the State.

In Tasmania myxomatosis was slow to gain momentum; and for nearly two years after the inoculation effort disease activity, though widespread, was unspectacular. During 1953–54 only two grade 1 outbreaks were recorded, each being restricted to a few

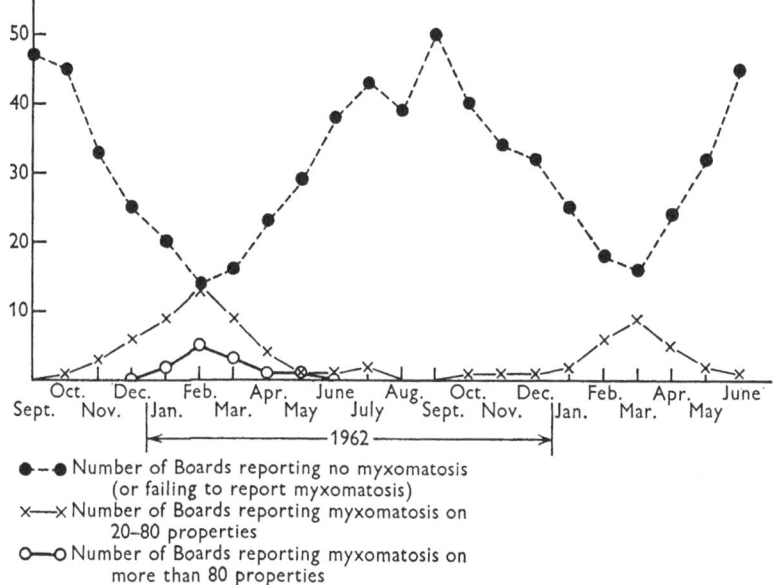

●--● Number of Boards reporting no myxomatosis (or failing to report myxomatosis)
×—× Number of Boards reporting myxomatosis on 20–80 properties
○—○ Number of Boards reporting myxomatosis on more than 80 properties

Fig. 25. Myxomatosis activity in New South Wales during the 1961–62 and 1962–63 seasons as revealed by monthly reports of Pastures Protection Boards (data from Fennessy, *in litt.*). Number of boards reporting myxomatosis on less than twenty properties not shown.

hundred acres. The first real impact of the disease on the rabbit population was felt in 1954–55, when effective kills were much more common. Myxomatosis in Tasmania has always tended to remain more localized than it has on the mainland. Winter activity and the limitation of outbreaks by netting fences have been reported on several occasions, suggesting transmission by ectoparasite exchange. The vector situation has never been investigated in the island State.

Tasmania pioneered the large-scale use of sodium fluoracetate ('1080') for rabbit control, and in that State the myxoma virus has had to compete with poison both for rabbit victims and for public acclaim. The poisoning campaign was pressed with great vigour, but it naturally took two or three years before even the more serious local infestations had all been dealt with. Myxomatosis was introduced before this had been achieved, and during the 1954–55 season it spread effectively and often spectacularly in dense local rabbit populations for the first and last time. There is a difference of opinion on the relative importance of poison and the virus in the amelioration of what was a very serious situation. The vermin-control authorities while recognizing the value of myxomatosis in helping to reduce rabbit numbers, give most credit to the direct-control effort; Tasmanian landholders, on the other hand, extol the benefit they derived from myxomatosis—possibly recalling the timely development of an outbreak which saved them from the necessity of implementing official instructions to poison.

This point has only been made to draw attention to the fact that myxomatosis can, in some circumstances, complicate the administration of rabbit control. If properly carried out, '1080' poisoning can be virtually guaranteed to reduce a rabbit population by over 90 %; and the effort involved in dealing with an infestation of one rabbit per acre is very little less than for a population five times as dense— and human nature being what it is, the job is less likely to be done with adequate attention to detail when it is not so obviously necessary. If inspectors' instructions to poison are set aside until outbreaks of myxomatosis have run their course, the result can be a serious dislocation of a district drive and a residual rabbit population which should, in wisdom, be dealt with, but which usually has to be left to breed up in the interest of good landholder–inspector relations. This problem, which stems from the facts that myxomatosis cannot be relied on now to reduce a rabbit population by more than 75 % and that the cost and labour of poisoning is not proportional to rabbit density, is by no means confined to Tasmania.

ECOLOGICAL EFFECTS OF MYXOMATOSIS IN AUSTRALIA

Although the great overall reduction in Australia's rabbit population as a result of myxomatosis must be classed as an ecological event of the first magnitude, we have practically no data of the kind that would enable us to assess its effects with precision. Thus we have no evidence of an increase in native marsupial herbivores which had suffered severely from competition with rabbits and domestic stock. The rabbit's main competitors, sheep and cattle, have certainly benefited greatly from the indirect effects of myxomatosis: it has been possible to increase stocking rates in many if not most areas; and in some places (e.g. parts of Tasmania) investment in improved pastures has become worth while, which it was not in the days of high rabbit numbers.

Immediately after the widespread eclipse of the rabbit during the initial spread of myxomatosis, there was some evidence of increased attacks by foxes on poultry and lambs; but reports of trouble from foxes are no more common today than they were before the advent of myxomatosis. It is reasonable to conclude that the status and reproductive success of some raptors which depended very largely on rabbits to feed their young in parts of their range (e.g. the wedge-tailed and little eagles, *Aquila audax* (Latham) and *Hieraaetus morphnoides* (Gould)) must have been affected by rabbit scarcity, but the possibility has not been investigated. A point made several times in this and earlier chapters is that the reduction in rabbit density has enhanced the efficacy of predation by the rabbit's natural enemies.

By far the most important effects of myxomatosis have been botanical, resulting from an easing of grazing and browsing pressure on pasture plants and the seedlings of shrubs and trees. The widespread and continuing regeneration of some of the latter, such as the mulga (*Acacia aneura*) and cypress pines (*Callitris* spp.) has been remarkable, at any rate in New South Wales (K. Myers, personal communication). Prior to myxomatosis, the Forestry Commission could only ensure regeneration in its *Callitris* reserves by the expenditure of considerable money and effort on rabbit control measures and netting fences. *Callitris* regeneration has been so dense on some properties in western New South Wales that landholders have begun to regard the seedlings almost as noxious weeds. This is the only parallel that Australia can provide to the chalk downlands of Britain, which could be regarded as a desirable vegetational disclimax produced and maintained primarily by rabbit grazing.

20-2

CHAPTER 17

MYXOMATOSIS IN EUROPE
1936–63

Western Europe is the ancestral home of *Oryctolagus cuniculus*, whence it spread, with Western European man, all over the world. Myxoma virus evolved in the Americas, and it was not until 1952 that it was successfully introduced into the wild rabbits of Europe. During the period 1936–38, however, several unsuccessful attempts were made to establish myxomatosis in Europe.

EARLY ATTEMPTS TO INTRODUCE MYXOMATOSIS

Following Sir Charles Martin's investigations in Cambridge (Martin, 1936) several attempts were made between 1936 and 1938 to introduce myxomatosis for the control of wild rabbits in different parts of Europe. Reports exist of the results of three such attempts, the essential features of which are summarized in Table 51. In the experiments on Skokholm, an island off the coast of Pembrokeshire (Lockley, 1940), myxomatosis was introduced on three successive years but failed to affect the size of the rabbit population, and did not become established. Lockley (1955) suggests that this fact, and the failure of myxomatosis to become established on Skokholm since 1953, was due to the absence of rabbit fleas. Martin did not seriously consider insect vector activity as a mechanism of transmission of myxomatosis.

On the Danish island of Vejrø (Schmit-Jensen, 1939) groups of about 150 rabbits were caught and inoculated with virulent virus, marked, and released again on one day in each of three successive years. The results each year were similar. About twice as many uninoculated rabbits died as the number of inoculated rabbits released. It was concluded that infection was confined to families, and myxomatosis did not persist for more than two or three months. As a method of rabbit control the trials were unsuccessful.

308

TABLE 51. *Attempts to introduce myxomatosis in Europe*

	Skokholm Island (England)	Vejro Island (Denmark)	Dufeke Estate (Sweden)	Maillebois Estate (France)	Heisker Islands (Scotland)
Time of introduction of virus	1936, late autumn; 1937, summer; 1938, spring	1936, autumn; 1937, late autumn; 1938, early spring	1938, spring	1952, summer	1952, summer; 1953, spring
Virus used	Martin's strain B *	Martin's strain B	Martin's strain B	Lausanne strain	Martin's strain B
Type of rabbit	Wild rabbits	Domestic rabbits run wild	Wild rabbits	Wild rabbits	Wild rabbits
Habitat	Bird-infested warrens	Underground warrens	In stone fences	Underground warrens	Underground warrens
Vector situation	Rabbit fleas absent	Not recorded	Not recorded	Both mosquitoes and rabbit fleas common	Rabbit fleas common 'biting insects a plague' but species not recorded
Method of introduction	Locally caught rabbits inoculated and released; 83, 55 and 7 on successive years	Locally caught rabbits inoculated and released; 148, 156 and 163 on successive years	18 locally caught wild rabbits inoculated and released	2 locally caught wild rabbits inoculated and released	Groups of a dozen infected laboratory rabbits were released
Result	Intrafamilial spread only, died out	Intrafamilial spread only, died out	Widespread infection over whole estate, then died out	Widespread infection, ultimately over Europe; established as enzootic disease	Sharp epizootic between March and April 1953, then died out
References	Lockley (1940, 1955)	Schmit-Jensen (1939)	Hvass & Schmit-Jensen (1939)	See chapter 17, text	Shanks *et al.* (1955)

* Martin's strain B later became the 'standard laboratory strain'.

On the Dufeke Estate in Skaane in southern Sweden, on the other hand, virus introduced on 10 May 1938 by the release of eighteen inoculated rabbits, caused an extensive epizootic all over the large estate. Virus must have over-wintered, for one infected rabbit was seen on 20 April 1939, but the disease then died out. No data were obtained on insect vectors, but Borg (1962) noted that when myxomatosis appeared in southern Sweden in 1962 (the first extension of the disease to Sweden from the great European epizootic) many ectoparasites were found on Swedish wild rabbits, including *Spilopsyllus cuniculi.*

None of these early attempts led to the establishment of myxomatosis as an enzootic disease amongst the wild rabbits of Europe. This occurred in 1952 after the Australian experience had shown that myxoma virus could indeed survive in nature in the absence of *Sylvilagus.*

SUCCESSFUL ESTABLISHMENT IN FRANCE IN 1952

On 14 June 1952, Dr P. F. Armand Delille inoculated two wild rabbits on his estate at Maillebois (Eure-et-Loir) with myxoma virus obtained from his friend Professor Hauduroy, of the Laboratoire de Bacteriologie, Lausanne, Switzerland. From this modest beginning, so different from the prolonged and frustrating experiences in Australia (chapter 16), myxomatosis spread over the whole of Europe, including Britain and Ireland, and into North Africa. It is now an established enzootic disease of the wild rabbits of Europe.

The virus used by Delille, designated Brazil/Campinas/1949/1 by Fenner & Marshall (1957) but often referred to as the 'Lausanne' strain, was obtained by Bouvier from an outbreak among domestic rabbits at Campinas, Brazil, in 1949 (Bouvier, 1954). This outbreak could only have originated from the disease enzootic in the tapeti (*Sylvilagus brasilensis*). After very few passages in domestic rabbits, the virus was deposited in the Virus Culture Collection at Lausanne, whence Dr Delille obtained it. He had been impressed by the results of myxomatosis in Australia, and wished to control the wild rabbits on his 600-acre country estate at Maillebois.

Thus the virus introduced into Europe differed from that used in Australia. The latter had undergone a very large number of serial

PLATE XV. Personalities concerned with myxomatosis in Europe. a, The late Dr P. F. Armand Delille; b, Dr H. Jacotot; c, Sir Christopher Andrewes; d, Mr H. V. Thompson; e, Medal presented to Dr Delille by French agriculturists and foresters, at a ceremony organized by M. Chavet, Secrétaire général de la Fédération nationale des Producteurs de bois et Reboiseurs français (from Siriez, 1960).

passages in domestic rabbits, over a period of nearly 40 years, before it was successfully introduced into the wild rabbit population of Australia, whereas the virus used in Europe had been subjected to very few passages in domestic rabbits. Both were of extremely high virulence for *Oryctolagus*, but the Lausanne strain produced a more florid disease, with very protuberant skin lesions and great swelling of the eyelids and face in the later stages. This florid picture is characteristic of the response of *Oryctolagus* to virus which had been maintained in *Sylvilagus brasiliensis* (chapter 15).

At about the same time as Dr Delille introduced myxoma virus into France, a much less publicized introduction was made in Britain. Following the successful use of myxomatosis in Australia, sheep farmers in Scotland requested that the virus should be tried in their rabbit-infested grazings. In order to test the possibilities of the virus Shanks, Sharman, Allan, Donald, Young & Marr (1955) introduced myxoma virus into the rabbit-infested Heisker Islands of the Hebrides. Several groups of infected laboratory rabbits were released there on eight occasions between July 1952 and May 1953 (Table 51). Although myxomatosis is thought to have spread to the wild rabbits and caused considerable mortality in 1953, it did not become enzootic, and by September 1954 the rabbit population was as large as it ever had been.

THE SPREAD OF MYXOMATOSIS THROUGH EUROPE

Within a month of introduction of the virus all the wild rabbits on Delille's estate had died, and cases of myxomatosis had also been seen amongst the wild rabbits at St Ange and Sfraze, two villages 8 and 45 km. distant from Maillebois. During 1952 cases were recognized in nine Départements. The conclusive demonstration that the new disease was indeed myxomatosis was made in October 1952 by Jacotot & Vallée (1953a), using material from a wild rabbit from Rambouillet, and the mystery of the origin of myxomatosis in Europe was solved when Dr Delille addressed the French Academy of Agriculture on 24 June 1953 (Delille, 1953).

Few cases were seen amongst the wild rabbits during the winter of 1952–53, but in the summer of 1953 violent and destructive outbreaks occurred in both wild and domestic *Oryctolagus*; and myxomatosis became a matter of great public interest. That year it spread to almost every part of France, although it was less common in mountainous districts. Brittany was spared for a long time, and the disease did not reach the Département of Finistère until 1957.

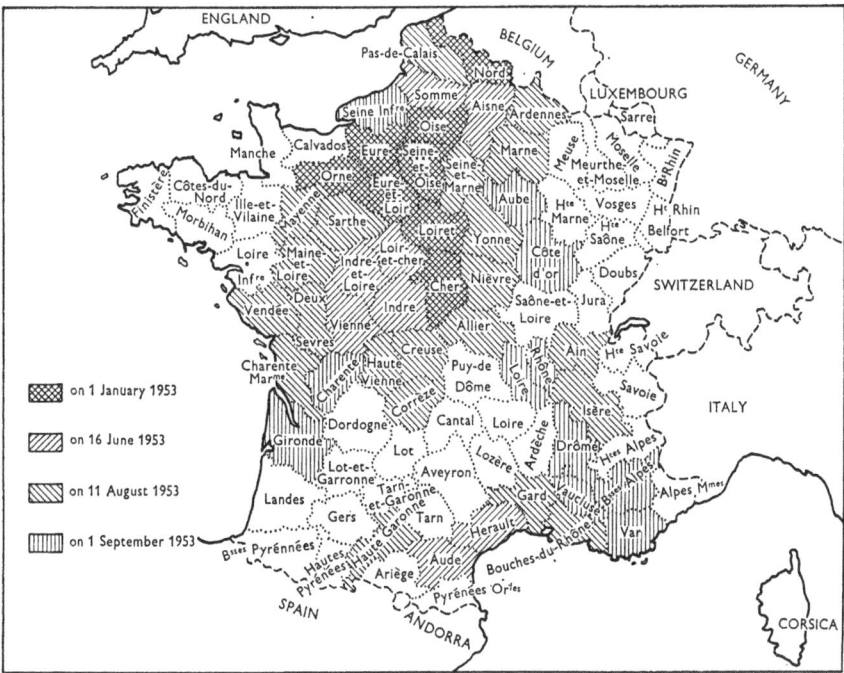

Fig. 26. Spread of myxomatosis in France in 1953, the year of major continental extension of the disease in Europe (data from Thompson, 1954).

As might be expected, myxomatosis took no account of national boundaries, and during 1953 cases occurred in Spain, Holland, Belgium, Germany and England. As in Australia, the exact mode of spread in the early stages can only be guessed at. Mosquito transmission was certainly important in continental Europe, though apparently not in Britain, and both deliberate and unintentional human movement of infected rabbits were also important factors. It is said that Delille was asked to supply sick rabbits in order to propagate the disease during the first outbreak on his estate at Maillebois, and deliberate spread by the introduction of sick rabbits may well have been responsible for setting up many new foci of infection. The practice by hunting clubs of restocking hunts in southern France from wild rabbits captured in the north also contributed, unwittingly, to the spread of the disease.

Subsequently, myxomatosis has spread to every part of Europe where rabbits occur, cases being first reported in Switzerland in 1954, Italy and Austria in 1955, Portugal in 1956, Denmark in 1960, and southern Sweden as recently as 1961 (Borg, 1962). The disease was carried to domestic rabbits in Algeria and Morocco in shipments of rabbits from France.

For two countries only, France and Britain, is information on myxomatosis sufficiently detailed to warrant further discussion here. Since both the importance of the rabbit (wild and domesticated) and the mode of spread of myxomatosis differed in these two countries, they will be discussed separately.

MYXOMATOSIS IN FRANCE

Both wild and domestic *Oryctolagus* are of considerable commercial importance in France. The French Game Act of 1844 protects wild rabbits as game; 'vermin' being any predators, such as fox, stoat, weasel and badger, which attack rabbits or the game birds. There is an open season for hunting rabbits between 1 September and early January, with possible extension (under exceptional circumstances) until 31 March.

In 1952, 1,850,000 shooting permits were issued, of which 80 % were held by hunters who were primarily rabbit hunters. The importance of adequate numbers of wild rabbits to these enthusiastic *chasseurs* can be readily appreciated, as well as their significance for such supporting industries as cartridge manufacture, makers of sporting clothes, etc. The chasseurs, through the major hunting organizations, the Conseil Superieur de la Chasse and the Saint Hubert Club de France, made prolonged and vigorous protests about the destruction of rabbits by myxomatosis, but all the measures undertaken on their behalf failed to preserve the rabbit as a game animal. The impact of myxomatosis on hunting can be appreciated from the figures shown in Tables 52 and 53, and the fact that a commentary in *The Times* of 28 August 1962 on hunting in France contained no mention of rabbits as game animals.

313

TABLE 52. *Numbers of rabbits killed annually on several hunts in different parts of France (data from Giban, 1956, and personal communication, 1963)*

	Hunt and area (in hectares)				
Year	Eure-et-Loir 805 ha	Charente 800 ha	Vienne 1,206 ha	Sologne (14 hunts) 8,900 ha	Le Mans (8 hunts) 5,600 ha
1946–47	326	985	1,270	—	—
1947–48	601	1,380	2,883	—	—
1948–49	1,251	1,205	1,595	23,917	8,436
1949–50	769	1,645	1,980	36,127	12,348
1950–51	1,875	1,325	3,574	44,999	10,666
1951–52	1,084	1,652	1,950	22,648	10,389
1952–53	1,052	1,840	1,715	36,986	13,540
1953–54	9	2,815	2,561	51	1,285
1954–55	128	2,405	0	204	13
1955–56	291	1,200	20	844	49
1956–57	4	340		843	106
1957–58	65	556	6	2,528	152
1958–59	34	542	1	1,364	102
1959–60	347	702	7	2,718	300
1960–61	261	875	54	3,768	533
1961–62	391	1,220	1,171	7,539	585

TABLE 53. *The reduction in numbers of rabbits killed in several hunts in France subsequent to the occurrence of myxomatosis (data from Giban, 1956)*

	Hunts		Average annual kill prior to myxomatosis	Relative no. of rabbits killed since myxomatosis (%)		
Département	No.	Area (ha)		1953–54	1954–55	1955–56
Orne	2	750	1,747	1·5	0·1	4·3
Sarthe	2	3,304	4,173	0·1	0·1	1·0
Seine-et-Marne	1	25	226	0	0	7·9
Seine-et-Oise	2	1,030	2,080	0	2·1	16·5
Var	1	3,389	4,500	0·6	0	4·5
Yonne	3	953	841	0	3·5	24·6
Aisne	9	22,844	22,634	2·9	1·5	3·8
Aisne	3	9,475	1,905	6·1	0·1	1·5
Indre	2	1,620	1,950	11·6	0·4	2·7
Marne	2	4,000	15,490	9·0	0·1	0·3
Oise	9	?	11,288	2·4	0·4	1·7
Pas-de-Calais	2	1,050	1,398	36·4	0	0·7
Sarthe	4	1,080	1,210	111·9	0·6	0·4
Ille-et-Vilaine	1	2,500	2,500	20	12·5	10
Gard	18	30,000	21,150	10·1	4·3	2·6

To other sections of the community, however, wild rabbits represented a major pest. In parts of France agricultural land had been converted to a desert by rabbits, and in some such areas farmers made their living by renting hunting grounds to the chasseurs. For the foresters, however, the wild rabbits were an unmitigated pest, and in State forests destruction of rabbits was permitted throughout the year. To the foresters the control of the wild rabbits by myxomatosis was a boon.

TABLE 54. *The incidence of myxomatosis amongst domestic rabbits in France (data from statistics of Office International des Epizooties)*

Figures for 1953–54 show total numbers of 'communes' (parishes) in which outbreaks were reported to have occurred during the month indicated. Figures for 1955–58 show numbers of communes in which new outbreaks were reported during the month indicated. Figures for 1959–62 show numbers of 'exploitations' (farms) on which new outbreaks were reported during the month indicated.

	1953	1954	1955	1956	1957	1958	1959	1960	1961	1962
January	—	5138	(376)	3	5	9	17	1	3	5
February	—	5085	10	3	7	6	24	1	5	4
March	—	4595	6	5	16	7	46	4	7	2
April	—	4579	21	14	15	5	26	3	6	6
May	—	4505	22	8	29	15	56	15	21	14
June	—	4504	57	20	54	18	31	20	43	5
July	1142	4670	193	43	322	159	64	46	104	8
August	3218	4379	418	223	307	638	145	116	202	—
September	4010	5184	323	237	200	566	119	181	178	—
October	5599	4107	218	298	90	180	162	114	140	—
November	5606	4079	99	53	45	69	24	52	44	—
December	4938	3888	13	11	14	29	14	9	8	—
Total	—	—	1380	918	1104	1701	728	562	761	—

The raising of domestic rabbits is practised on a very large scale in France, both commercially and in backyard 'clapiers'. Before myxomatosis some 140 million domestic rabbits were produced and consumed annually, and many retired workers were largely dependent upon rabbit raising for their livelihood. Domestic rabbits were infected as soon as the disease spread amongst the wild rabbits, in 1953; and in 1954 it was estimated that 30–40% of the domestic rabbit population was destroyed. Numerous cases have occurred amongst domestic rabbits each year (see Table 54), but recently, with the great reduction in the size of the wild rabbit reservoir of myxomatosis, outbreaks have become much less common. Figure 27 summarizes the importance of rabbits in France, before the advent of myxomatosis.

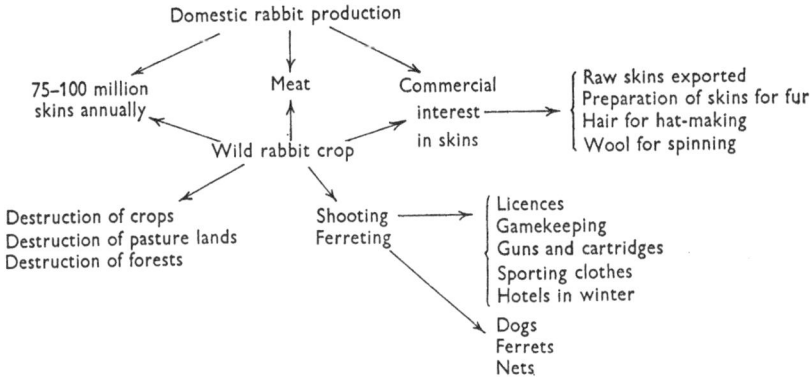

Fig. 27. The place of the rabbit in France (modified from Barthélémy, 1953)

Mode of spread of myxomatosis in France

In the early days of the French epizootic fantastic stories circulated concerning the mode of spread of myxomatosis, and some of these are reflected in the regulations introduced in an attempt to control the disease (see next section). The consumption by domestic rabbits of grass which might have been contaminated by myxomatous wild rabbits was regarded as dangerous, and motor cars which ran over infected rabbits were blamed for the movement of the disease to new districts. There were discussions in the French press about the height to which rabbits could jump, for it was held that the disease may have escaped from Delille's walled estate in this way.

Adequate field investigations have not been made, but there is now general agreement that in France, as elsewhere, myxomatosis is primarily spread by insect vectors. Both mosquitoes and the rabbit flea are probably important. Although no investigations on its activity have been carried out in France, the flea may well be largely responsible for the maintenance of infection through the winter. Mosquitoes certainly play a major role in the summer epizootics, and are usually responsible for transferring the virus from wild to domesticated rabbits.

As pointed out in chapter 11, any mosquito is a potential vector of

316

myxomatosis. Jacotot, Toumanoff, Vallée & Virat (1954) showed that *Anopheles labranchiae atroparvus* was an efficient vector in the laboratory, and the rabbit constitutes a favoured source for its blood meals. Roubaud (quoted by Jacotot, Toumanoff *et al.* (1954)) suggested that this mosquito was the most important vector in France, especially among domestic rabbits. Among wild rabbits he thought that in addition to *Anopheles atroparvus* various *Aedes* mosquitoes, *Anopheles claviger*, and especially *Anopheles plumbeus* were probably important vectors.

During 1952 and 1953 myxomatosis was introduced into many parts of France by the transfer of infected rabbits, either deliberately, or unwittingly, as when some of the southern hunts were restocked with rabbits from northern France, or when infected domestic rabbits were shipped from one place to another whilst incubating the disease. Deliberate movement of infected rabbits by farmers and foresters, though illegal, is still important in setting up new foci of infection where the wild rabbit population has built up.

Official action on myxomatosis

No immediate official action was taken when myxomatosis was first recognized, on 20 October 1952. After the rapid spread of the disease during the following spring, however, a decree was published (Ministry of Agriculture, 27 May 1953) making myxomatosis a notifiable disease, and prohibiting the movement of rabbits in an infected area, with appropriate measures for the isolation and disinfection of domestic rabbitries (Radot & Lépine, 1953).

Dr Delille was pleased with the success of his experiment of June 1952, and read a paper to the French Academy of Agriculture, on 24 June 1953 on 'A new means of controlling rabbits'. His action, however, met with a mixed, predominantly hostile reception, and in 1956 legal action was taken against him, by a plaintiff who claimed that following the introduction of myxomatosis, he had lost the possibility of shooting rabbits on his land. The Court of Appeal gave judgement against Delille, but said that the damage was slight, as it was offset by the increased yield of the plaintiff's crops. It awarded only 5000 francs damages, instead of the million francs claimed. The ambivalent French attitude towards myxomatosis is symbolized by the fact that shortly after this the National Federation

317

of Foresters presented Dr Delille with a medal in recognition of his valuable service to the forestry industry (Siriez, 1960).

Effects of myxomatosis on the wild rabbit population

The spread of myxomatosis in 1953 was erratic, so that although the mortality rate in many areas was enormous, many pockets of uninfected rabbits remained even in Départements in which infection was widespread. In subsequent years most of the pockets have been affected, but the disease has never been common in the higher country of the Massif Central. By the end of 1954 it was estimated that 90 % of the wild rabbits in France had been killed by myxomatosis.

Vigorous efforts were made by hunter's organizations to minimize the effects of myxomatosis, to no avail. Prohibition of the movements of rabbits was of little importance when the regulation enforcing this was introduced, for the disease was already very widely spread so that numerous foci existed for subsequent dissemination of the virus by mosquitoes.

In 1953 attempts were made to create 'sanitary barriers' around favoured hunts by inoculating large numbers of wild rabbits in the surrounding areas with Shope's fibroma virus, but as might have been expected these were unsuccessful. Other measures tried were the introduction of cottontail rabbits from America, both as game animals and with the hope that they might be crossed with *Oryctolagus* to produce hybrids resistant to myxomatosis. Experimentally, in crosses between *Oryctolagus* and *Sylvilagus*, in both directions, fertilization was achieved but the embryos failed to develop. Having heard that the Australian wild rabbits were showing some degree of genetic resistance to myxomatosis (see chapter 14) some French landowners even made attempts to repopulate their hunts with these animals.

In spite of all efforts, myxomatosis has persisted as a lethal disease of the wild rabbits of Europe, as it did in Australia. For the present the rabbit has ceased to be a game animal in many parts of France, its place having been taken by hares, partridges, and pheasants. In scattered localities there has been no myxomatosis for several years, and here the rabbit is again common enough to be hunted. Tests on rabbits from several such areas show them to be fully susceptible to myxomatosis (A. Lucas, personal communication, 1962).

In the Sologne, which was a great rabbit hunting area for hunters from Paris, myxomatosis almost exterminated the rabbits and the large private hunts there were repopulated with hares and pheasants. Since these are better sport, the hunters of the Sologne now wish to keep down rabbits, but in the south of France where less wealthy hunters shoot on common lands the loss of the rabbit is still deplored. Whether its numbers may eventually rise so that it can again become the prime target for the chasseurs' bullets depends upon the processes of natural selection, operating through changes in the virulence of the virus, and changes in the genetic resistance of the rabbit.

Effects of myxomatosis on domestic rabbits

Outbreaks of myxomatosis among domestic rabbits were very common in 1953 and 1954. Although a number still occur each year (Table 54) outbreaks are now less common, since the size of the reservoir of infection amongst the wild rabbits has become so much smaller.

Fibroma vaccine was used on a very large scale some years ago (tens of millions of doses annually), and is now used on a smaller scale, primarily to protect breeding animals. It was moderately effective, protecting some 70 % of vaccinated animals from death; but its continued popularity probably depends upon the comparative rarity of challenge, since myxomatosis is now much less common among domestic rabbits.

At present domestic rabbits are raised on a somewhat smaller scale than before the advent of myxomatosis, and rabbit products fetch correspondingly higher prices. For coats and trimmings, 'lapin' has been replaced by muskrat, and fur felt hats (which used to be made from wild rabbit skins) have been replaced by hats made of other materials.

Attenuated strains of myxomatosis virus in France

In France, as in Australia, attenuated strains of myxoma virus have been found in naturally infected rabbits. The first attenuated strain recognized was that designated 'Loiret 55' (France/Loiret/4–55/1). Both Jacotot, Vallée & Virat (1955e) and Fenner & Marshall (1957) showed that this strain was attenuated, with a mean survival time of 33 days under standard conditions of test. This strain and seven others subsequently examined by Jacotot *et al.* (1956), produced protuberant nodular lesions rather than flat skin lesions found with the attenuated Australian strains (see chapter 15).

We have been able to examine a few French strains obtained from the environs of Paris in 1962–63 (see Table 38). All were attenuated, some of them strikingly so (Grade IV virulence). In all cases the skin lesions were only slightly raised, rather than protuberant, as those produced by the Lausanne and Loiret 55 strains had been.

Enhanced genetic resistance amongst French wild rabbits

Although claims have been made (e.g. Giban, 1959) that wild rabbits are becoming more resistant to myxomatosis, no real evidence for this exists and where supposedly resistant rabbits have been tested they have proved to be fully susceptible (A. Lucas, personal communication, 1962). From the Australian experience it would be safe to predict that when attenuation of the virus allows enough survivors for selective breeding, there will be an increase of genetic resistance amongst the wild rabbits. Since infected domestic rabbits are destroyed, no such selection for resistance will occur amongst domestic rabbits, and any modification of the disease seen in these animals must be ascribed to alterations in the virulence of the virus.

Ecological effects of myxomatosis

The drastic reduction in the numbers of rabbits in parts of France, where they were previously very abundant, and their continued rarity since about 1953, had dramatic ecological effects, due to the sudden removal of a major grazing animal, and for some predators to the sudden removal of an important component of their diet. Some of the changes have been described at the sixth technical meeting of the International Union for Conservation of Nature and Natural Resources at Edinburgh in June 1956 (Terre et la Vie, 1956), and by Siriez (1957, 1960).

The ultimate effects on the vegetation will in part depend upon the continued scarcity of rabbits, but within a few years of the epizootics of myxomatosis the vegetational landscape had greatly changed in those areas where the rabbit density had been high, as in the Sologne. This change is merely a reflexion of the previous intensity of rabbit grazing, which was selective. In forests rabbits ate young seedlings of oak, beech and hornbeam, for example, but spared chestnuts, birches and aspen (Morel, 1956); and where the density of rabbits had been high for 50 years this quite changed the aspect of the forests. Indeed Morel considers that it was primarily rabbit grazing that had altered forests of lofty oaks, with thicket undergrowth, into the birch and aspen moors of the Sologne. Attempted re-afforestation was rendered impossible by rabbits, and their sudden removal had an immediate and spectacular effect. The first result in the degraded forests was a luxuriant growth of grasses (e.g. *Molinia coerulea*); this was followed by an increase of broom (*Sarothamnus scoparius*) in the understory, and then by the widespread appearance of oak and beech seedlings. Similar vegetational regeneration is occurring elsewhere, as in the Bouches du Rhône, the Camargue, and even the limestone wastelands of the Gâtinais.

The sharp decrease in rabbits in Europe had obvious and rapid effects on the ecology of two predators, the fox and the common buzzard (*Buteo buteo*). Foxes have adapted by altering their diet—to birds and fish in the Camargue, to field-mice in Seine-et-Marne.

The disappearance of the rabbit has eliminated interspecific competition with the hare, but has increased predator pressure on the latter. There appears to have been a substantial increase in the numbers of hares, especially in certain habitats such as the edges of woods. Siriez (1960) quotes estimates of the increase which vary between twofold and tenfold in different areas. There has also been a substantial importation of hares from Alsace, Hungary and Yugoslavia, for hunting. Rothschild (1958) also notes increases in the numbers of hares on three estates. The impact of the increased number of hares on agriculture has not been of any significance, but they have been troublesome to foresters and orchardists because of their habit of ringbarking trees, especially when snow lies on the ground.

MYXOMATOSIS IN BRITAIN

Shortly after Martin's (1936) investigations of myxomatosis at Cambridge, on behalf of the Australian Council of Scientific and Industrial Research, an attempt was made to establish myxomatosis on the island of Skokholm (Lockley, 1940). The disease spread a little, but soon died out.

There was no further active investigation of myxomatosis until the epizootics in Australia in 1950–51, after which experimental introductions were made in the Hebrides in 1952–53 (see Table 51). The proximity of the great epizootics in France in 1953 made it inevitable that myxomatosis would spread to England, and the disease probably reached England from France in August or September 1953, the actual means of entry being unknown. The first confirmed case was obtained from Bough Beech, near Edenbridge, Kent, in October, 1953. The virus from this outbreak (England/Kent/10–53/1) was found by Fenner & Marshall (1957) to be indistinguishable from the Lausanne strain; so that myxomatosis in England may be regarded as having been initiated, a year later than in France, by essentially the same virus.

The spread of myxomatosis in the first outbreak has been reported in detail by Armour & Thompson (1955). As soon as the disease had been recognized attempts were made to contain or eradicate it, by local extermination of rabbits. These attempts failed, and there was a slow centrifugal spread and the occasional establishment of new foci several miles from Bough Beech. By November 1954 diseased rabbits had been seen at a total of 408 separate sites, in the counties of Kent, Surrey, and East Sussex; and outbreaks in more distant areas (Cornwall, Norfolk and Radnor) had occurred as early as May of that year, and were very extensive by December (see Fig. 28). Thus myxomatosis had survived through the English winter, and became established in so many localities that a year after its entry it had to be regarded as an enzootic disease of the British wild rabbit. By the end of 1955 there were few areas of Britain unaffected by the disease, and at this time it was calculated that over 90 % of the wild rabbits on the island had died. There were, of course, small pockets of rabbits which escaped infection and continued to breed.

Myxomatosis has remained an enzootic disease throughout England, Scotland, and Wales, and since 1958 cases have occurred in most counties each month of the year. Temporary freedom from myxomatosis has led to local increases of rabbits, into which myxomatosis has then been introduced, usually by the deliberate (though illegal) release of infected rabbits. The overall rabbit population remains low compared with the numbers present in 1952.

KEY
- Outbreaks before 1 April 1954
- Spread up to 31 Dec 1954
- Spread up to 31 December 1955

Fig. 28. The spread of myxomatosis in England, September 1953 to December 1955 (data from Thompson, 1956, with the permission of H.M. Stationery Office. Crown copyright).

21-2

There are still a few scattered areas in Great Britain where myxomatosis has apparently not occurred. It is interesting to note that the island of Skokholm is one of these, although rabbits were infected on the neighbouring island of Skomer, and to realize that rabbit fleas are absent from the rabbits of Skokholm but not from those of Skomer (Lockley, 1955). In general, however, myxomatosis in Britain, as in Australia and continental Europe, is now co-extensive with the wild rabbit population.

The rabbit in Britain

The importance of the rabbit in Britain has been well described by Thompson & Worden (1956), the position being quite different from that obtaining in France. The wild rabbit is now not a game animal, and official policy since the Second World War has regarded the rabbit as a major animal pest (to agriculture and forestry), to be combatted by all available 'humanitarian' means. There is, on the other hand, much sentimental affection for the rabbit, inspired by the appealing appearance of the young animals, and many 'nursery rhyme' associations.

Domestic rabbit breeding is not nearly as important as in France, but is a growing industry. About 12,000,000 domestic rabbits are produced and consumed annually in England, mainly by large-scale commercial breeders (Thompson & Worden, 1956).

Official action on myxomatosis

On 3 November 1953 an Advisory Committee on Myxomatosis was appointed by the Minister of Agriculture and the Secretary of State for Scotland, and it has presented two reports (*Reports*, 1954, 1955). Out of respect for public sentiment, the committee recommended that myxomatosis should not be deliberately spread, but should be allowed to run its own course. It also recommended that where epizootics of myxomatosis did occur energetic follow-up measures should be taken to destroy any surviving rabbits.

The widespread occurrence of myxomatosis in the summer of 1954 led to heated public debate on the ethics of exploiting the disease as a means of rabbit control. Great Britain is a densely populated island, wild rabbits were then common, and rabbits with advanced myxomatosis present a horrible aspect, especially when infected with the Lausanne strain of virus. The animal lovers of Britain were instrumental in obtaining the insertion of a clause into the Pests Act, 1954, which virtually banned deliberate spreading of myxomatosis in Britain by any means, an action which led to the

STOP this
heartless traffic
in MYXOMATOSIS!

STOP the deliberate spreading of MYXOMATOSIS! Victims of this horrible disease blind, misshapen, tormented — are being caught for sale as carriers, to be let loose in infection-free areas. Effective rabbit-control can be maintained by humane methods : myxomatosis kills only after intense, prolonged pain and misery. Nothing can justify this callous encouragement of animal suffering, and the R.S.P.C.A. appeals for your moral and material support in demanding an immediate legal ban. **Volunteers in infected areas, who must be expert shots, apply please,** to the Chief Secretary, R.S.P.C.A. (Dept. T.) 105 Jermyn Street, London, S.W.1 or to the nearest R.S.P.C.A. Inspector.

Remember the **R S P C A**

Fig. 29. Advertisement inserted in newspapers in Britain in 1954, by the Royal Society for the Prevention of Cruelty to Animals.

resignation of two members of the Advisory Committee on Myxomatosis. This legislation was reinforced by the publication in the national newspapers of the advertisement shown in Fig. 29.

The Pests Act, 1954, provides that any person who knowingly uses or permits the use of a rabbit infected with myxomatosis to spread it among uninfected rabbits thereby commits an offence and is liable to a maximum fine of £20 for the first offence and £50 for a subsequent offence. Inoculation of rabbits with virus was already illegal. This provision of the Act has proved impossible to police and there is no doubt that deliberate spread by means of infected rabbits continues. Up to December 1962 there had been two prosecutions and convictions, on this count.

Full use is being made of the provision of the Pests Acts, 1954, which relate to rabbit clearance areas, and a large number of Rabbit Clearance Societies have been set up, some 46 % of all agricultural and wooded land in England and Wales now being serviced by these societies, of which there are now 670 (Newsletter, 1964).

Mode of spread in Britain

Myxomatosis in Britain has shown very interesting epidemiological differences from what was observed in France and in Australia. The seasonal incidence has been much less pronounced (Table 55) although since rabbits are much more common in summer, and more readily observed then, many more cases occur and are seen during

TABLE 55. *The seasonal incidence of myxomatosis in wild rabbits in England and Wales in 1961*

		Tests on infected rabbits at Weybridge†	
Month	No. of counties* affected (out of 61)	Antigen and antibody	Antigen (or lesions) only
January	52	2	10
February	48	4	8
March	53	0	7
April	51	4	15
May	52	2	7
June	52	11	18
July	55	14	43
August	57	6	17
September	57	8	18
October	57	13	23
November	58	11	27
December	58	4	7

* Data from monthly reports of Rabbit Clearance Societies, most of which note the occurrence of some new outbreak each month of the year.

† Gel-diffusion tests with lesion material and blood (Mansi & Thomas, 1958) 'antigen + antibody' has been interpreted as indicating the occurrence of attenuated virus strain, 'antigen (or lesions) only' as indicating that the infection studied was caused by highly virulent virus (but see chapter 13).

the summer months. Outbreaks have often appeared to be restricted by mechanical barriers to rabbit movement, such as fences, and uninfected pockets of rabbits are often found adjacent to areas severely affected. Cases among domestic rabbits have been very rare.

All these facts suggest a different mode of transmission in England from that found in France, where mosquitoes cause summer epizootics and long distance movement of the disease, and the rabbit flea is probably important in maintaining infection through the winter months. In chapter 11 evidence was presented that the rabbit flea (*Spilopsyllus cuniculi*) was an efficient vector of myxomatosis, and British investigators believe that it has been much the most important vector in Britain (Andrewes, Thompson & Mansi, 1959). Woodland mosquitoes, although effective vectors in the laboratory, have proved to be of minor importance in the field (Muirhead-Thompson, 1956a) although *Anopheles atroparvus* may play a more important role in some coastal areas. All the epidemiological facts described earlier are explicable if the flea is the major vector in Britain, and this mode of transmission has important implications in its selective effect on strains of differing virulence.

The effect of myxomatosis on the wild rabbit population

By the end of 1955 it was estimated that the wild rabbit population of Britain had been reduced to 10 % of its previous level (Andrewes *et al.* 1959). Myxomatosis was less common between 1956 and 1960 although the incidence relative to the reduced rabbit population was still considerable. The energetic efforts of the Rabbit Clearance Societies have maintained many areas relatively free of rabbits. Elsewhere there was a slow increase in rabbit numbers, but in 1961 and 1962 myxomatosis appeared to be more common, although the wild rabbit population is now appreciably higher than in 1955.

Attenuated strains of myxoma virus in Britain

As in Australia and in France, attenuated strains of myxoma virus appeared in England about a year after the introduction of the virus. Fenner & Marshall (1957) reported that two strains recovered respectively from a diseased rabbit in Sussex in September 1954, and from a pool of *Anopheles atroparvus* mosquitoes in October 1954 (Andrewes *et al.* 1956), were both slightly attenuated, with mean survival times in laboratory rabbits of twenty-two and nineteen days instead of the usual twelve days. The most interesting attenuated strain recovered in England is that designated England/Nottingham/4–55/1 by Fenner & Marshall (1957), which proved to contain a mixture of particles of high and low virulence (for details see chapter 13). This 'strain' was recovered from Sherwood Forest in Nottinghamshire in April 1955 (Hudson & Mansi, 1955; Hudson *et al.* 1955; Fenner & Marshall, 1955). Myxomatosis first appeared in this area in September 1954, and in spite of high rabbit and flea densities it was less destructive than expected. On one estate, for example, half the surviving rabbits tested serologically were immune, and four strains recovered from the area produced a mild infection characterized by the development of large well demarcated nodules in the skin (see Pl. XIII, fig. 7). A few miles to the south, fully virulent myxoma virus was producing a heavy mortality.

Since 1957 attempts have been made by the Central Veterinary Laboratory, Weybridge, to classify strains of myxoma virus obtained from naturally infected rabbits as 'typical' (i.e. fully virulent) or 'atypical' (attenuated), on the basis of Mansi's gel diffusion precipitation test (Mansi & Thomas, 1958). The sort of results obtained are shown in Table 55, cases being registered as 'typical' if antigen only is found in the tumours and/or blood, and 'atypical' if antigen and antibody are found in any combination of infected tissue and blood.

As explained in chapter 13, Mansi's method, although ingenious, is open to different interpretations from that just proposed, and recent experiments by Chapple *et al.* (1963) have shown that even cases scored as 'typical' may be due to attenuated strains of myxoma virus. The only reliable method of assessing the virulence of strains of myxoma virus, in terms of their lethality for rabbits, appears to be that adopted by Fenner & Marshall (1957), that is the inoculation of rabbits of standard genetic resistance (fully susceptible laboratory rabbits) with small doses of virus, observations being made of the clinical course of the disease, the mean survival time, and the case-mortality rate. Using such methods we have recently completed the examination of 222 strains from 80 counties of Britain, collected between October and December 1962 (Fenner & Chapple, 1965). The results are summarized in Table 56 and Fig. 30.

328

In Britain, as in Australia and in France, moderately virulent strains (Grade III virulence) are now predominant. The overall proportion of more virulent strains has remained higher than in Australia and in France, probably because fleas and not mosquitoes were the main vectors (see page 342). In spite of the fact that there was only one introduction of highly virulent myxoma virus of the Lausanne type, in September 1953, and that inoculation of the virulent strain has been illegal, no less than 22 % of the strains examined were

TABLE 56. *The virulence of strains of myxoma virus from Britain, and the types of lesion produced*

Figures expressed as percentages (data from Fenner & Chapple, 1965).

	Virulence Grade					
	I	II	III A	III B	IV	V
Mean survival time (days)...	< 13	14–16	17–22	23–28	29–50	—
1953–54	100	—	—	—	—	—
1962	4·1	17·6	38·8	24·8	14·4	0·5

	Type of lesion		
	Protuberant	Raised	Flat
1953–54	100	—	—
1962	20	35	45

highly virulent (Grades I or II). This was not merely due to persistence of the original virulent strain, however, for only 30 % of these Grade I–II strains had the protuberant type of skin lesion which characterized the Lausanne strain and the early English isolates. Nevertheless, with flea transmission in Britain as with mosquito transmission in Australia, moderately virulent strains (Grade III) have become the predominant group, comprising 63 % of the total, although the majority were of the more virulent subgroup (III A).

Strains with protuberant, raised, and flat skin lesions occurred in all groups, and over half of all those examined had flat skin lesions similar to those once thought to be characteristic of Australian strains. These predominated in the more attenuated strains (Grades IV and V), suggesting that the 'Nottingham attenuated' type of virus (see page 228) had not spread widely in Britain.

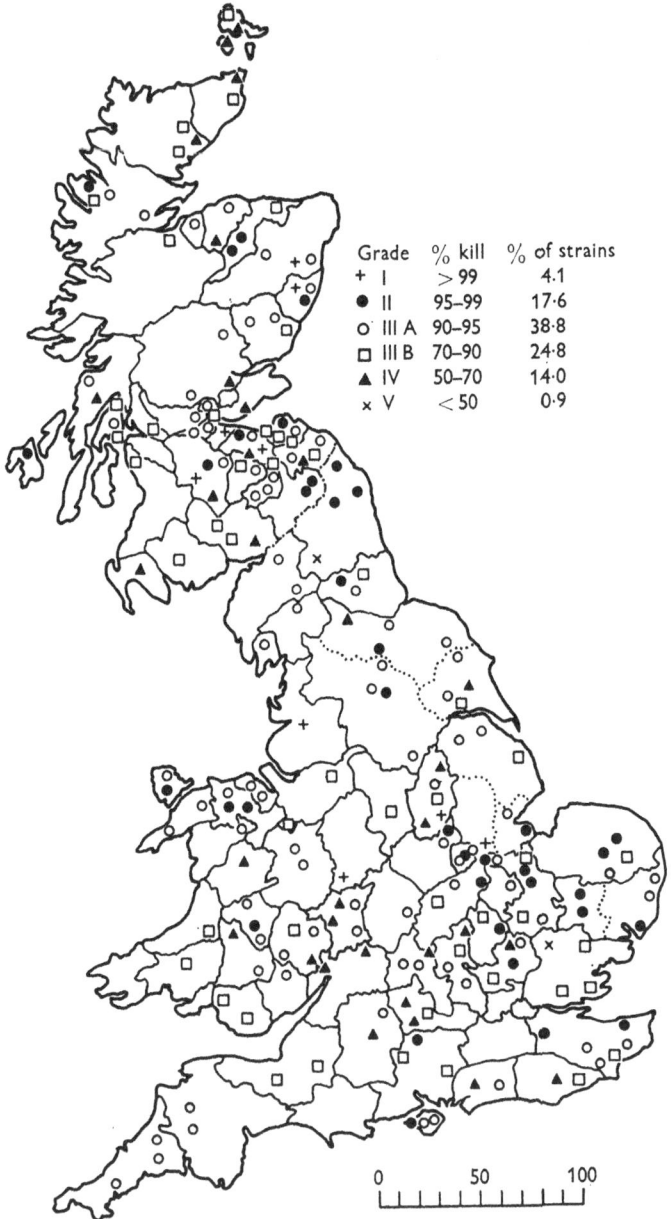

Grade	% kill	% of strains
+ I	> 99	4.1
● II	95–99	17.6
○ III A	90–95	38.8
□ III B	70–90	24.8
▲ IV	50–70	14.0
× V	< 50	0.9

Fig. 30. Map of Britain showing the distribution and virulence of strains of myxoma virus collected in October–November 1962 (from Fenner & Chapple, 1965).

330

Enhanced genetic resistance among English wild rabbits

There is as yet no evidence of the possible emergence of rabbits of enhanced genetic resistance in England. The irregular incidence of myxomatosis, which means that rabbits in a particular area may be exposed to infection only at intervals of three or four years, would tend to minimize the selective pressure for genetic resistance, and the continued somewhat higher proportion of virulent strains of myxoma virus would leave a smaller proportion of the rabbits which are infected available for breeding. The emergence of resistant rabbits, if it does occur, may therefore be much slower than was observed in those areas of Australia where annual epizootics of infection with slightly attenuated strains of virus imposed a very effective selection for genetic resistance (see chapter 14). This is an indirect effect of flea rather than mosquito transmission.

Ecological effects of myxomatosis

The effects on the vegetation of the sudden removal of rabbits from areas where the population density had long been high was much the same as in France. Thomas made an admirable quantitative study of the vegetational changes on the chalk downlands of southern England, his observations starting before the rabbits were affected by myxomatosis. Fixed transects about five miles long, on seven areas of downland, were examined twice a year until 1962 (Thomas, 1960; and personal communication, 1962). On every site there was a dramatic change in the vegetation. The turf increased in height and there was a spectacular increase in the abundance of spring flowers. There was an increase in woody plants, both shrubs (brambles, gorse and hawthorn) and latterly ash and oak seedlings. There seems little doubt that in the absence of heavy grazing the chalk downland will ultimately revert to woodland. On the reserves of the Nature Conservancy it was only the rabbit which maintained the downland as a treeless turf. Similar effects have been observed in other habitats, for example sand-dunes (Ranwell, 1960).

There appears to have been a slight increase in hares after the advent of myxomatosis, in most places less than that seen in France, although on Rothschild's property there was a very substantial

331

increase, due in large part to the extermination of foxes in 1959 (Rothschild, 1961). During 1961, when there were numerous hares but very few rabbits at Ashton Wold, Rothschild (1963) recorded flea counts (*Spilopsyllus cuniculi*) of twenty to twenty-five on the hares. Two female fleas were collected from a pregnant doe, one of which had pigmented corpora lutea. Rothschild believes that the degree of flea infestation of hares, in the absence of anything except trivial numbers of rabbits, indicates that *S. cuniculi* can successfully parasitize and maintain a life cycle on the hare, although it is not a wholly satisfactory host.

Southern & Watson (1941) showed that rabbits, when they were abundant, provided 50 % or more of the food of British foxes. With the virtual disappearance of this prey item, the fox was forced to change its hunting habits, paying much more attention to small rodents such as voles (*Microtus agrestis*) and wood mice(*Apodemus* spp.). H. N. Southern (personal communication) had the opportunity of observing at first hand this change in the foxes' hunting behaviour. In his study area near Oxford he saw myxomatosis take effect, then observed the signs of intense predation by foxes on wood mice. He was able to measure the effect of this on the wood mouse population, and also the effect of the reduction in mouse numbers on the breeding success of the tawny owls (*Strix aluco*) inhabiting the area. Lever, Armour & Thompson (1957) examined a series of fox stomachs in 1955, some obtained from districts in which myxomatosis had reduced rabbits to scarcity and some from areas to which the disease had not spread. They confirmed Southern & Watson's assessment of the importance of rabbits in the foxes' normal diet, and showed that where rabbits were no longer readily available voles replaced them as a major food item. These authors found evidence of only a slight increase in the killing of sheep and lambs by foxes, immediately after myxomatosis, but perhaps a significant increase in attacks on poultry. Reports from keepers indicated increased fox predation on game birds, especially during the rearing season.

The buzzard (*Buteo buteo*), like the fox, is a general predator for which in Britain the rabbit is a very important food item—especially young rabbits, and when the birds have nestlings to feed. Moore (1956) has reported the effects of the reduction in rabbit numbers on the breeding of the buzzard, which in recent years has become quite common in the west and south-west of England. The survey in 1955 included some areas in which rabbits had not been affected by myxomatosis; and the successful breeding in these areas provided an extraordinary contrast with the complete failure of the nesting pairs to rear young in those areas where the disease had already taken effect. As the buzzard maintains quite high population levels in some parts of Europe that are rabbit-free, Moore attributed the breeding failure not to the scarcity of rabbits as such, but to the sudden reduction of the total prey biomass for which several predator species competed. Later observations (Moore, 1957) indicated that the buzzard was in process of stabilizing its population at a new level, somewhat lower than the one ruling prior to myxomatosis.

SUMMARY

Several attempts were made to introduce myxomatosis into populations of wild *Oryctolagus* in different parts of Europe in the period 1936–38. None was successful in establishing the enzootic disease. This only occurred in June 1952 after Dr A. Delille inoculated two rabbits on his estate near Paris with a strain of virus recently recovered from Brazil. From here the disease spread through much of Europe by the end of 1953, and by 1961 myxomatosis had been recognized in almost every country of Europe where wild *Oryctolagus* occurs.

The wild rabbit was most common, and the domestic rabbit industry most important, in France and in Britain, which suffered most severely from the new disease. Enormous mortalities of wild rabbits occurred in both countries, and in France (but not in Britain) the domestic rabbit population was severely affected. The reduction in wild rabbits, which used to be the most important game animal in France, has had important consequences for the *chasseurs* as well as for farmers and foresters.

The European virus was highly virulent and produced a very florid disease in rabbits. Attenuated strains were recognized as early as 1954. They have not entirely replaced the highly virulent strain as they did in Australia, but co-exist with it, although now much more common. This difference in behaviour is probably due to the different vector situation in Europe, where in Britain at least the

333

rabbit flea is the major vector. In France mosquitoes are also very important.

The sudden disappearance of such a common grazing animal as the rabbit had dramatic ecological effects. Some areas of moor and downland are reverting to woodlands, hares have increased in numbers, and foxes and buzzards have altered their eating and breeding habits.

CHAPTER 18

THE CONTINUING EVOLUTION OF MYXOMATOSIS

It was J. B. S. Haldane (1949) who first explicitly propounded the view that infectious diseases must have played an important part in the evolution of animals and plants. Myxomatosis in wild populations of *Oryctolagus cuniculus* has provided for the first time an opportunity to study this process in a large mammal. Concurrently, the myxoma virus has encountered a new situation, and new selective forces have resulted in rapid evolutionary changes in the virus also.

In this chapter we will try to integrate observations which have been described in several of the earlier chapters so as to provide a comprehensive picture of evolutionary aspects of myxomatosis. We will also indulge in some speculations about the evolutionary history of viruses in general and the myxoma-fibroma viruses in particular, and venture some prophecies concerning future trends in myxomatosis of *Oryctolagus*, the virulence of myxoma virus, and the genetic resistance of *Oryctolagus*.

EVOLUTIONARY HISTORY OF THE MYXOMA-FIBROMA SUBGROUP OF THE POXVIRUSES

When one considers the size and structure of animal viruses in general one cannot but be struck by the unique nature of the poxviruses. Compared with the large number of animal viruses which use RNA as their genetic material (the 'riboviruses' of Cooper, 1961) there are comparatively few viruses of vertebrates with DNA as their genetic material ('deoxyriboviruses'). Among the latter the poxviruses are sharply distinguished from the others by two characteristics which may be related. They are the only deoxyviruses of vertebrates which multiply exclusively in the cytoplasm. The virions are also much larger and much more complex than those of the other deoxyviruses, and they have very much more genetic material (see Table 57).

335

Whatever may have been the origin of the icosahedral deoxyviruses we would postulate that the poxviruses represent another, and perhaps a unique, evolutionary development. There are persuasive reasons for regarding most viruses as having originated from components of the cells of bacteria or higher organisms (Luria, 1958), but the parasitic degeneration hypothesis of Green (1935) and Laidlaw (1938) seems more likely to account for the poxviruses.

TABLE 57. *The major groups of deoxyviruses which infect vertebrates, the amount of their genetic material and the molar dissymmetry of its nucleotides, and the structure of the virions*

Virus group	Genetic material		Structure of the virion
	Molecular weight ($\times 10^6$)	Ratio: $\dfrac{\text{adenine and thymine}}{\text{guanine and cytosine}}$	
Poxvirus	180	1·70	Complex, with several membranes closely applied to the nucleocapsid
Adenovirus	10	0·77	Icosahedron with 252 capsomeres; no envelope
Papovavirus	3·5	1·08	Icosahedron with 42 capsomeres, no envelope
Herpesvirus	50–60	0·36	Icosahedron with 162 capsomeres, usually enclosed within a loosely applied envelope
Rabbit cells	—	1·28	
KB cell	—	1·38	
Chick embryo	—	1·35	

The structure and chemical complexity of poxvirions, their common internal antigen, and the phenomenon of non-genetic reactivation suggest that all poxviruses had a common origin. Subsequently, there has been some morphological differentiation, as yet examined adequately only in a few members of the group. But vaccinia and myxoma are morphologically indistinguishable, even by negative staining and thin section electron microscopy.

The recent recovery of a virus of the myxoma-fibroma subgroup from fibromas of *Lepus europaeus* in southern France and northern Italy, and the identification of this hare fibroma with the 'fibrosarcoma' of hares of Dungern & Coca (1909), and the hare fibroma of Mello (1929), allows of some speculation on the antiquity of the subgroup.

The fossil history of the Leporidae is very incomplete, but recent studies of teeth from Pliocene and Pleistocene deposits suggests that the three genera of the rabbit family which are of concern in the story of myxomatosis (*Sylvilagus*, *Oryctolagus*, and *Lepus*), arose during the early Pleistocene from a common North American ancestral stock (Hibbard, 1963). There are several species of *Sylvilagus* in North America and one in South America, *Oryctolagus* has never spread beyond Europe and North Africa (except for its recent wide distribution by man), and different species of hares (*Lepus*) occur in Europe, Asia, Africa, and North America. Fossil *Oryctolagus* are found in Pleistocene deposits in Europe, fossil *Sylvilagus* in the Pleistocene in North and South America, and fossil *Lepus* in the Pleistocene of Eurasia and North America (Dice, 1929).

The common ancestor of the hare fibroma virus and of the various fibroma viruses of the American *Sylvilagus* must have antedated the dispersal of *Lepus*, and the movement of *Sylvilagus* into South America during the Pleistocene, and must therefore date back probably into the early Pleistocene.

MYXOMA-FIBROMA VIRUSES IN THE AMERICAS

In earlier chapters (7 and 15) we pointed out that there were at least three climax associations of *Sylvilagus* and a virus of the myxoma-fibroma group in the Americas. In eastern and mid-western U.S.A., over an area corresponding approximately to the distribution *S. floridanus mearnsi*, viruses have been recovered from naturally infected cottontails which appear to be indistinguishable from Shope's rabbit fibroma virus, originally recovered in New Jersey. Minor differences may exist between strains obtained from different geographic areas, for comparisons have been limited to the pathology in *Sylvilagus* and *Oryctolagus* and cross immunity studies in *Oryctolagus*.

In California, *Sylvilagus bachmani* is probably enzootically infected with a poxvirus over the whole of its geographical range. In its natural host this produces a fibroma very like that produced in *S. floridanus* by Shope's fibroma virus, and it would not be inappropriate to call this agent the Marshall–Regnery fibroma. In *Oryctolagus*, however, this virus causes an acute infection, often associated with neurological signs, which is rapidly lethal, with few external signs. In the occasional rabbit which survives from as long as twelve days swelling of the muco-cutaneous junctions and eyelids appear which resemble the signs of myxomatosis in *Oryctolagus*, and our anthropocentric habit of regarding disease in man or his domestic animals as the yardstick has led to this virus being designated Californian myxoma virus. It can be distinguished in many of its characteristics from both Shope's fibroma virus and classical (Brazilian) myxoma virus, and the comparative studies of Marshall & Regnery, described in chapters 7 and 15, show that there is probably a highly evolved climax association between Californian myxoma virus and *S. bachmani*. We have not been able to distinguish between strains of Californian myxoma virus recovered from areas as far apart as San Diego and San Francisco, and at various times between 1949 and 1959, by animal pathogenicity studies, plaque or pock character, or gel-diffusion tests.

The classical myxoma viruses were derived from *Oryctolagus* which had been naturally infected in Brazil (Moses (1911), the standard laboratory strain; Bouvier (1954), the Lausanne strain). Aragão showed that the natural host of this virus in Brazil is the tapeti, *S. brasiliensis*, in which infection produces a fibroma analogous to Shope's fibroma in *S. floridanus*. An appropriate eponym would be Aragão's fibroma. Transferred to *Oryctolagus* this virus causes classical myxomatosis, a florid disease of great severity, almost always lethal.

Myxoma virus occurs over a large part of the wide range of *S. brasiliensis*. Strains recovered from areas as far apart as Panama and Colombia, Brazil and Argentina, and at various times between 1898 and 1960, are all equally lethal for *Oryctolagus*, and produce the same florid symptoms. Our preliminary investigations, however, show that there are minor antigenic and other differences between strains from Panama and Colombia, and those from Brazil.

In American leporids, therefore, there exist at least three viruses of the fibroma-myxoma subgroup, each of which is highly adapted to one species of *Sylvilagus*, and probably could not survive as a natural infection in other species of *Sylvilagus* or in *Lepus*. Each is mechanically transmitted by insect vectors which feed through the fibroma. These associations are old and stable, and the virus and the leporids may have evolved together. In keeping with this we find that all isolates of each of these viruses behave in a similar way in a novel host, *Oryctolagus cuniculus*, which is by an accident of its genetic constitution highly susceptible to lethal infection with two of them. This stability is not the consequence of lack of mutability of the virus, however, but must be ascribed to the stabilizing effect of natural selection in these particular situations. When it was introduced into populations of wild *Oryctolagus* in Australia and Europe, Brazilian myxoma virus was exposed to quite new selective forces, and it underwent rapid changes in the characteristic of the greatest importance in ensuring its continued transmission (and therefore its survival in nature), namely its virulence for *Oryctolagus*.

MYXOMA VIRUS IN *ORYCTOLAGUS CUNICULUS*

Myxomatosis due to Brazilian myxoma virus became established as an enzootic-epizootic infection of wild *Oryctolagus* in Australia in 1950, in Europe in 1952, and possibly in Chile since 1954. It is worth considering separately the evolution of the virus in this novel situation in three geographical areas, Australia, Britain and France. The reasons for such separate consideration are twofold, primarily because three different mechanisms of insect transmission have been involved in these three localities, and secondarily because different

TABLE 58. *Characteristics of enzootic-epizootic myxomatosis of* Oryctolagus *in Australia, Britain and France*

Country	Strain of virus originally introduced	First natural spread	Dominant vectors
Australia	Standard laboratory strain, originally from Moses (1911); re-introduced annually by widespread inoculation campaigns	December 1950	Mosquitoes in violent summer epizootics. The rabbit flea does not occur in Australia
France	Lausanne (Bouvier, 1954)	June 1952	Mosquitoes in violent summer epizootics. The rabbit flea is common
Britain	Virulent derivative of Lausanne	September 1953	Mosquitoes are relatively unimportant. Rabbit flea is responsible for transmission throughout the year

22-2

strains of virus were used in Europe and Australia (see Table 58). Relevant information is most complete for Australia, least for France.

The evolution of myxoma virus in Australia

The uniformly high kills which characterized the early epizootics of myxomatosis in Australia reflected the relatively homogeneous nature of the virus introduced into the wild rabbit population, at least for the genes concerned with virulence for $O.$ cuniculus. Case mortality rates in large populations of the order of 99·8 % would not have occurred if mutants of diminished virulence had been at all common, for in most cases mixed infections of rabbits with attenuated and virulent myxoma viruses prolong the survival time, and may lead to recovery of the infected rabbit. Yet after the first winter, strains of demonstrably diminished virulence for Oryctolagus were common enough to be found in a very superficial survey; and at Lake Urana, where intensive observations were made, none of eight isolates from wild caught mosquitoes was as virulent as the virus which had caused mortalities of over 99 % in large populations of rabbits a year before.

In subsequent years natural isolates of still lower virulence were found, but no highly attenuated viruses like neuromyxoma have ever been recovered from natural cases. The situation a decade after its introduction was that many mutant strains of varying virulence for $O.$ cuniculus were established as enzootic infections in various local populations of rabbits, and these spread with varying efficiency and over variable distances each summer, depending upon the vector incidence, rabbit numbers, and other factors. How can we explain this lability in a virus which in South America had shown such great uniformity and stability in just this character, namely its virulence for Oryctolagus?

In $S.$ brasiliensis myxoma virus survives as a mosquito-transmitted agent producing trivial tumours, the annual breeding of its host providing an annual crop of new susceptible animals. Persistence of an infectious tumour, in at least a few animals, is a necessary and sufficient reason for survival of the virus. In Oryctolagus, on the other hand, this virus causes an infection which, although it produces many infectious tumours, kills the host within five days of the onset of infectivity. In a large population of susceptible rabbits, and with a large population of an efficient vector like Anopheles annulipes, this is enough to ensure almost complete spread of the virus through the population. Over-wintering, however, involves difficulties in survival for which the rapidly lethal viruses are ill-fitted. There are few vectors, and susceptible rabbits are rare. A virus which causes a disease which is infectious for only five days therefore fails to survive. In the absence of survival in another vertebrate or invertebrate host, long persistence of infectious lesions is an essential condition for successful survival through the winter, and this is precisely the situation produced by the attenuated strains of virus. So far no viruses have appeared in *Oryctolagus* in Australia which cause long persistence of solitary or scattered tumours. The very attenuated viruses which have been recovered after laboratory manipulations cause a mild generalized disease from which the rabbit rapidly recovers, so that it ceases to be infectious due to healing of the lesions somewhat more rapidly than the highly virulent virus removes its infected host by killing it.

That this is a reasonable explanation for the appearance of moderately attenuated strains of virus after the winter is borne out by the advantage shown by such strains in direct competition with highly virulent strains during summer epizootics at Lake Urana and Merricumbene.

Selection on the basis of transmissibility is probably the correct explanation for the replacement of the highly virulent strain of myxoma virus by others of diminished virulence. However, it does not explain the emergence, in individual rabbits, of a viral population comprising principally the attenuated virus, and this event must have occurred repeatedly in many parts of Australia, and must still occur. To explain this we can only speculate, although an experimental study of the problem is feasible. A possible explanation is that provided by Lwoff & Lwoff (1960). They suppose that attenuated strains of myxoma virus, like attenuated strains of poliovirus type 1, are less thermoresistant and less psychrosensitive (in their replication) than virulent strains. The growth of poliovirus at low temperatures selects attenuated 'cold' variants, and the same may hold for myxoma virus. Hence the lowered temperatures experienced during the winter months would select for attenuated strains. Having arisen, they would be maintained and would spread by the operation of the mechanisms we have described earlier.

The evolution of myxoma virus in Britain

The virus introduced into England in September 1953 was a highly virulent derivative of the Lausanne strain, released in France fifteen months earlier. Its spread over the whole of Britain was slow, compared with the spread of myxomatosis over south-eastern Australia by mosquitoes in 1950–51, and was effected by man and by the rabbit flea, which continue to be the major vectors of myxomatosis. Regular inoculation of rabbits with virulent virus has not been attempted.

TABLE 59. *Comparison of the virulence of field strains of myxoma virus in Australia and Britain several years after its introduction (figures indicate percentages)*

	Virulence Grade					
	I	II	III A	III B	IV	V
Estimated case mortality rate (%)	> 99	95–99	90–95	70–90	50–70	< 50
Australia 1950–51	100	—	—	—	—	—
1958–59	0	25	29	27	14	5
1963–64*	0	0·3	26·0	34·0	31·3	8·3
Britain 1953	100	—	—	—	—	—
1962	4·1	17·6	38·8	24·8	14·4	0·5

* From Fenner & Woodroofe (unpublished observations).

A survey of 222 strains of virus from 80 counties of Britain nine years after the introduction of the virus revealed a somewhat different situation from that found in Australia (see Table 59). Several strains were recovered which were just as virulent as the Lausanne strain, and produced similar lesions. Most of the rest were of Grade II or high Grade III virulence. Since the winters in Britain are much colder than in Australia selection of the postulated somewhat attenuated 'cold' variants might be expected to be even more pronounced. Yet the virulence level remains somewhat higher than in Australia. Possible explanations for this are the greater mutability of the standard laboratory strain, which was released in Australia, or the effect of the different vector in the two countries, namely the rabbit flea in England and the mosquito in Australia. Tests on viruses obtained from France (see next section), and laboratory studies with the Lausanne virus, lead us to favour the latter explanation.

Although Mead-Briggs found that fleas are more mobile than was previously suspected, infected fleas would much more frequently leave a dead animal than a live one. Hence flea transmission leads to selection for lethality. At the low temperatures experienced during winter in Britain even moderately attenuated strains would often cause lethal infections, so that one would not expect flea transmission to select absolutely for highly virulent strains. But it would select against strains which allowed the infected rabbit to survive.

Evolution of myxoma virus in France

Both flea and mosquito transmission occur in France, and summer mosquito-transmitted epizootics are a feature of the epidemiology. Our admittedly inadequate tests of virus strains recovered from France have revealed more attenuated strains there than were obtained in England. If we can accept the data at their face value, and not as the result of inadequate and biassed sampling, this result emphasizes the overriding importance of the mode of transmission in determining the relative frequency of strains of virus which differ in their virulence for *Oryctolagus*.

Comment

Under static conditions of natural selection, myxoma viruses in the Americas have remained stable for many years, and over a wide geographic area, in their virulence for a novel host, *Oryctolagus cuniculus*, which is exquisitely susceptible to lethal infection by them. When introduced into large natural populations of *Oryctolagus*, however, they were subjected to quite different forces of natural selection. In this new situation, in Australia, maintenance of high virulence would have led to the local eradication of rabbits, and the disappearance of the virus. This did not happen. Mechanical transmission by mosquitoes selected for viruses which would allow survival of rabbits in an infectious state for a prolonged period and attenuated strains of virus appeared after the first winter season. Strains of moderate virulence have now completely replaced the original highly virulent strain.

In England the rabbit flea is the principal vector. Infective fleas are much more likely to leave rabbits which die from acute myxomatosis than those which die after a prolonged illness, or recover, and this leads to selection for lethal strains. Tests of many virus strains recovered nine years after the introduction of the virus showed that the proportion of British strains which caused death in most infected rabbits was higher than in Australia, although most of them produced disease characterized by a longer survival time than was found with the virus originally introduced.

Tests of these attenuated strains in *S. brasiliensis* would be of great interest. It might be expected that the attenuated strains would fail to produce infectious tumours in this host, thus explaining the absence of such strains in recoveries of virus made in South America.

Only limited investigations have been made on possible changes in characteristics other than virulence for *Oryctolagus*. All the strains examined in rabbits have been titrated on the chorioallantoic membrane, but pocks distinctly different from those produced by the original virulent strains have been detected in only two instances.

W. R. Sobey (personal communication, 1962) compared the soluble antigens of the prototype attenuated Australian strains with those of the original virulent strain, and found no differences between them. Further experiments of this type, with a large number of strains, would be interesting, in view of the differences which we have found in different geographic isolates from South America.

EVOLUTIONARY CHANGES IN *ORYCTOLAGUS CUNICULUS* DUE TO MYXOMATOSIS

Myxomatosis in Australia has provided a unique opportunity for the investigation of the influence of an infectious disease on the evolution of a mammal. It is worth recalling just how suitable the situation was.

(*a*) The wild rabbit population of the mainland of Australia (but not that of Tasmania) derives from two dozen wild rabbits introduced in the 'Lightning' shipment of 1859, with negligible additions to the gene pool by escaped domestic rabbits, and no other introductions of wild rabbits. It is thus likely to be more homogeneous genetically than populations of wild rabbits in Europe.

(*b*) The rabbit is large enough to make population counts feasible, although difficult. Attempts to set aside a laboratory colony of wild rabbits before they had been exposed to myxomatosis failed, but it was shown that laboratory rabbits behaved, in their reaction to both virulent and attenuated strains of myxoma virus, in a manner very similar to genetically unselected wild rabbits. They thus constitute a control animal with which to compare progressive changes in the genetic resistance of wild rabbits.

(c) Myxoma virus produced disease only in the rabbit, and there was therefore no pressure to control or limit its spread. Modern techniques of storage at low temperature made it possible to keep stocks of virus for years with no loss in titre, and no further passage.

(d) The rabbit was so important a pest in Australia that adequate finances to study its reaction to myxomatosis were available.

In chapter 14 we described the results of tests on wild rabbits, which were confirmed by laboratory studies on selection for resistance to myxomatosis. Both showed that under the rigorous selective stress of the regular exposure of each generation to strains of virus which killed 90 % of unselected rabbits there was a relatively rapid increase in genetic resistance, such that in a period of seven years the case mortality rate fell from 90 to 25 %. Tests on laboratory bred rabbits showed that the heritability of resistance (h^2) was about 0·30.

There are few or no comparable experiences on the selection of wild animals for resistance to an infectious disease, although there is now an extensive literature on the genetics of resistance of mice and chickens to several bacterial and viral pathogens (Gowen, 1963; Hutt, 1963).

One of the few recorded examples of differences in the innate resistance of different populations of a single species of a wild mammal to a natural infection is plague in rats (Anonymous, 1912; Sokhey & Chitre, 1937). Wild rats were captured from many cities in India in which the recent experience of plague differed greatly. After maintenance for two weeks in the laboratory they were inoculated with a standard dose of plague bacilli. The mortality rates were inversely proportional to the recent experience of plague, and varied from 91 % for rats from cities with no plague for the previous thirty years to 10 % for rats from cities with severe plague up to two years before the capture of the rats. Further studies by Habbu (1960) showed that the resistance of Bombay rats remained at a high level from 1931 until 1959, a period during which Bombay was free of plague. Thus genetic resistance acquired during the plague years had persisted without substantial change for at least 30 years.

Intermittent epizootics of plague appear to have had no effect on the innate resistance of gerbils (*Tatera brantsi*), as far as can be recognized by observations on mortality rates in the field. Plague has been enzootic in the gerbils of northern Orange Free State, with epizootic episodes, since 1922. The observations of Davis (1953) showed that the gerbil population was still extremely susceptible to plague in 1940, and the same author noted that reports by officers of the anti-plague staff of the Union Health Department showed that the same pattern of temporary increase followed by population decline occurred throughout the next decade.

Very little work has been done to see whether there have been changes in characteristics of rabbits other than genetic resistance. The view is widespread in Britain that a much larger proportion lives above ground, rather than in burrows, than before 1953 (Anonymous 1963); but it is impossible to obtain precise data.

INTERACTIONS BETWEEN VIRUS VIRULENCE AND HOST RESISTANCE

The nature of the disease produced in the rabbit by myxoma virus, in relation to mechanical transmission by insects, is the material upon which natural selection for the virus operates. Too rapidly lethal a disease, or a disease with few infectious skin lesions, like Californian myxomatosis in *Oryctolagus*, does not provide adequate opportunity for mosquitoes to become infected. Neither does too mild a generalized infection, like neuromyxoma in genetically unselected rabbits. Two forms of disease offer the maximum opportunity for mosquito transmission. The form which has been important in myxomatosis in *Oryctolagus* in Australia, up to the present, is a severe generalized infection in which the lesions remain infective for mosquitoes for a prolonged period. A disease like fibroma or myxoma in the natural *Sylvilagus* hosts, that is a localized tumour which does not seriously interfere with the health of the rabbit, but does remain infectious for mosquitoes for many months, would also constitute an efficient reservoir of infection.

So far there has been no evidence that lesions of the latter type will develop in *Oryctolagus*. The picture with moderately virulent strains of virus in genetically resistant rabbits is like that of neuromyxoma in genetically unselected rabbits, that is a mild form of generalized myxomatosis with slight development and rapid healing of the skin lesions. But the possibility that with further selection for resistance of the wild rabbit a fibroma-like lesion will appear cannot be excluded.

Whether or not such a situation develops, we can envisage an interaction of genetic resistance and virus virulence (each measured by using the appropriate controls, a stored preparation of virus, and the laboratory rabbit respectively), such that the commonest form of myxomatosis will remain, for a long time, similar to that produced in laboratory rabbits by strains of Grade III virulence. In genetically resistant rabbits such a syndrome would be caused by viruses which would be classified by tests in laboratory rabbits as Grade I, and it is not inconceivable that selection of the virus, in genetically resistant rabbits, could lead to the emergence and perhaps the eventual dominance of 'hypervirulent' strains.

Since all available evidence shows that resistance to virulent strains operates through the same mechanism as resistance to attenuated strains the long-term effect of such interactions as we have envisaged will be continued selection for resistance. Apart from unexpected developments, like the widespread occurrence of increased resistance transmitted maternally, by congenital infection with an interfering virus, two possible climax associations can be visualized.

(a) A disease like fibroma may develop in genetically selected Australian wild rabbits. If virus from such lesions is transferred to genetically unselected rabbits it would behave as a very virulent myxoma virus. This is the climax association which has developed in *Sylvilagus* in South America and California, but there is no evidence of any such trend in *Oryctolagus* in Australia or Europe.

(b) There will not be continued progressive selection for genetic resistance, because after a certain degree of resistance has been acquired no further changes in the virus are possible which would enable it to produce the type of generalized disease suited to mosquito transmission. We could then envisage a climax association in which myxomatosis still caused a moderately severe disease with an appreciable mortality, much as smallpox does in human communities. The reproductive capacity of the rabbit is such that this sort of disease need not seriously interfere with its population size.

BIBLIOGRAPHY AND CITATION INDEX

*Numbers in parentheses following each reference indicate the
text pages where the reference has been cited*

ABEL, P. (1962). Multiplicity reactivation and marker rescue with vaccinia virus. *Virology*, **17**, 511. (69)

AHLSTRÖM, C. G. (1938). The histology of the infectious fibroma in rabbits. *J. Path. Bact.* **46**, 461. (106)

AHLSTRÖM, C. G. (1940a). On the anatomical character of the infectious myxoma of rabbits. *Acta path. microbiol. scand.* **18**, 377. (74, 106, 107)

AHLSTRÖM, C. G. (1940b). The reaction of tarred rabbits to the myxoma virus. *Acta path. microbiol. scand.* **18**, 394. (106, 112)

AHLSTRÖM, C. G. & ANDREWES, C. H. (1938). Fibroma virus infection in tarred rabbits. *J. Path. Bact.* **47**, 65. (97, 106, 111, 150)

ALLAN, R. M. & SHANKS, P. L. (1955). Rabbit fleas on wild rabbits and the transmission of myxomatosis. *Nature, Lond.*, **175**, 692. (155)

ANDERSON, S. G. & HAMILTON, J. (1949). The epidemiology of primary herpes simplex infection. *Med. J. Aust.* **1**, 308. (119)

ANDREWES, C. H. (1936). A change in rabbit fibroma virus suggesting mutation. I. Experiments on domestic rabbits. *J. exp. Med.* **63**, 157. (64)

ANDREWES, C. H. (1951). Viruses and Linnaeus. *Acta path. microbiol. scand.* **28**, 211. (61)

ANDREWES, C. H. & HARISIJADES, S. (1955). Propagation of myxoma virus in one-day old mice. *Brit. J. exp. Path.* **36**, 18. (80, 87)

ANDREWES, C. H. & HORSTMANN, D. M. (1949). The susceptibility of viruses to ethyl ether. *J. gen. Microbiol.* **3**, 290. (71)

ANDREWES, C. H., MUIRHEAD-THOMPSON, R. C. & STEVENSON, J. P. (1956). Laboratory studies of *Anopheles atroparvus* in relation to myxomatosis. *J. Hyg., Camb.*, **54**, 478. (159, 228, 327, Table 29)

ANDREWES, C. H., THOMPSON, H. V. & MANSI, W. (1959). Myxomatosis: present position and future prospects in Great Britain. *Nature, Lond.*, **184**, 1179. (327)

ANONYMOUS (1912). Sixth report on plague investigations in India. *J. Hyg., Camb.*, **12**, Plague Supplement 1, 229. (345)

ANONYMOUS (1942). A mechanical device for the spread of disease agents amongst rabbits. *J. Coun. sci. industr. Res. Aust.* **15**, 82. (273)

ANONYMOUS (1951). Murray Valley encephalitis. *Med. J. Aust.* **1**, 526. (281)

ANONYMOUS (1955). Myxomatosis in hares. *Vet. Rec.* **67**, 455. (89, 92)

BIBLIOGRAPHY

ANONYMOUS (1959). *Annual Report on the Operations of 'The Stock Routes and Rural Lands Protection Acts, 1944 to 1951' for the year* 1958–59. Brisbane: Government Printer. (289)

ANONYMOUS (1963). Differences in rabbits since myxomatosis. *The Times*, 11 May 1963, p. 11. (346)

APPLEYARD, G., WESTWOOD, J. C. N. & ZWARTOUW, H. T. (1962). The toxic effect of rabbitpox virus in tissue culture. *Virology*, **18**, 159. (66)

ARAGÃO, H. DE B. (1920). Transmissão do virus do myxoma dos coelhos pelas pulgas. *Brazil-med.* **34**, 753. (155)

ARAGÃO, H. DE B. (1927). Myxoma of rabbits. *Mem. Inst. Osw. Cruz*, **20**, 237. (1, 61, 65, 86)

ARAGÃO, H. DE B. (1942). Sensibilidade do coelho do mato ao virus do mixoma; transmissão pelo "Aedes scapularis" e pelo "Stegomyia". *Brazil-med.* **56**, 207. (3)

ARAGÃO, H. DE B. (1943). O virus do mixoma no coelho do mato (*Sylvilagus minensis*), sua transmissão pelos *Aedes scapularis* e *aegypti*. *Mem. Inst. Osw. Cruz*, **38**, 93. (3, 88, 90, 145, 147, 154, 155, 252, 260)

ARMOUR, C. J. & THOMPSON, H. V. (1955). Spread of myxomatosis in the first outbreak in Great Britain. *Ann. appl. Biol.* **43**, 511. (322)

AUSTIN, A. W. (1955). The rabbit in Australia. *Field*, **205**, 1034. (17)

BANG, F. B., LEVY, E. & GEY, G. O. (1951). Some observations on host-cell-virus relationships in fowl pox. II. The inclusion produced by the virus on the chick chorioallantoic membrane. *J. Immunol.* **66**, 329. (65)

BARNARD, J. E. & ELFORD, W. J. (1931). The causative organism in infectious ectromelia. *Proc. roy. Soc.* B, **109**, 360. (108)

BARTHÉLÉMY, F. (1953). *Cahiers français d'Information*, **237**, 4. Paris. (316)

BENJAMIN, B. & RIVERS, T. M. (1931). Regeneration of virus myxomatosum (Sanarelli) in the presence of cells of exudates surviving *in vitro*. *Proc. Soc. exp. Biol., N.Y.*, **28**, 791. (73)

BERRY, G. P. & DEDRICK, H. M. (1936). A method for changing the virus of rabbit fibroma (Shope) into that of infectious myxomatosis (Sanarelli). *J. Bact.* **31**, 50. (69)

BERRY, G. P. & LICHTY, J. A., JR. (1936). Immunological and serological evidence of a close relationship between the viruses of rabbit fibroma (Shope) and infectious myxomatosis (Sanarelli). *J. Bact.* **31**, 49. (128)

BORG, K. (1962). Om myxomatos. *Medlemsbl. Sveriges Vet.-Förb.* **14**, 89. (9, 310, 313)

BORG, K. & BAKOS, K. (1963). Dissemination of myxomatosis by birds. *Nord. VetMed.* **15**, 159. (169)

BOSWELL, F. W. (1947). Electron microscope studies of virus elementary bodies. *Brit. J. exp. Path.* **28**, 253. (108)

BOUVIER, G. (1954). Quelques remarques sur la myxomatose. *Bull. Off. int. Epiz.* **46**, 76. (310, 338, 339)

BRAMBELL, F. W. R. (1942). Intra-uterine mortality in the wild rabbit, *Oryctolagus cuniculus* (L.). *Proc. roy. Soc.* B, **130**, 462. (42)

BRAMBELL, F. W. R. (1944). The reproduction of the wild rabbit *Oryctolagus cuniculus* (L.). *Proc. zool. Soc. Lond.* **114**, 1. (35, 40, 42, 116)

BRAMBELL, F. W. R., HEMMINGS, W. A. & HENDERSON, M. (1951). *Antibodies and Embryos*. University of London, The Athlone Press. (130)

BRERETON, J. LE GAY (1953). Initial spread of myxomatosis in Australia. *Nature, Lond.*, **172**, 108. (201)

BRONSON, L. H. & PARKER, R. F. (1943). The inactivation of the virus of infectious myxomatosis by heat. *J. Bact.* **45**, 177. (70)

BROWN, A., MAYYASI, S. A. & OFFICER, J. E. (1959). The 'toxic' activity of vaccinia virus in tissue culture. *J. infect. Dis.* **104**, 193. (69)

BROWN, P. W., ALLAN, R. M. & SHANKS, P. L. (1956). Rabbits and myxomatosis in the north-east of Scotland. *Scot. Agric.* **35**, 204. (167)

BUIST, J. B. (1886). The life history of the microorganisms associated with variola and vaccinia. *Proc. roy. Soc. Edinb.* **13**, 603. (65)

BULL, L. B. & DICKINSON, C. G. (1937). The specificity of the virus of rabbit myxomatosis. *J. Coun. sci. industr. Res. Aust.* **10**, 291. (6, 86, 87, 91).

BULL, L. B. & MULES, M. W. (1944). An investigation of *Myxomatosis cuniculi* with special reference to the possible use of the disease to control rabbit populations in Australia. *J. Coun. sci. industr. Res. Aust.* **17**, 79. (6, 153, 155, 166, 200, 217, 273, 276, 283)

BULL, P. C. (1956). Some facts and theories on the ecology of the wild rabbit. *N.Z. Sci. Rev.* **14**, 51. (45, 67)

BURNET, F. M. (1955). *The Principles of Animal Virology*, p. 436. New York: Academic Press, Inc. (215)

CAIRNS, J. (1960). The initiation of vaccinia infection. *Virology*, **11**, 603. (68, 73, 75, 108)

CALABY, J. H. (1954). Unpublished report quoted by Marshall, Regnery & Grodhaus (1963). (265)

CALABY, J. H., GOODING, C. D. & TOMLINSON, A. R. (1960). Myxomatosis in Western Australia. *C.S.I.R.O. Wildl. Res.* **5**, 89. (190, 194, 200, 290)

CHAMBERLAIN, R. W. & SUDIA, W. D. (1961). Mechanism of transmission of viruses by mosquitoes. *Annu. Rev. Entomol.* **6**, 371. (158)

CHAPPLE, P. J. & BOWEN, E. T. W. (1963). A note on two attenuated strains of myxoma virus isolated in Great Britain. *J. Hyg., Camb.*, **61**, 161. (228)

CHAPPLE, P. J., BOWEN, E. T. W. & LEWIS, N. D. (1963). Some observations on the use of the Ouchterlony gel diffusion technique in the study of myxomatosis. *J. Hyg., Camb.*, **61**, 373. (238, 328)

CHAPPLE, P. J. & LEWIS, N. D. (1964). An outbreak of myxomatosis caused by a moderately attenuated strain of myxoma virus. *J. Hyg., Camb.*, **62**, 423. (167)

351

BIBLIOGRAPHY

CHAPPLE, P. J. & MUIRHEAD-THOMPSON, R. C. (1964). Effect of varying the site of intradermal inoculation of myxoma virus on the course of the disease. *J. comp. Path.* **74**, 366. (212)

CHAPPLE, P. J. & WESTWOOD, J. C. N. (1963). Electron microscopy of myxoma virus. *Nature, Lond.,* **199**, 199. (65)

CHAPRONIERE, D. M. (1956). The effect of myxoma virus on cultures of rabbit tissues. *Virology,* **2**, 599. (73)

CHAPRONIERE, D. M. & ANDREWES, C. H. (1957). Cultivation of rabbit myxoma and fibroma viruses in tissues of non-susceptible hosts. *Virology,* **4**, 351. (90, 92, 93, 98)

CHAPRONIERE, D. M. & ANDREWES, C. H. (1958). Factors involved in the susceptibility of tissues of various species to myxoma virus. *Virology,* **5**, 120. (90, 92)

CLARINGBOLD, P. J. & SOBEY, W. R. (1955). The biological assay of myxoma virus. *Brit. J. exp. Path.* **36**, 573. (80, 81)

CLEMMESEN, J. (1939). The influence of roentgen radiation on immunity to Shope fibroma virus. *Amer. J. Cancer,* **35**, 378. (106, 111, 127, 150)

COLLINS, J. J. (1955). Myxomatosis in the common hare—*Lepus europaeus. Irish Vet. J.* **9**, 268. (92)

CONSTANTIN, T., FEBVRE, H. & HAREL, J. (1956). Cycle de multiplication du virus du fibrome de Shope *in vitro* (souche OA). *C. R. Soc. Biol., Paris,* **150**, 347. (76)

COOPER, P. D. (1961). A chemical basis for the classification of animal viruses. *Nature, Lond.,* **190**, 302. (335)

DALES, S. & SIMINOVITCH, L. (1961). The development of vaccinia virus in Earle's L strain cells as examined by electron microscopy. *J. biophys. biochem. Cytol.* **10**, 475. (83)

DALMAT, H. T. (1958a). Effects of X-rays and chemical carcinogens on infectivity of domestic rabbit fibromas for arthropods. *J. infect. Dis.* **102**, 153. (150)

DALMAT, H. T. (1958b). Immunity of rabbits to Shope's fibroma virus. *J. infect. Dis.* **102**, 179. (127, 258)

DALMAT, H. T. (1958c). Passage of Shope's rabbit fibroma virus through one-day-old mice. *Proc. Soc. exp. Biol., N.Y.,* **97**, 219. (81)

DALMAT, H. T. (1959). Arthropod transmission of rabbit fibromatosis (Shope). *J. Hyg., Camb.,* **57**, 1. (146, 147, 148, 150, 156, 163, Table 29)

DALMAT, H. T. & STANTON, M. F. (1959). A comparative study of the Shope fibroma in rabbits in relation to transmissibility by mosquitoes. *J. nat. Cancer Inst.* **22**, 593. (108, 109, 150, 157)

DAVIS, D. H. S. (1953). Plague in South Africa: a study of the epizootic cycle in gerbils (*Tatera brantsi*) in the northern Orange Free State. *J. Hyg., Camb.,* **51**, 427. (346)

DAWSON, I. M. & MCFARLANE, A. S. (1948). Structure of an animal virus. *Nature, Lond.,* **161**, 464. (65)

352

BIBLIOGRAPHY

Day, M. F., Fenner, F., Woodroofe, G. M. & McIntyre, G. A. (1956). Further studies on the mechanism of mosquito transmission of myxomatosis in the European rabbit. *J. Hyg., Camb.*, **54**, 258. (70, 150, 157, 162, 164, 165, 166, 227, Table 29)

Delille, P. F. A. (1953). Une methode nouvelle permettant à l'agriculture de lutter efficacement contre la pullulation du lapin. *C. R. Acad. agric. Fr.* **39**, 638. (2, 9, 311)

De Maeyer, E. & De Maeyer-Guignard, J. (1964). Inhibition by 3-methylcholanthrene of interferon formation in rat-embryo cells infected with Sidnbis virus. *J. nat. Cancer Inst.* **32**, 1317. (112)

Dice, L. R. (1929). The phylogeny of the Leporidae, with description of a new genus. *J. Mammal.* **10**, 340. (337)

Dobrotworsky, N. V. (1954). The genus *Theobaldia* (Diptera, Culicidae) in Victoria. *Proc. Linn. Soc. N.S.W.* **79**, 65. (195)

Douglas, G. W. (1958). Myxomatosis in Victoria. *J. Dep. Agric. Vict.* **56**, 779. (292)

Douglas, G. W. (1962). The Glenfield strain of myxoma virus. Its use in Victoria. *J. Dep. Agric. Vict.* **60**, 511. (236, 292)

Douglas, G. W. & Tighe, F. G. (1965). Observations on the innate resistance of the rabbit (*Oryctolagus cuniculus*), in Victoria, Australia, to selected strains of myxomatosis. To be published. (243, 246, 247)

Downie, A. W., McCarthy, K., MacDonald, A., MacCallum, F. O. & Macrae, A. D. (1953). Virus and virus antigen in the blood of smallpox patients. Their significance in early diagnosis and prognosis. *Lancet*, ii, 164. (103)

Dungern & Coca (1909). Ueber Hasensarkome, die in Kaninchen wachsen und über das Wesen der Geschwulstimmunität. *Z. ImmunForsch.* **2**, 391. (99, 336)

Duran-Reynals, F. (1940). Production of degenerative inflammatory or neoplastic effects in the newborn rabbit by the Shope fibroma virus. *Yale J. Biol. Med.* **13**, 99. (106, 111)

Duran-Reynals, F. (1945). Immunological factors that influence the neoplastic effects of the rabbit fibroma virus. *Cancer Res.* **5**, 25. (97, 106, 111, 127, 128)

Dyce, A. L. (1961). Transmission of myxomatosis on the spines of thistles, *Cirsium vulgare* (savi) Ten. *C.S.I.R.O. Wildl. Res.* **6**, 88. (169)

Dyce, A. L. & Lee, D. J. (1962). Blood-sucking flies (Diptera) and myxomatosis transmission in a mountain environment in New South Wales. II. Comparison of the use of man and rabbit as bait animals in evaluating vectors of myxomatosis. *Aust. J. Zool.* **10**, 84. (195, 198, 199)

Easterbrook, K. B. (1961). The multiplication of vaccinia virus in suspended KB cells. *Virology*, **15**, 404. (75, 76)

Edlinger, E. & Harel, J. (1956). Propriétés antigéniques du virus fibromateux de Shope au cours du cycle de multiplication virale. Appari-

23   353   F&RM

tion de l'antigen 'soluble' dans la phase d'éclipse. *C. R. Soc. Biol., Paris*, **150**, 867. (76)

ENDERS, J. F. (1960). A consideration of the mechanisms of resistance to viral infection based on recent studies of the agents of measles and poliomyelitis. *Trans. Coll. Phycns Philad.* **28**, 68. (239)

ENGLISH, M. I., MACKERRAS, I. M. & DYCE, A. L. (1957). Notes on the morphology and biology of a new species of *Chalybosoma* (Diptera, Tabanidae). *Proc. Linn. Soc. N.S.W.* **82**, 289. (Pl. XI)

EPSTEIN, B., REISSIG, M. & DE ROBERTIS, E. (1952). Studies by electron microscopy of thin sections of infectious myxomatosis in rabbits. *J. exp. Med.* **96**, 347. (83, 109)

EPSTEIN, M. A. (1958). An investigation into the purifying effect of a fluorocarbon on vaccinia virus. *Brit. J. exp. Path.* **39**, 436. (66)

FARRANT, J. L. & FENNER, F. (1953). A comparison of the morphology of vaccinia and myxoma virus. *Aust. J. exp. Biol. med. Sci.* **31**, 121. (65, Pl. II)

FEBVRE, H. (1962). The Shope fibroma virus of rabbits. '*Tumours induced by Viruses: Ultrastructural Studies.*' Ed. A. J. Dalton & F. Hagenau. New York and London: Academic Press, Inc. (82)

FENNER, F. (1953*a*). Classification of myxoma and fibroma viruses. *Nature, Lond.*, **171**, 562. (61)

FENNER, F. (1953*b*). Changes in the mortality-rate due to myxomatosis in the Australian wild rabbit. *Nature, Lond.*, **172**, 228. (210, 243)

FENNER, F. (1958). The biological characters of several strains of vaccinia, cowpox and rabbitpox viruses. *Virology*, **5**, 502. (64)

FENNER, F. (1962). Interactions between poxviruses. The Leeuwenhoek Lecture. *Proc. roy. Soc.* B, **156**, 388. (62, 69, 72)

FENNER, F. (1965). Viruses of the myxoma-fibroma subgroup of the poxviruses. II. Comparison of soluble antigens by gel diffusion tests, and a general discussion of the subgroup. *Aust. J. exp. Biol. med. Sci.* **43**, 143. (63, 64, 67, 260, 262, 263)

FENNER, F. & BURNET, F. M. (1957). A short description of the pox-virus group (vaccinia and related viruses). *Virology*, **4**, 305. (61)

FENNER, F. & CHAPPLE, P. J. (1965). Evolutionary changes in myxoma virus in Britain. An examination of 222 naturally occurring strains obtained from 80 counties during the period October–November 1962. *J. Hyg., Camb.* **63**, 175. (123, 229, 230, 328, 329, 330)

FENNER, F. & DAY, M. F. (1952). Contrasting methods of transmission of animal viruses by mosquitoes. *Nature, Lond.*, **170**, 204. (170)

FENNER, F., DAY, M. F. & WOODROOFE, G. M. (1952). The mechanism of the transmission of myxomatosis in the European rabbit (*Oryctolagus cuniculus*) by the mosquito *Aëdes aegypti. Aust. J. exp. Biol. med. Sci.* **30**, 139. (156, 159, 163, Table 29)

FENNER, F., DAY, M. F. & WOODROOFE, G. M. (1956). Epidemiological consequences of the mechanical transmission of myxomatosis by mosquitoes. *J. Hyg., Camb.*, **54**, 284. (155, 156, 161, 163, 165, 219, 220, 296, Table 29)

FENNER, F., MARSHALL, I. D. & WOODROOFE, G. M. (1953). Studies in the epidemiology of infectious myxomatosis of rabbits. I. Recovery of Australian wild rabbits (*Oryctolagus cuniculus*) from myxomatosis under field conditions. *J. Hyg., Camb.*, **51**, 225. (120, 124, 125)

FENNER, F., POOLE, W. E., MARSHALL, I. D. & DYCE, A. L. (1957). Studies in the epidemiology of infectious myxomatosis of rabbits. VI. The experimental introduction of the European strain of myxoma virus into Australian wild rabbit populations. *J. Hyg., Camb.*, **55**, 192. (294, 295)

FENNER, F. & MARSHALL, I. D. (1954). Passive immunity in myxomatosis of the European rabbit (*Oryctolagus cuniculus*): the protection conferred on kittens born by immune does. *J. Hyg., Camb.*, **52**, 321. (110, 126, 130, 131, 132, 244)

FENNER, F. & MARSHALL, I. D. (1955). Occurrence of attenuated strains of myxoma virus in Europe. *Nature, Lond.*, **176**, 782. (210, 227, 328)

FENNER, F. & MARSHALL, I. D. (1957). A comparison of the virulence for European rabbits (*Oryctolagus cuniculus*) of strains of myxoma virus recovered in the field in Australia, Europe and America. *J. Hyg., Camb.*, **55**, 149. (4, 9, 78, 79, 80, 93, 110, 125, 126, 153, 210, 211, 212, 213, 217, 222, 226, 230, 231, 232, 236, 254, 256, 310, 319, 322, 327, 328)

FENNER, F. & McINTYRE, G. A. (1956). Infectivity titrations of myxoma virus in the rabbit and the developing chick embryo. *J. Hyg., Camb.*, **54**, 246. (78, 81)

FENNER, F. & WOODROOFE, G. M. (1953). The pathogenesis of infectious myxomatosis: The mechanism of infection and the immunological response in the European rabbit (*Oryctolagus cuniculus*). *Brit. J. exp. Path.* **34**, 400. (103, 104, 121, 122, 152, 156, 168)

FENNER, F. & WOODROOFE, G. M. (1954). Protection of laboratory rabbits against myxomatosis by vaccination with fibroma virus. *Aust. J. exp. Biol. med. Sci.* **32**, 653. (125, 127, 128)

FENNER, F. & WOODROOFE, G. M. (1960). The reactivation of pox-viruses. II. The range of reactivating viruses. *Virology*, **11**, 185. (62, 80)

FILSHIE, B. K. (1964). Observations with the electron microscope of myxoma virus on mosquito mouthparts. *Aust. J. biol. Sci.* **17**, 903. (158)

FISHER, E. R. (1953). The nature and staining reactions of the fibroma-cell inclusions of the Shope fibroma of the rabbit. *J. nat. Cancer Inst.* **14**, 355. (108, 109, 157)

FRITH, H. J. (1964). Mobility of the red kangaroo, *Megaleia rufa. C.S.I.R.O. Wildl. Res.* **9**, 1. (Pl. XI)

GALASSO, G. J. & SHARP, D. G. (1963). Homologous inhibition, toxicity, and multiplicity reactivation with ultraviolet-irradiated vaccinia virus. *J. Bact.* **85**, 1309. (69)

GAYLORD, W. H. & MELNICK, J. L. (1953). Intracellular forms of pox-viruses as shown by the electron microscope (vaccinia, ectromelia, molluscum contagiosum). *J. exp. Med.* **98**, 157. (108)

GIBAN, J. (1956). Repercussion de la myxomatose sur les populations de lapin de garenne en France. *Terre et la Vie*, **103**, 179. (314)

GIBAN, J. (1959). *The Development of Myxomatosis in France. Report of the International Conference on Harmful Mammals and their Control*, p. 68. Europe and Mediterranean Plant Protection Organization, Paris. (320)

GINDER, D. R. & FRIEDEWALD, W. F. (1951). Effect of Semliki Forest virus on rabbit fibroma. *Proc. Soc. exp. Biol.*, *N.Y.*, **77**, 272. (114, 142)

GINDER, D. R. & FRIEDEWALD, W. F. (1952). Effect of Semliki Forest virus on rabbit myxoma. *Proc. Soc. exp. Biol.*, *N.Y.*, **79**, 615. (114, 142)

GISPEN, R. (1952). Silver impregnation of smallpox elementary bodies after treatment with xylol. *Leeuwenhoek ned. Tijdschr.* **18**, 107. (65)

GOODSON, J. K. (1963). M.Sc. thesis, University of Wisconsin. (75, 76)

GOWEN, J. W. (1963). In *Genetic Selection in Man*, ed. W. J. Schull. Ann Arbor: The University of Michigan Press. (345)

GREEN, R. G. (1935). On the nature of filterable viruses. *Science*, **82**, 443. (336)

GRODHAUS, G., REGNERY, D. C. & MARSHALL, I. D. (1963). Studies in the epidemiology of myxomatosis in California. II. The experimental transmission of myxomatosis in brush rabbits (*Sylvilagus bachmani*) by several species of mosquitoes. *Amer. J. Hyg.* **77**, 205. (95, 147, 148, 267, 268)

GUO, K. D. (1937). Ueber die Immunitätsbeziehungen zwischen dem Shopeschen Fibroma-Virus, dem Myxomatose-Virus und dem Neurolapine-Virus bei Kaninchen. *Zbl. Bakt.* (*Abt.* 1), **139**, 308. (64)

HAAGEN, E. & DU, D. H. (1938). Weitere Untersuchungen über das Verhalten des Kaninchenmyxomvirus *in vitro*. *Zbl. Bakt.* (*Abt.* 1), **143**, 23. (77, 78, 234)

HABBU, M. K. (1960). The present position of plague in India. *Ind. J. Path. Bact.* **3**, 123. (345)

HALDANE, J. B. S. (1949). Disease and evolution. *Ric. sci.* **19**, 68. (335)

HANAFUSA, H. (1962). Factors involved in the initiation of multiplication of vaccinia virus. *Cold Spr. Harb. Symp. quant. Biol.* **27**, 209. (69)

HANAFUSA, T., HANAFUSA, H. & KAMAHORA, J. (1959). Transformation phenomena in pox group viruses. II. Transformation between several members of pox group. *Biken's J.* **2**, 85. (62)

HANSARD (1951). *Parliamentary Debates of the Commonwealth of Australia*, **212**, 171; **214**, 523. (281)

HARADA, K. & MATUMOTO, M. (1962). Antigenic relationship between mammalian and avian poxviruses as revealed by complement fixation reaction. *Jap. J. exp. Med.* **32**, 369. (64)

BIBLIOGRAPHY

HAREL, J. (1956). Rôle de la résistance naturelle dans l'évolution des tumeurs provoquées par le virus fibromateux de Shope (souche OA) chez le lapereau. Action de la cortisone. Transmission à la progéniture de l'immunité acquise par la mère. *C. R. Soc. Biol., Paris*, **150**, 351. (130)

HAREL, J. & CONSTANTIN, T. (1954). Sur la malignité des tumeurs provoquées par le virus fibromateux de Shope chez le lapin nouveau-né et le lapin adulte traité par des doses massives de cortisone. *Bull. Cancer*, **41**, 482. (97, 106, 111, 112)

HAYWARD, J. S. (1961). The ability of the wild rabbit to survive conditions of water restriction. *C.S.I.R.O. Wildl. Res.* **6**, 160. (35, 39)

HENDERSON, D. W. (1952). An apparatus for the study of airborne infection. *J. Hyg., Camb.*, **50**, 53. (153)

HENLE, W. & HENLE, G. (1947). The effect of ultraviolet irradiation on various properties of influenza viruses. *J. exp. Med.* **85**, 347. (69)

HERMAN, C. M., KILHAM, L. & WARBACH, O. (1956). Incidence of Shope's rabbit fibroma in cottontails at the Patuxent Research Refuge. *J. Wildlife Mgmt*, **20**, 85. (95, 98, 259)

HERMAN, C. M. & REILLY, J. R. (1955). Skin tumors on squirrels. *J. Wildlife Mgmt*, **19**, 402. (64, 98, 99)

HERSHKOVITZ, P. (1950). Mammals of Northern Colombia. Preliminary report No. 6: Rabbits (Leporidae) with notes on the classification and distribution of the South American forms. *Proc. U.S. nat. Mus.* **100**, 327. (12, 14)

HERZBERG, K. & THELEN, A. (1938). Über den Nachweis und den Vermehrungsvorgang des Virus des Shopeschen Kaninchenfibroms. *Virchows Arch.* **303**, 81. (65)

HIBBARD, C. W. (1963). The origin of the Ps pattern of *Sylvilagus, Caprolagus, Oryctolagus* and *Lepus. J. Mammal.* **44**, 1. (337)

HINZE, H. C. & WALKER, D. L. (1964). Response of cultured rabbit cells to infection with the Shope fibroma virus. I. Proliferation and morphological alteration of the infected cells. *J. Bact.* **88**, 1185. (77)

HOBBS, J. R. (1931). The occurrence of natural and acquired immunity to infectious myxomatosis of rabbits. *Science*, **73**, 94. (88, 89)

HOFFSTADT, R. E. & PILCHER, K. S. (1938). The use of the chorioallantoic membrane of the developing chick embryo as a medium in the study of virus myxomatosum. *J. Bact.* **35**, 353. (77)

HOLLAND, C. W. (1923). Rabbits and their introduction into Australia. *Qd Nat.* **4**, 7. (17)

HOULIHAN, R. B. & DERRICK, W. A. (1945). Infectious myxomatosis in malnourished rabbits. *Science*, **101**, 364. (142)

HUDSON, J. R. & MANSI, W. (1955). Attenuated strains of myxoma virus in England. *Vet. Rec.* **67**, 746. (231, 328)

HUDSON, J. R., THOMPSON, H. V. & MANSI, W. (1955). Myxoma virus in Britain. *Nature, Lond.*, **176**, 783. (210, 328)

357

HURST, E. W. (1937a). Myxoma and the Shope fibroma. I. The histology of myxoma. *Brit. J. exp. Path.* **18**, 1. (74, 106, 108)

HURST, E. W. (1937b). Myxoma and the Shope fibroma. II. The effect of intracerebral passage on the myxoma virus. *Brit. J. exp. Path.* **18**, 15. (106, 107, 210, 221, 234)

HURST, E. W. (1937c). Myxoma and the Shope fibroma. III. Miscellaneous observations bearing on the relationship between myxoma, neuromyxoma and fibroma viruses. *Brit. J. exp. Path.* **18**, 23. (97, 111, 128, 233)

HURST, E. W. (1938a). Myxoma and the Shope fibroma. IV. The histology of fibroma. *Aust. J. exp. Biol. med. Sci.* **16**, 53. (106)

HURST, E. W. (1938b). Myxoma and the Shope fibroma. V. Myxoma in the fibroma-immune rabbit, with a summary of present knowledge of the relationship between myxoma and fibroma viruses. *Aust. J. exp. Biol. med. Sci.* **16**, 205. (106, 107, 128)

HUTT, F. B. (1963). In *Genetic Selection in Man*, ed. W. J. Schull. Ann Arbor: The University of Michigan Press. (345)

HVASS, J. & SCHMIT-JENSEN, H. O. (1939). Report of a visit to the Dufeke estate in Skaane (Sweden) on 27 April 1939, for the purpose of studying the effects of infection by myxomatosis among wild rabbits. Report to Danish Ministry for Agriculture and Fisheries. (See also Notini *et al.* 1952.) (309)

HYDE, K. E. (1936). The relationship between the viruses of infectious myxoma and the Shope fibroma of rabbits. *Amer. J. Hyg.* **23**, 278. (91, 96)

HYDE, R. R. (1936). Immunity to virus myxomatosis as affected by the portal of entry. *Amer. J. Hyg.* **23**, 425. (128, 130)

HYDE, R. R. (1939a). The pathogenesis of infectious myxomatosis (Sanarelli) as modified by certain immunizing agents. *Amer. J. Hyg.* (B), **30**, 37. (123)

HYDE, R. R. (1939b). Infectious myxomatosis of rabbits (Sanarelli) versus the fibroma virus (Shope) with especial reference to the time interval in the establishment of concomitant immunity. *Amer. J. Hyg.* (B), **30**, 47. (128)

HYDE, R. R. & GARDNER, R. E. (1933). Infectious myxoma of rabbits. *Amer. J. Hyg.* **17**, 446. (86, 89, 90)

HYDE, R. R. & GARDNER, R. E. (1939). Transmission experiments with the fibroma (Shope) and myxoma (Sanarelli) viruses. *Amer. J. Hyg.* (B), **30**, 57. (130)

JACOTOT, H., LEVADITI, J., VALLÉE, A. & VIRAT, B. (1954). Un cas de généralisation du fibrome infectieux de Shope. *Ann. Inst. Pasteur*, **86**, 247. (97)

JACOTOT, H., TOUMANOFF, C., VALLÉE, A. & VIRAT, B. (1954). Transmission expérimentale de la myxomatose au lapin par *Anopheles maculipennis atroparvus* et *A. stephensi*. *Ann. Inst. Pasteur*, **87**, 477. (155, 317)

JACOTOT, H. & VALLÉE, A. (1953a). Un foyer de myxome infectieux chez des lapins de garenne dans la région de Rambouillet. *Ann. Inst. Pasteur*, **84**, 448. (9, 311)

JACOTOT, H. & VALLÉE, A. (1953b). Essais infructueux de vaccination contre la myxomatose des lapins par anavirus tissulaire. *Ann. Inst. Pasteur*, **85**, 133. (124)

JACOTOT, H., VALLÉE, A. & VIRAT, B. (1954a). Sur un cas de myxomatose chez le lièvre. *Ann. Inst. Pasteur*, **86**, 105. (91, 92, Pl. VI)

JACOTOT, H., VALLÉE, A. & VIRAT, B. (1954b). L'immunité contre la myxomatose des lapins est-elle transmissible de la mère à ses laperaux? *Bull. Acad. vét. Fr.* **27**, 465. (130)

JACOTOT, H., VALLÉE, A. & VIRAT, B. (1955a). Étude sur la transmission expérimentale de la myxomatose au lièvre. *Ann. Inst. Pasteur*, **88**, 1. (92)

JACOTOT, H., VALLÉE, A. & VIRAT, B. (1955b). Stabilité relative du pouvoir pathogène du virus du myxome infectieux. *Ann. Inst. Pasteur*, **88**, 234. (235)

JACOTOT, H., VALLÉE, A. & VIRAT, B. (1955c). Considérations sur la durée et le mécanisme de l'immunité engendrée par le virus du fibrome de Shope contre le virus du myxome de Sanarelli. *Ann. Inst. Pasteur*, **88**, 381. (128, 129)

JACOTOT, H., VALLÉE, A. & VIRAT, B. (1955d). Étude d'une souche atténuée de virus du myxome (Uriarra 111 d'Australie). *Ann. Inst. Pasteur*, **89**, 8. (220, 235)

JACOTOT, H., VALLÉE, A. & VIRAT, B. (1955e). Apparition en France d'un mutant naturellement atténué du virus de Sanarelli. *Ann. Inst. Pasteur*, **89**, 361. (319)

JACOTOT, H., VALLÉE, A. & VIRAT, B. (1955f). Sur la conservation et la destruction dans les peaux du virus de la myxomatose des lapins. *Ann. Inst. Pasteur*, **89**, 290. (Table 29)

JACOTOT, H., VALLÉE, A. & VIRAT, B. (1956). Étude de quelques souches française de virus atténuée du myxome infectieux. *Ann. Inst. Pasteur*, **90**, 779. (235, 320)

JACOTOT, H., VALLÉE, A. & VIRAT, B. (1957). Incidence de la quantité de virus inoculée sur l'évolution de la myxomatose expérimentale. *Ann. Inst. Pasteur*, **92**, 262. (212)

JACOTOT, H., VALLÉE, A. & VIRAT, B. (1958). Sur l'immunisation contre le virus du myxome infectieux par inoculation de virus du fibrome de Shope. *Ann. Inst. Pasteur*, **94**, 282. (129)

JACOTOT, H., VALLÉE, A. & VIRAT, B. (1962). Influence de la cortisone sur l'immunisation du lapin contre la myxomatose par inoculation de virus du fibrome. *Ann. Inst. Pasteur*, **103**, 285. (112, 129)

JOKLIK, W. K. (1962a). The purification of four strains of poxvirus. *Virology*, **18**, 9. (65, 68)

JOKLIK, W. K. (1962b). Some properties of poxvirus deoxyribonucleic acid. *J. Mol. Biol.* **5**, 265. (68)

JOKLIK, W. K. (1962c). The multiplication of poxvirus DNA. *Cold Spr. Harb. Symp. quant. Biol.* **27**, 199. (69, 72)

JOKLIK, W. K., ABEL, P. & HOLMES, I. H. (1960). Reactivation of poxviruses by a non-genetic mechanism. *Nature, Lond.*, **186**, 992. (69)

KATO, S. & CUTTING, W. (1959). A study of the inclusion bodies of rabbit myxoma and fibroma virus and a consideration of the relationship between all pox virus inclusion bodies. *Stanf. med. Bull.* **17**, 34. (109)

KATO, S., TAKAHASHI, M., KAMEYAMA, S. & KAMAHORA, J. (1959). A study on the morphological and cyto-immunological relationship between the inclusions of variola, cowpox, rabbitpox, vaccinia (variola origin) and vaccinia IHD and a consideration of the term 'Guarnieri body'. *Biken's J.* **2**, 353. (108)

KATO, S., TAKAHASHI, M., KAMEYAMA, S., MORITA, K. & KAMAHORA, J. (1959). Studies on the carrier culture of rabbit fibroma and myxoma virus. *Biken's J.* **2**, 30. (76)

KATO, S., TAKAHASHI, M., MIYAMOTO, H. & KAMAHORA, J. (1963). Shope fibroma and rabbit myxoma viruses. I. Autoradiographic and cytoimmunological studies on 'B' type inclusions. *Biken's J.* **6**, 127. (68)

KEJDANA, S. (1955). Myxomatosis in hares. *Méd. vét., Varsovie,* **11**, 136. (92)

KESSEL, J. F., FISK, R. T. & PROUTY, C. C. (1934). Studies with the Californian strain of the virus of infectious myxomatosis. *Proc. Fifth Pacific Sci. Congr.* **4**, 2927. (4, 257)

KESSEL, J. F., PROUTY, C. C. & MEYER, J. W. (1931). Occurrence of infectious myxomatosis in southern California. *Proc. Soc. exp. Biol., N.Y.*, **28**, 413. (2, 4, 265)

KIDDLE, MARGARET (1961). *Men of Yesterday. A Social History of the Western District of Victoria,* p. 573. Melbourne University Press. (17)

KILBOURNE, E. D., SMART, M. K. & POKORNY, B. A. (1961). Inhibition by cortisone of the synthesis and action of interferon. *Nature, Lond.*, **190**, 650. (112)

KILHAM, L. (1955). Metastasizing viral fibromas of grey squirrels: pathogenesis and mosquito transmission. *Amer. J. Hyg.* **61**, 55. (99)

KILHAM, L. (1958). Fibroma-myxoma virus transformation in different types of tissue culture. *J. nat. Cancer Inst.* **20**, 729. (93)

KILHAM, L. (1959). Relation of thermoresistance to virulence among fibroma and myxoma viruses. *Virology,* **9**, 486. (141, 240)

KILHAM, L. & DALMAT, H. T. (1955). Host-virus-mosquito relations of Shope fibromas in cottontail rabbits. *Amer. J. Hyg.* **61**, 45. (96, 146, 147, 148, 258, Table 29)

KILHAM, L. & FISHER, E. R. (1954). Pathogenesis of fibromas in cottontail rabbits. *Amer. J. Hyg.* **59**, 104. (96, 146)

KILHAM, L., HERMAN, C. M. & FISHER, E. R. (1953). Naturally occurring fibromas of grey squirrels related to Shope's rabbit fibroma. *Proc. Soc. exp. Biol., N.Y.*, **82**, 298. (64, 98, Pl. VI)

BIBLIOGRAPHY

KILHAM, L., LERNER, E., HIATT, C. & SHACK, J. (1958). Properties of myxoma virus transforming agent. *Proc. Soc. exp. Biol., N.Y.*, **98**, 689. (71)

KILHAM, L. & WOKE, P. A. (1953). Laboratory transmission of fibromas (Shope) in cottontail rabbits by means of fleas and mosquitoes. *Proc. Soc. exp. Biol., N.Y.*, **83**, 296. (146, 147, Table 29)

KIRSCHSTEIN, R. L., RABSON, A. S. & KILHAM, L. (1958). Pulmonary lesions produced by fibroma viruses in squirrels and rabbits. *Cancer Res.* **18**, 1340. (99)

KLEIN, J. M. & MARKS, E. N. (1960). Australian mosquitoes described by Macquart. I. Species in the Paris Museum, *Aëdes (Finlaya) alboannulatus* (Macquart), *Aëdes (Finlaya) rubrithorax* (Macquart), *Aëdes (Ochlerotatus) albirostris* (Macquart). New synonymy and a new species from New Zealand. *Proc. Linn. Soc. N.S.W.* **85**, 107. (177)

LAFENÈTRE, H., CORTEZ, A., RIOUX, J. A., PAGES, A., VOLLHARDT, Y. & QUATREFAGES, H. (1960). Enzootie de tumeurs cutanées chez le lièvre. *Bull. Acad. vét. Fr.* **33**, 379. (99, 100)

LAIDLAW, P. P. (1938). *Virus Diseases and Viruses.* Rede Lecture, University of Cambridge. Cambridge University Press. (336)

LEDINGHAM, J. C. G. (1937). Studies on the serological interrelationships of the rabbit viruses, myxomatosis (Sanarelli, 1898), and fibroma (Shope, 1932). *Brit. J. exp. Path.* **18**, 436. (64, 119)

LEE, D. J., CLINTON, K. J. & O'GOWER, A. K. (1954). The blood sources of some Australian mosquitoes. *Aust. J. biol. Sci.* **7**, 282. (174, 182, 184)

LEE, D. J., DYCE, A. L. & O'GOWER, A. K. (1957). Blood-sucking flies (Diptera) and myxomatosis transmission in a mountain environment in New South Wales. *Aust. J. Zool.* **5**, 355. (175, 180, 189, 197, 205)

LEINATI, L., MANDELLI, G. & CARRARA, O. (1959). Lesioni cutanee nodulari nelle lepri della pianura padana. *Atti. Soc. ital. Sci. vet.* **13**, 429. (99, 100)

LEINATI, L., MANDELLI, G., CARRARA, O., CILLI, V., CASTRUCCI, G. & SCATOZZA, F. (1961). Ricerche anatomo-istopatologiche e virologiche sulla malattia cutanea nodulare delle lepri padane. *Boll. Ist. sieroter. Milano*, **40**, 295. (99, 100)

LEOPOLD, A. S. (1959). *Wildlife of Mexico: the Game Birds and Mammals.* University of California Press. (14)

LERNER, I. M. (1950). *Population Genetics and Animal Improvement.* Cambridge University Press. (249)

LEVER, R. A., ARMOUR, C. J. & THOMPSON, H. V. (1957). Myxomatosis and the fox. *Agriculture, Lond.* **64**, 105. (332)

LEWIS, M. R. & GARDNER, R. E. (1932). A simple method for studying the cytology of infectious myxoma of the rabbit. *Amer. J. Path.* **8**, 583. (65)

LINES, E. W. L. (1952). Some natural factors affecting rabbit populations in Australia. *S. Aust. Nat.* **27**, 28. (40, 202, 288)

361

LIPSCHÜTZ, B. (1927). Untersuchungen über die Aetiologic der Myxom-krankheit des Kaninchens: vorlaufe Mitteilung. *Wien. klin. Wschr.* **40**, 1101. (65)

LLOYD, B. J., JR. & KAHLER, H. (1955). Electron microscopy of the virus of rabbit fibroma. *J. nat. Cancer Inst.* **15**, 991. (65)

LOCKLEY, R. M. (1940). Some experiments in rabbit control. *Nature, Lond.*, **145**, 767. (308, 309, 321)

LOCKLEY, R. M. (1954). The European rabbit-flea, *Spilopsyllus cuniculi*, as a vector of myxomatosis in Britain. *Vet. Rec.* **66**, 434. (155, 167)

LOCKLEY, R. M. (1955). Failure of myxomatosis on Skokholm Island. *Nature, Lond.*, **175**, 906. (308, 309, 324)

LUCAS, A., BOULEY, G., QUINCHON, C. & TOUCAS, L. (1953). La myxomatose du lièvre. *Bull. Off. int. Epiz.* **39**, 770. (91)

LURIA, S. E. (1958). Viruses as infective genetic materials. *Immunity & Virus Infection*, ed. V. A. Najjar, p. 188. New York: John Wiley and Sons. (336)

LUSH, D. (1937). The virus of infectious myxomatosis of rabbits on the chorioallantoic membrane of the developing egg. *Aust. J. exp. Biol. med. Sci.* **15**, 131. (77, 81, 87, 120, 234)

LUSH, D. (1939). The serological relationship of myxoma and Shope's fibroma viruses. *Aust. J. exp. Biol. med. Sci.* **17**, 85. (119, 121)

LWOFF, A. (1957). The concept of virus. *J. gen. Microbiol.* **17**, 239. (60)

LWOFF, A. (1959). Factors influencing the evolution of viral diseases at the cellular level and in the organism. *Bact. Rev.* **23**, 109. (134)

LWOFF, A. & LWOFF, M. (1960). Sur les facteurs du développement viral et leur rôle dans l'évolution de l'infection. *Ann. Inst. Pasteur*, **98**, 173. (134, 142, 240, 341)

LWOFF, A., ANDERSON, T. F. & JACOB, F. (1959). Remarques sur les caractéristiques de la particle virale infectieuse. *Ann. Inst. Pasteur*, **97**, 281. (64)

MAGALLON, P., BAZIN, O. & BAZIN, J. (1953). La myxomatose du lièvre. *Bull. Off. int. Epiz.* **39**, 765. (91)

MANSI, W. (1957a). Serological investigation of myxoma and fibroma viruses. I. Complement-fixation test. *J. comp. Path.* **67**, 208. (121)

MANSI, W. (1957b). The study of some viruses by the plate gel diffusion precipitin test. *J. comp. Path.* **67**, 297. (67, 121)

MANSI, W. & THOMAS, V. (1958). Serological investigation of myxoma and fibroma viruses. II. The gel diffusion precipitin test. *J. comp. Path.* **68**, 188. (67, 103, 104, 121, 123, 237, 326, 328)

MARSHALL, I. D. (1959). The influence of ambient temperature on the course of myxomatosis in rabbits. *J. Hyg., Camb.*, **57**, 484. (136, 137, 138, 141, 212, Pl. VIII)

MARSHALL, I. D. (1961). Myxomatosis investigations carried out in Central and South America. Report to the Australian Wool Research Fund Committee. (3, 252, 253, 261, 263, 266)

MARSHALL, I. D. & DOUGLAS, G. W. (1961). Studies in the epidemiology of infectious myxomatosis of rabbits. VIII. Further observations on changes in the innate resistance of Australian wild rabbits exposed to myxomatosis. *J. Hyg., Camb.*, **59**, 117. (141, 143, 213, 243, 245, 246, 247)

MARSHALL, I. D., DYCE, A. L., POOLE, W. E. & FENNER, F. (1955). Studies in the epidemiology of infectious myxomatosis of rabbits. IV. Observations of disease behaviour in two localities near the northern limit of rabbit infestation in Australia, May 1952 to April 1953. *J. Hyg., Camb.*, **53**, 12. (115, 116, 204)

MARSHALL, I. D. & FENNER, F. (1958). Studies in the epidemiology of infectious myxomatosis of rabbits. V. Changes in the innate resistance of Australian wild rabbits exposed to myxomatosis. *J. Hyg., Camb.*, **56**, 288. (212, 243, 245, 246)

MARSHALL, I. D. & FENNER, F. (1960). Studies in the epidemiology of infectious myxomatosis of rabbits. VII. The virulence of strains of myxoma virus recovered from Australian wild rabbits between 1951 and 1959. *J. Hyg., Camb.*, **58**, 485. (223, 224)

MARSHALL, I. D. & REGNERY, D. C. (1960). Myxomatosis in a Californian brush rabbit (*Sylvilagus bachmani*). *Nature, Lond.*, **188**, 73. (4, 94, 145, 148, 252, 257, 265)

MARSHALL, I. D. & REGNERY, D. C. (1963). Studies in the epidemiology of myxomatosis in California. III. The response of brush rabbits (*Sylvilagus bachmani*) to infection with exotic and enzootic strains of myxoma virus, and the relative infectivity of the tumors for mosquitoes. *Amer. J. Hyg.* **77**, 213. (88, 91, 149, 265)

MARSHALL, I. D., REGNERY, D. C. & GRODHAUS, G. (1963). Studies in the epidemiology of myxomatosis in California. I. Observations on two outbreaks of myxomatosis in coastal California and the recovery of myxoma virus from a brush rabbit (*Sylvilagus bachmani*). *Amer. J. Hyg.* **77**, 195. (4, 93, 127, 211, 254)

MARTIN, C. J. (1936). Observations on *Myxomatosis cuniculi* (Sanarelli) made with a view to the use of the virus in the control of rabbit plagues. *Bull. Coun. sci. industr. Res. Aust.* No. 96. (6, 86, 130, 153, 217, 321)

MATSUMOTO, S. (1958). Electron microscope studies of ectromelia replication. *Annu. Report Inst. Virus Research, Kyoto Univ.* Ser. A, **1**, 151. (108)

McKEE, C. M. (1939). Immunization against infectious myxomatosis with heat-inactivated virus in conjunction with the type III pneumococcus. *Amer. J. Hyg.* (B), **29**, 165. (123)

McKERCHER, D. G. (1952). Infectious myxomatosis. I. Vaccination. II. Antibiotic therapy. *Amer. J. vet. Res.* **13**, 425. (124)

McKERCHER, D. G. & SAITO, J. (1964). An attenuated live virus vaccine for myxomatosis. *Nature, Lond.*, **202**, 933. (77, 130, 235, 239)

MEAD-BRIGGS, A. R. (1964). Some experiments concerning the interchange of rabbit-fleas, *Spilopsyllus cuniculi* (Dale), between living rabbit hosts. *J. anim. Ecol.* **33**, 13. (168)

BIBLIOGRAPHY

MEAD-BRIGGS, A. R. & RUDGE, A. J. B. (1960). Breeding of the rabbit-flea, *Spilopsyllus cuniculi* (Dale): requirement of a 'factor' from a pregnant rabbit for ovarian maturation. *Nature, Lond.*, **187**, 1136. (168)

MELLO, U. (1929). Di una affezione neoplasica ad andamenio epizootico nelle lepri. *Ann. Staz. sper. lotta mal inf. bestiame Piem.-Lig.* **2**, 47. (99, 337)

MILES, J. A. R., FOWLER, M. C. & HOWES, D. W. (1951). Isolation of a virus from encephalitis in South Australia: a preliminary report. *Med. J. Aust.* **1**, 799. (281)

MIMS, C. A. (1960). Intracerebral injections and the growth of viruses in the mouse brain. *Brit. J. exp. Path.* **41**, 52. (81)

MIMS, C. A. (1964). Aspects of the pathogenesis of virus diseases. *Bact. Rev.* **28**, 30. (87, 93, 103, 105)

MOORE, M. S. & WALKER, D. L. (1962). Accentuation of plaques of myxoma and fibroma viruses by immune serum. *Proc. Soc. exp. Biol., N.Y.*, **111**, 493. (75, Pl. IV)

MOORE, N. W. (1956). Rabbits, buzzards and hares. Two studies on the indirect effects of myxomatosis. *Terre et la Vie*, **103**, 220. (332)

MOORE, N. W. (1957). The past and present status of the buzzard in the British Isles. *Brit. Birds*, **50**, 173. (333)

MOREL, A. (1956). Influence de l'épidémie de myxomatose sur la flore française. *Terre et la Vie*, **103**, 226. (321)

MOSES, A. (1911). O virus do mixoma dos coelhos. *Mem. Inst. Osw. Cruz*, **3**, 46. (1, 3, 6, 9, 86, 217, 338, 339)

MOULDER, J. W. (1962). *The Biochemistry of Intracellular Parasitism.* The University of Chicago Press. (60)

MUIRHEAD-THOMPSON, R. C. (1956a). The part played by woodland mosquitoes of the genus *Aëdes* in the transmission of myxomatosis in England. *J. Hyg., Camb.*, **54**, 461. (155, 327)

MUIRHEAD-THOMPSON, R. C. (1956b). Field studies of the role of *Anopheles atroparvus* in the transmission of myxomatosis in England. *J. Hyg., Camb.*, **54**, 472. (155)

MUIRHEAD-THOMPSON, R. C. (1956c). Report to the Scientific Sub-committee of the Myxomatosis Advisory Committee, Great Britain. (168)

MYERS, K. (1954). Studies in the epidemiology of infectious myxomatosis of rabbits. II. Field experiments, August–November 1950, and the first epizootic of myxomatosis in the riverine plain of south-eastern Australia. *J. Hyg., Camb.*, **52**, 47. (8, 155, 175, 274, 275)

MYERS, K. (1956). Methods of sampling winged insects feeding on the rabbit *Oryctolagus cuniculus* (L.). *C.S.I.R.O. Wildl. Res.* **1**, 45. (174, 185, 188)

MYERS, K. (1957). Some observations on the use of sight counts in estimating populations of the rabbit, *Oryctolagus cuniculus* (L.). *C.S.I.R.O. Wildl. Res.* **2**, 170. (36)

BIBLIOGRAPHY

MYERS, K. (1958). Further observations on the use of field enclosures for the study of the wild rabbit *Oryctolagus cuniculus* (L.). *C.S.I.R.O. Wildl. Res.* **3**, 40. (30)

MYERS, K. (1962). A survey of myxomatosis and rabbit infection trends in the Eastern Riverina, New South Wales, 1951–1960. *C.S.I.R.O. Wildl. Res.* **7**, 1. (56)

MYERS, K., MARSHALL, I. D. & FENNER, F. (1954). Studies in the epidemiology of infectious myxomatosis of rabbits. III. Observations on two succeeding epizootics in Australian wild rabbits on the Riverine plain of south-eastern Australia, 1951–1953. *J. Hyg., Camb.*, **52**, 337. (155, 175, 184, 192, 218, 285, Pl. X)

MYERS, K. & MYKYTOWYCZ, R. (1958). Social behaviour in the wild rabbit. *Nature, Lond.*, **181**, 1515. (48)

MYERS, K. & POOLE, W. E. (1959). A study of the biology of the wild rabbit, *Oryctolagus cuniculus* (L.), in confined populations. I. The effects of density on home range and the formation of breeding groups. *C.S.I.R.O. Wildl. Res.* **4**, 14. (48)

MYERS, K. & POOLE, W. E. (1961). A study of the biology of the wild rabbit, *Oryctolagus cuniculus* (L.), in confined populations. II. The effects of season and population increase on behaviour. *C.S.I.R.O. Wildl. Res.* **6**, 1. (48, 54)

MYERS, K. & POOLE, W. E. (1962). A study of the biology of the wild rabbit, *Oryctolagus cuniculus* (L.) in confined populations. III. Reproduction. *Aust. J. Zool.* **10**, 225. (41, 43, 48)

MYKYTOWYCZ, R. (1953). An attenuated strain of the myxomatosis virus recovered from the field. *Nature, Lond.*, **172**, 448. (210, 220)

MYKYTOWYCZ, R. (1956). The effect of season and mode of transmission on the severity of myxomatosis due to an attenuated strain of the virus. *Aust. J. exp. Biol. med. Sci.* **34**, 121. (139, 143, 211, 212, 220, 230)

MYKYTOWYCZ, R. (1957). The transmission of myxomatosis by *Simulium melatum* Wharton (Diptera: Simuliidae). *C.S.I.R.O. Wildl. Res.* **2**, 1. (155, 197)

MYKYTOWYCZ, R. (1958a). Social behaviour of an experimental colony of wild rabbits, *Oryctolagus cuniculus* (L.). I. Establishment of the colony. *C.S.I.R.O. Wildl. Res.* **3**, 7. (48)

MYKYTOWYCZ, R. (1958b). Contact transmission of infectious myxomatosis of the rabbit *Oryctolagus cuniculus* (L.). *C.S.I.R.O. Wildl. Res.* **3**, 1. (152, 153, 155, 200)

MYKYTOWYCZ, R. (1959a). Social behaviour of an experimental colony of wild rabbits *Oryctolagus cuniculus* (L.). II. First breeding season. *C.S.I.R.O. Wildl. Res.* **4**, 1. (43, 48)

MYKYTOWYCZ, R. (1959b). Effect of infection with myxomatosis on the endoparasites of rabbits. *Nature, Lond.*, **183**, 555. (143)

MYKYTOWYCZ, R. (1960). Social behaviour of an experimental colony of wild rabbits *Oryctolagus cuniculus* (L.). III. Second breeding season. *C.S.I.R.O. Wildl. Res.* 5, 1. (40, 43, 48, 51)

MYKYTOWYCZ, R. (1961). Social behaviour of an experimental colony of wild rabbits, *Oryctolagus cuniculus* (L.). IV. Conclusion: Outbreak of myxomatosis; third breeding season and starvation. *C.S.I.R.O. Wildl. Res.* 6, 142. (38, 169)

MYKYTOWYCZ, R. (1962a). Territorial function of chin gland secretion in the rabbit *Oryctolagus cuniculus* (L.). *Nature, Lond.*, 193, 799. (49)

MYKYTOWYCZ, R. (1962b). Epidemiology of coccidiosis (*Eimeria* spp.) in an experimental colony of the Australian wild rabbit, *Oryctolagus cuniculus* (L.). *Parasitology*, 52, 375. (59)

NAGINGTON, J. & HORNE, R. W. (1962). Morphological studies of orf and vaccinia viruses. *Virology*, 16, 248. (65, 66)

NAGINGTON, J., PLOWRIGHT, W. & HORNE, R. W. (1962). The morphology of bovine papular stomatitis virus. *Virology*, 17, 361. (66)

NEWSLETTER (1964). Quarterly Newsletter of the Federation of Rabbit Clearance Societies Limited, Summer 1964. (326)

NOTINI, G., FORSELIUS, S., BRAMFORD, S. & MELLSTRÖM, B. (1952). The rabbit problem in the Gotland Island. *K. Skogshösk. Skr.* 9, 95, 104. (358)

OZAKI, Y. & HIGASHI, N. (1959). Studies on the growth of viruses of ectromelia and vaccinia in strain L cells and HeLa cells. *Ann. Rep. Inst. Virus Res., Kyoto Univ.*, Ser. B, 2, 65. (83)

PADGETT, B. L., MOORE, M. S. & WALKER, D. L. (1962). Plaque assays for myxoma and fibroma viruses and differentiation of the viruses by plaque form. *Virology*, 17, 462. (74, 82)

PADGETT, B. L. & WALKER, D. L. (1962). Single growth cycle of myxoma virus in RK cells. *Bact. Proc.* 1, 133. (75, 76)

PADGETT, B. L., WRIGHT, M. J., JAYNE, A. & WALKER, D. L. (1964). Electron microscopic structure of myxoma virus and some reactivable derivatives. *J. Bact.* 87, 454. (66, Pl. II)

PARKER, R. F. (1940). Studies of the infectious unit of myxoma. *J. exp. Med.* 71, 439. (64, 81, 135)

PARKER, R. F. & BRONSON, L. H. (1941). Neutralization of virus of myxoma by specific immune serum. *J. Immunol.* 40, 147. (119)

PARKER, R. F. & THOMPSON, R. L. (1942). The effect of external temperature on the course of infectious myxomatosis of rabbits. *J. exp. Med.* 75, 567. (134, 135, 138)

PASTEUR, L. (1888). Sur la destruction des lapins en Australie et dans la Nouvelle-Zélande. *Ann. Inst. Pasteur*, 2, 1. (272)

PETERS, D. (1960). Struktur und Entwicklung der Pockenviren. *Fourth International Conference on Electron Microscopy Berlin*, 1958, 2, 551. (65, 66)

PLACIDI, L. (1957). Le hérisson (*Aethechinus algirus algirus*) n'est pas réceptif au virus de Sanarelli (myxome infectieuse du lapin). *Bull. Acad. vét. Fr.* 30, 281. (87)

366

BIBLIOGRAPHY

PLOTZ, H. (1932). Culture du virus myxomatosum (Sanarelli) en présence de cellules vivantes. *C. R. Soc. Biol., Paris*, **109**, 1327. (73)

POOLE, W. E. (1960). Breeding of the wild rabbit, *Oryctolagus cuniculus* (L.), in relation to the environment. *C.S.I.R.O. Wildl. Res.* **5**, 21. (115, 116)

PÖTZ, L. (1957). Elektronenmikroskopische Untersuchungen der Kaninchenhaut bei infectktiöser Myxomatose. *Beitr. path. Anat.* **118**, 1. (83, 109)

RADOT, C. & LÉPINE, P. (1953). *La Myxomatose*. Paris: Flammarion. (317)

RANWELL, D. S. (1960). Newborough Warren, Anglesey. III. Changes in the vegetation on parts of the dune system after the loss of rabbits by myxomatosis. *J. Ecol.* **48**, 385. (331)

RATCLIFFE, F. N., MYERS, K., FENNESSY, B. V. & CALABY, J. H. (1952). Myxomatosis in Australia. A step towards the biological control of the rabbit. *Nature, Lond.*, **170**, 7. (38, 184, 192, 202)

REID, P. A. (1953). Some economic results of myxomatosis. *Quart. Rev. agric. Econ.* **6**, 93. (27)

REISNER, A. H., SOBEY, W. R. & CONOLLY, D. (1963). Differences among the soluble antigens of myxoma viruses originating in Brazil and California. *Virology*, **20**, 539. (67, Pl. III)

RENDTORFF, R. C. & WILCOX, A. (1957). The role of nematodes as an entry for viruses of Shope's fibromas and papillomas of rabbits. *J. infect. Dis.* **100**, 119. (146)

RENZONI, A. & CASTRUCCI, G. (1960). Ricerche istochimiche sugli inclusi citoplasmatici nel fibroma di Shope del coniglio. *Boll. Ist. sieroter. Milano*, **39**, 378. (109)

Report (1954). *Myxomatosis. Report of the Advisory Committee on Myxomatosis of the Ministry of Agriculture and Fisheries.* London: H.M.S.O. (324)

Report (1955). *Myxomatosis. Second Report of the Advisory Committee on Myxomatosis of the Ministry of Agriculture and Fisheries.* London: H.M.S.O. (325)

RHODES, A. J. (1938). The effect of intracerebral passage on the virus of infectious myxomatosis of rabbits. *J. Path. Bact.* **46**, 217. (107, 235)

RIVERS, T. M. (1930). Infectious myxomatosis of rabbits. Observations on the pathological changes induced by virus myxomatosum (Sanarelli). *J. exp. Med.* **51**, 965. (74, 108, 217)

RIVERS, T. M. & WARD, S. M. (1937). Infectious myxomatosis of rabbits. Preparation of elementary bodies and studies of serologically active materials associated with the disease. *J. exp. Med.* **66**, 1. (66, 119, 120)

RIVERS, T. M., WARD, S. M. & SMADEL, J. E. (1939). Infectious myxomatosis of rabbits. Studies of a soluble antigen associated with the disease. *J. exp. Med.* **69**, 31. (66, 120)

RODRÍGUEZ LOUSTAU, J. A., QUEVEDO, J. M., JR., TORRE, E. J. & RIZZO, H. R. (1955). Mixomatosis. Estudio de un foco-tentativas de inmunización. *Gac. vet., B. Aires*, **17**, 172. (368)

367

RONDLE, C. J. M. & DUMBELL, K. R. (1962). Antigens of cowpox virus. *J. Hyg., Camb.*, **60**, 41. (66)

ROOYEN, C. E. VAN (1937). Elementary (Paschen) bodies in infectious myxomatosis of the rabbit (virus myxomatosum Sanarelli). *Zbl. Bakt.* (*Abt.* 1), **139**, 130. (65, 152)

ROSENBUSCH, F. (1919). *Mem. Soc. Rural Arg.* p. 26, quoted by Rodríguez Loustau et al. (1955). (263)

ROTHSCHILD, M. (1953). Notes on the European rabbit flea. Report to the Myxomatosis Advisory Committee, 6 December 1953. (166, 167)

ROTHSCHILD, M. (1958). A further note on the increase of hares (*Lepus europaeus*) in France. *Proc. zool. Soc. Lond.* **131**, 328. (321)

ROTHSCHILD, M. (1961). Increase of hares at Ashton Wold. *Proc. zool. Soc. Lond.* **137**, 634. (332)

ROTHSCHILD, M. (1963). A rise in the flea-index on the hare (*Lepus europaeus* Pallas) with relevant notes on the fox (*Vulpes vulpes* (L.)) and wood-pigeon (*Columba palumbus* L.) at Ashton, Peterborough. *Proc. zool. Soc. Lond.* **140**, 341. (332)

ROWE, B., MANSI, W. & HUDSON, J. R. (1956). The use of fibroma virus (Shope) for the protection of rabbits against myxomatosis. *J. comp. Path.* **66**, 290. (129)

RUBIN, H., CORNELIUS, A. & FANSHIER, L. (1961). The pattern of congenital transmission of an avian leukosis virus. *Proc. nat. Acad. Sci., Wash.*, **47**, 1058. (144)

RUIZ-GOMEZ, J. & ISAACS, A. (1963a). Optimal temperature for growth and sensitivity to interferon among different viruses. *Virology*, **19**, 1. (239)

RUIZ-GOMEZ, J. & ISAACS, A. (1963b). Interferon production by different viruses. *Virology*, **19**, 8. (239)

RUSKA, H. & KAUSCHE, G. A. (1943). Ueber Form, Grössenverteilung und Struktur einiger Virus-Elementarkörper. *Zbl. Bakt.* (*Abt.* 1), **150**, 311. (65)

SANARELLI, G. (1898). Das myxomatogene Virus. Beitrag zum Stadium der Krankheitserreger ausserhalb des Sichtbaren. *Zbl. Bakt.* (*Abt.* 1), **23**, 865. (1, 87)

SCHMIT-JENSEN, H. O. (1939). Summary of the experiments carried out in Vejrø by the State Veterinary Serum Laboratory on the extermination of rabbits by myxomatosis virus. *Report to Danish Ministry for Agriculture and Fisheries.* (308, 309)

SCHWERDT, P. R. & SCHWERDT, C. E. (1962). A plaque assay for myxoma virus infectivity. *Proc. Soc. exp. Biol., N.Y.*, **109**, 717. (74)

SCOTT, ERNEST (1936). *A Short History of Australia*, 6th ed. Oxford University Press. (20)

SHAFFER, J. G. (1941). Antigenic relationship of infectious myxoma and fibroma viruses of the rabbit. *Amer. J. Hyg.* (B), **34**, 102. (64, 121)

368

BIBLIOGRAPHY

SHANKS, P. O., SHARMAN, G. A. M., ALLAN, R., DONALD, L. G., YOUNG, S. & MARR, T. G. (1955). Experiments with myxomatosis in the Hebrides. *Brit. Vet. J.* **111**, 25. (155, 309, 311)

SHOPE, R. E. (1932). A transmissible tumor-like condition in rabbits. A filtrable virus causing a tumor-like condition in rabbits and its relationship to virus myxomatosum. *J. exp. Med.* **56**, 793. (63, 95, 108, 128)

SHOPE, R. E. (1936a). Infectious fibroma of rabbits. III. The serial transmission of virus myxomatosum in cottontail rabbits, and cross-immunity tests with the fibroma virus. *J. exp. Med.* **63**, 33. (63, 91, 128)

SHOPE, R. E. (1936b). Infectious fibroma of rabbits. IV. The infection with virus myxomatosum of rabbits recovered from fibroma. *J. exp. Med.* **63**, 43. (91)

SHOPE, R. E. (1938). Protection of rabbits against naturally acquired infectious myxomatosis by previous infection with fibroma virus. *Proc. Soc. exp. Biol., N.Y.*, **38**, 86. (128)

SHOPE, R. E. (1949). The spread of viruses from infected to susceptible hosts. *The Diplomate*, **21**, 235. (146)

SHOPE, R. E., MANGOLD, R., MACNAMARA, L. G. & DUMBELL, K. R. (1958). An infectious cutaneous fibroma of the Virginia whitetailed deer (*Odocoileus virginianus*). *J. exp. Med.* **108**, 797. (64)

SIMMONS, J. S. (1934). The virus of infectious myxomatosis. Thesis, George Washington University. (265)

SIRIEZ, H. (1957). *La myxomatose, moyen de lutte biologique contre le lapin, rongeur nuisible.* Paris: Editions Sep. (320)

SIRIEZ, H. (1960). *Lapins et myxomatose. L'evolution de la maladie de 1956 à 1960 et quelques compléments à une précédente étude.* Paris: Editions Sep. (318, 320, 321)

SMADEL, J. E. & HOAGLAND, C. L. (1942). Elementary bodies of vaccinia. *Bact. Rev.* **6**, 79. (66, 68)

SMADEL, J. E., RIVERS, T. M. & HOAGLAND, C. L. (1942). Nucleoprotein antigen of vaccine virus. I. A new antigen from elementary bodies of vaccinia. *Arch. Path.* **34**, 275. (61)

SMADEL, J. E., WARD, S. M. & RIVERS, T. M. (1940). Infectious myxomatosis of rabbits. II. Demonstration of a second soluble antigen associated with the disease. *J. exp. Med.* **72**, 129. (66)

SMITH, M. H. D. (1948). Propagation of rabbit fibroma virus in the embryonated egg. *Proc. Soc. exp. Biol., N.Y.*, **69**, 136. (79)

SMITH, M. H. D. (1952). The Berry–Dedrick transformation of fibroma into myxoma in the rabbit. *Ann. N.Y. Acad. Sci.* **54**, 1141. (106, 112)

SMITH, W., ANDREWES, C. H. & LAIDLAW, P. P. (1933). A virus obtained from influenza patients. *Lancet*, ii, 66. (85)

SOBEY, W. R. (1960). Myxomatosis: the virulence of the virus and its relation to genetic resistance in the rabbit. *Aust. J. Sci.* **23**, 53. (236, 297)

SOBEY, W. R. & TURNBULL, K. (1956). Fertility in rabbits recovering from myxomatosis. *Aust. J. biol. Sci.* **9**, 455. (115, 248)

SOKHEY, S. S. & CHITRE, R. G. B. D. (1937). L'immunité des rats sauvage de l'Inde vis-à-vis de la peste. *Bull. Off. int. Hyg. publ.* **29**, 2093. (345)

SOUTHERN, H. N. (1940). The ecology and population dynamics of the wild rabbit (*Oryctolagus cuniculus*). *Ann. appl. Biol.* **27**, 509. (44)

SOUTHERN, H. N. (1948). Sexual and aggressive behaviour in the wild rabbit. *Behaviour*, **1**, 173. (48)

SOUTHERN, H. N. & WATSON, J. S. (1941). Summer food of the red fox (*Vulpes vulpes*) in Great Britain: a preliminary report. *J. anim. Ecol.* **10**, 1. (332)

SPLENDORE, A. (1909). Ueber das Virus myxomatosum der Kaninchen. *Zbl. Bakt. (Abt.* 1), **48**, 300. (1)

SPRUNT, D. H. (1932). Infectious myxomatosis (Sanarelli) in pregnant rabbits. *J. exp. Med.* **56**, 601. (112)

SYVERTON, J. T. & BERRY, G. P. (1947). The superinfection of rabbit papilloma (Shope) by extraneous viruses. *J. exp. Med.* **86**, 131. (114)

TAKAHASHI, M., KAMEYAMA, S., KATO, S. & KAMAHORA, J. (1959). The immunological relationship of the pox virus group. *Biken's J.* **2**, 27. (64)

TEIXEIRA, J. DE CASTRO & SMADEL, J. E. (1941). Further studies on the serological reactions of the soluble antigens of infectious myxomatosis. *J. Bact.* **42**, 591. (66, 119)

THOMAS, A. S. (1960). Changes in vegetation since the advent of myxomatosis. *J. Ecol.* **48**, 287. (331)

THOMAS, O. (1901). On mammals obtained by Mr Alphonse Robert on the Rio Jordão, S.W. Minas Geraes. *Ann. Mag. Nat. Hist.* Ser. 7, **8**, 526. (14)

THOMAS, O. (1913). Notes on S. American Leporidae. *Ann. Mag. Nat. Hist.* Ser. 8, **11**, 209. (14)

THOMPSON, H. V. (1954). The rabbit disease: myxomatosis. *Ann. appl. Biol.* **41**, 358. (9, 312)

THOMPSON, H. V. (1956). Myxomatosis: a survey. *Agriculture, Lond.*, **63**, 51. (323)

THOMPSON, H. V. & WORDEN, A. N. (1956). *The Rabbit.* London: Collins. (11, 17, 18, 324)

THOMPSON, R. L. (1938). The influence of temperature upon proliferation of infectious fibroma and infectious myxoma viruses in vivo. *J. infect. Dis.* **62**, 307. (134)

THOMPSON, R. L. & COATES, M. S. (1942). The effect of temperature upon the growth and survival of myxoma, herpes and vaccinia viruses in tissue culture. *J. infect. Dis.* **71**, 83. (141, 240)

TORRES, S. (1936). Transmissão da mixomatose dos coelhos pelo *Culex quinquefasciatus. Bol. Soc. Bras. Med. vet.* **6**, 4. (154, 155)

TRAUB, E. (1939). Epidemiology of lymphocytic choriomeningitis in a mouse stock observed for four years. *J. exp. Med.* **69**, 801. (144)

BIBLIOGRAPHY

VAIL, E. L. & McKENNY, F. D. (1943). *Diseases of Domestic Rabbits.* Conserv. Bull. No. 31, Fish and Wildlife Service, U.S. Dept. of the Interior. (4, 267)

VERNA, J. E. & EYLAR, O. R. (1962). Rabbit fibroma virus plaque assay and *in vitro* studies. *Virology*, **18**, 266. (74)

WATSON, J. S. (1957). Reproduction of the wild rabbit, *Oryctolagus cuniculus* (L.) in Hawke's Bay, New Zealand. *N.Z. J. Sci. Tech.* **38**, 451. (40, 44)

WATSON, J. S. (1961). Feral rabbit populations on Pacific Islands. *Pacif. Sci.* **15**, 591. (18)

WHITE, H. C. (1929). Observations on rabbit myxoma. *N.S.W. Dept. Agric. Vet. Res. Rep. No.* 5, 1927–1928, p. 45. (5)

WHITTY, B. T. (1955). Myxomatosis in the common hare—*Lepus europaeus. Irish Vet. J.* **9**, 267. (92)

WODZICKI, K. A. (1950). *Introduced Mammals of New Zealand.* D.S.I.R. Bull. No. 98, Wellington, N.Z., 250 pp. (18)

WOODROOFE, G. M. & FENNER, F. (1960). Genetic studies with mammalian poxviruses. IV. Hybridization between several different poxviruses. *Virology*, **12**, 272. (63)

WOODROOFE, G. M. & FENNER, F. (1962). Serological relationships within the poxvirus group: an antigen common to all members of the group. *Virology*, **16**, 334. (61, 62, 64, 82, 120)

WOODROOFE, G. M. & FENNER, F. (1965). Viruses of the myxoma-fibroma subgroup of the poxviruses. I. Plaque production in cultured cells, plaque-reduction tests, and cross-protection tests in rabbits. *Aust. J. exp. Biol. med. Sci.* **43**, 123. (63, 75, 82, 99, 100, 120, 229, Pl. IV)

YUILL, T. M. & HANSON, R. P. (1964). Infection of suckling cottontail rabbits with Shope's fibroma virus. *Proc. Soc. exp. Biol., N.Y.*, **117**, 376. (96)

ZEUNER, F. E. (1963). *A History of Domesticated Animals.* London: Hutchinson. (15, 16)

Map of south-eastern Australia, showing the location of main myxomatosis study areas.

SUBJECT INDEX

INDEX

INDEX

377

INDEX

379